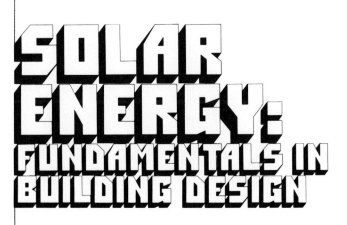

SOLAR ENERGY: FUNDAMENTALS IN BUILDING DESIGN

SOLAR ENERGY: FUNDAMENTALS IN BUILDING DESIGN

BRUCE ANDERSON
TOTAL ENVIRONMENTAL ACTION, INC.
HARRISVILLE, NEW HAMPSHIRE

McGraw-Hill Book Company

New York St. Louis San Francisco Auckland Bogotá
Düsseldorf Johannesburg London Madrid Mexico
Montreal New Delhi Panama Paris São Paulo
Singapore Sydney Tokyo Toronto

34567890 VHVH 7854321098

The editors for this book were Jeremy Robinson and Angela Dorenkamp. The designer was Paula Kursh. Illustrations were drawn by Robert Conti and Mark Frey. It was set in Baskerville by University Graphics, Inc., and it was produced by Total Environmental Action Press.

Printed and bound by Von Hoffmann Press, Inc.

Library of Congress Cataloging in Publication Data

Anderson, Bruce N
    Solar energy: fundamentals in building design.

    Bibliography: p.
    Includes index.
    1. Solar heating. 2. Architecture and solar radiation. I. Title.
TH7413.A5 1977   697'.78   76-45467

ISBN 0-07-001751-4

CHARLES HAVERSTICK

CONTENTS

To Nina Winters
my patient and understanding friend

PREFACE

Somehow I just can't help thinking that solar energy is going to make it this time. Twenty years ago public interest in solar peaked due to rising fuel prices and predicted energy shortages, then waned due to solar's relatively high cost and to intense Federal interest in nuclear. Energy prices are still low relative to the cost of solar, and tenacious Federal support for nuclear continues unabated; but this round, fortunately, the American public knows better. We know that energy prices are being held down artificially and that they could skyrocket when political forces see it to be in their best interest. We know enough not to be so easily seduced by the nuclear hard sell. We are at last realizing that the world's pantry of non-renewable natural resources does indeed have a bottom shelf; that we had better be damned frugal with what little remains; and that we had better turn to less conventional sources to provide for our reduced needs.

Energy conservation in building design is a natural, easily identifiable, and extremely important aspect of this frugality. The use of solar energy to provide us comfort in our energy-conserving buildings is the next logical step in this effort.

Having been involved over the last several years in dozens of energy conserving solar projects, I have become convinced of this one-two sequence: energy conservation first; solar energy next. Perhaps even more importantly, I have become convinced of step *three:* wise energy use by the occupants. Although not covered in detail here in this book, careful energy use may in fact be *the* most important step in using less energy. Unless we are willing to make an effort in our own daily lives to consume less, all of our efforts to conserve energy and to utilize solar energy will be in vain. Energy frugality in daily living is a must, of course, in using coal, oil, gas, or nuclear, too. Wise energy use is so important, in fact, that occupants of an energy-leaky building can use less energy than wasteful occupants of an energy-conserving building of the same size.

The process of becoming "book" is always an arduous one. *Solar Energy: Fundamentals in Building Design* is no exception. It had its origin in 1972, when, in the fall of that year, I wrote my thesis, *Solar Energy and Shelter Design,* for my Master's Degree in Architecture from MIT.

In February 1974, when I signed a contract with the publishing arm of Arthur D. Little, Inc. to expand on my thesis, I had no way of knowing that they would quietly leave the publishing field a year later, just one week after I had delivered my 1300-page manuscript, *Solar Energy in Building Design.* Fortunately, by then I had my own research, design and education firm, Total Environmental Action, Inc. (TEA). The book's publication would not be thwarted; during the next year we printed and sold unbound copies of the raw beast to an appreciative group of people who were happy that so much source material on solar energy had been developed into useable form.

Dr. Stephen Feldman, a good friend and co-principal investigator on a major National Science Foundation grant, crystallized our efforts to produce a publication. As we proceeded into the project, McGraw-Hill presented a contract, and here we are. Dr. Feldman continued assisting, heroically condensing the first five chapters of the original manuscript into Part I.

Simultaneously with the production of this book, two other friends joined forces to finance and produce a drastically revised softcover version of the original manuscript, *Solar Energy in Building Design,* for the purpose of reaching a general readership audience. Published by Cheshire Books, *The Solar Home Book: Heating, Cooling and Designing with the Sun* has, in fact, sentences which may be identical with those in this book.

The origin of the inspiration, the interest, the information, and the assistance for this book is clouded in an array of experiences, interactions, events, and long hours of work involving not only myself but scores of other people. My friends, relatives, and associates have been extraordinarily supportive with their time, their emotions, and their understanding of the load which this endeavor placed on me.

I owe particular gratitude to the dedicated professors at MIT who prepared me for this work and who assisted me in the writing of my thesis. A bit more elusive, but just as important, is my gratitude to the many dedicated people, too numerous to mention, who have given themselves to the field of solar energy. Their perseverance and courage when other men's eyes were blind and minds were closed has truly been an inspiration. In addition to this inspiration is their

The Christian Science Monitor

"We've got to look for a new source of energy!"

contribution to a broad and firm foundation of knowledge upon which this book is based and upon which the future of the utilization of solar energy now stands.

Thank you, also, to the innumerable people and organizations who freely shared their information, much of which is reproduced or referenced in these pages.

Many friends assisted me immeasurably with the many routine, but important, tasks of putting all this material into communicable form. In particular, my warmest thanks to my good friend Nina Winters as well as to Gretchen Poisson, Missy Michal, Suzanne Coonley and Carol Scully. Special gratitude also goes to many of my staff and associates: Francie Harrison, Susie Withol, Hilda Wetherbee, Michelle Artese, Karen Tolman, and in particular, to Charles Michal for significant contributions to the Resource Section and the Appendix, and to Douglas Mahone who wrote Part IV as well as several Resource Sections.

Particular credit is due on the production of this version: Angela Dorenkamp, editor; Rob Conti, illustrator, and his assistant, Mark Frey; Paula Kursh, designer. These four people performed exceptional work and put in long and tedious hours.

And finally, to Jeremy Robinson, Senior Editor at McGraw-Hill, who followed my work in the field of solar energy with confidence for three years before the first book came off the press, my appreciation.

It is my hope that the information in this book will be used wisely for the betterment of the human condition, to enable all of us to discover who we are, and why, in the context of one world with limited resources. The proper use of solar energy is just one important ingredient toward the realization of this goal.

Bruce Anderson
Harrisville, NH

PART I

Solar Energy:
A Context For Its Proper Use

The importance of energy's role in the world economy was made evident by the oil embargo of 1973. Since then, the expansion and diffusion of ideas on the development and utilization of alternate energy sources have received added impetus. Not since the era of Sputnik, however, has this technically-oriented society been faced with such an arduous challenge, a challenge which can be successfully met only by adopting a holistic approach toward energy and environmental management. This book is based on such an approach.

Many researchers view solar energy use narrowly, e.g., as a substitute for fossil fuels. Thus solar energy collectors are often added to a structure which is designed to appeal to a "contemporary" or "modern" clientele. These dwellings may have extra insulation, heavy drapes at the windows, or even triple glazing, but the occupants, who are not otherwise aware that their heat is derived from the sun, continue other ecologically destructive aspects of their lives. When solar systems engineers fail to realize that behavioral and aesthetic aspects of habitation are important to this design, a lack of occupant participation often results, necessitating the installation of high-level technology control systems which feature less reliability and lower overall efficiency.

Making the transition to widespread use of solar energy requires more than attaching a collector to a structure and judging its success by the resultant reduction in fuel bills. This common approach of many engineers, designers, and social scientists toward the conservation and utilization of energy must be altered, and new attitudes toward home-house-shelter-building design must be fostered. Indeed, the architectural significance of the integration of ecological technologies into shelter design lies far deeper than the attempt to synthesize, for example, a huge solar heating collector and its accompanying huge storage tank into an architectural whole, or to use alternative, mechanical accessories. Inextricably interwoven into the broader understanding of the inter-relationship between ecology and architectural design must be an awareness of the man-nature interface and the inter-relationship of natural systems.

An intense sensitivity to this man-nature interaction is, in fact, a pre-requisite in the application of ecological principles to building form. Since our beginnings, people have searched for and sometimes succeeded in discovering our role as members of Earth's community, our special and powerful place in the ecosystem. Art and architecture have provided some of the most sensitive manifestations of this search, each contribution proving the possibility and even the necessity for synthesis.

CHAPTER I·A

An Introduction To Solar Energy

Contemporary man takes pride in his new consciousness, in his discovery of the abundant energy which strikes our planet a mere eight and a half minutes after leaving a giant furnace 93,000,000 miles away. Since their beginning, living things have found ingenious ways of taking advantage of, working with, and protecting themselves from this powerful source of energy which drives the forces of the endless life-death-life cycle. The green color of plants tempers the intense rays of the sun with those of less intensity. Man himself has responded emotionally and spiritually to the sun, fearing it and worshipping it, and designing his habitations with respect for the sun's power, generosity, and cruelty.

In his *Memorabilia* (III, viii, 8–14), Xenophon records some teachings of Socrates (470–399 B.C.) regarding dwellings:

> Again his dictum about houses . . . was a lesson in the art of building houses as they ought to be. He approached the problem thus: 'When one means to have the right sort of house, must he contrive to make it as pleasant to live in and as useful as can be?' And this being admitted, 'Is it pleasant,' he asked, 'to have it cool in summer and warm in winter?' And when they agreed with this also, 'Now in houses with a south aspect, the sun's rays penetrate into the porticoes in winter, but in summer the path of the sun is right over our heads and above the roof, so that there is shade. If, then, this is the best arrangement, we should build the south side loftier to get the winter sun and the north side lower to keep out the cold winds. To put it shortly, the house in which the owner can find a pleasant retreat at all seasons and can store his belongings safely is presumably at once the pleasantest and the most beautiful.'

Man brought conscious sophistication and ingenuity to this interaction with the sun. Since pre-history, the sun has dried and preserved man's food. It has evaporated the waters of the ocean to yield salt. Since man began to reason, he has recognized the sun as the motive power behind every natural phenomenon and as a primary determinant of every man-made phenomenon. Various commentators believe that the Great Pyramid of Egypt, one of man's greatest engineering feats, was built as a stairway to the sun. Some twenty-five hundred years ago, virgin priestesses in the temples of Vesta focused metal cones and kindled sacred fires with the sun's rays. And in 212 B.C., the Greek physicist Archimedes used a huge, concave metallic mirror (in the form of hundreds of polished shields, all reflecting on the same ship) to burn the Roman fleet which attacked Syracuse.

A solar collector for producing mechanical energy (55 horse-power). Created by Frank Shuman and C. V. Boys in Meadi, Egypt, 1913, for pumping water. (Photo by Dr. C. G. Abbott, Smithsonian Institution and provided courtesy of Martin Marietta Corporation.)

A printing press driven by solar energy in Paris in 1878.

In addition to the thousands of ways in which the sun's energy has been used by both nature and man throughout time—to grow food, to see by, to get a sun tan, to dry clothes—it has also been deliberately harnessed to perform a number of other "chores." Solar energy is used to heat and cool buildings, to heat water and swimming pools; to power refrigerators; and to operate engines, pumps and sewage treatment plants. It powers cars, ovens, water stills, furnaces, distillation equipment, crop dryers, and sludge dryers. Powered by solar energy, wind is used to generate electricity and mechanical power, and solar-converted electricity is used both on earth and in space. Stoves and cars run on solar-made methane gas, power plants operate on organic trash, and sewage plants produce methane gas. The sun powered evaporation/rain cycle, in combination with gravity, powers machines and electric turbines. Solar electrolizers convert water to clean hydrogen gas (a fuel). Farrington Daniels' *The Direct Use of the Sun's Energy* is an excellent study of the many and varied uses of solar energy.

None of these uses, however, nor those discussed in this text, can be comprehended without a knowledge of the basic principles of solar energy. Most of the energy we receive from the sun comes in the form of light, a short-wave radiation, not all of which is visible to the human eye. When this radiation strikes a solid or liquid, it is absorbed and transformed into heat energy; the material becomes warm and stores the heat, conducts it to surrounding materials (air, water, other solids or liquids), or reradiates it to other materials of lower temperature. This reradiation is a long-wave radiation.

Glass easily transmits short-wave radiation, which means that it poses little interference to incoming solar energy, but it is a very poor transmitter of long-wave radiation. Once the sun's energy has passed through glass windows and has been absorbed by some material inside, the heat will not be reradiated back outside. Glass, therefore, acts as a heat trap, a phenomenon which has been recognized for some time in the construction of greenhouses, which can get quite warm on sunny days, even in the middle of winter; this has come to be known, in fact, as the "greenhouse effect." Solar collectors for home heating, usually called *flat plate collectors,* almost always have one or more glass *covers,* although various plastic and other transparent materials are often used instead of glass.

Beneath the cover plate, collectors commonly have another plate which absorbs the sun's rays hitting it. This *absorber plate* is usually made of copper, aluminum, steel, or another suitable material and is usually coated with a substance—black paint or one of the more sophisticated *selective coatings* available—that will help it absorb the most heat, rather than reflect or reradiate it. Once the heat is absorbed, it can be picked up and used. The glass cover plates help to reduce the loss of heat through the front while insulation reduces heat loss through the back.

From the absorber plate, heat is transferred by conduction to a transfer fluid, usually a liquid or air, which flows by the absorber plate, often with the help of a pump or blower. The liquids (water or a non-freezing fluid such as ethylene glycol) flow over the black surface or through tubes incorporated into the absorber plate. If air is used, it is blown across the surface of the absorber plate, which should have many small irregular surfaces with which the air can come in contact.

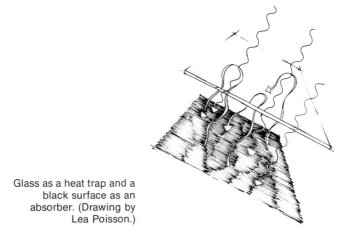

Glass as a heat trap and a black surface as an absorber. (Drawing by Lea Poisson.)

In some cases, it is possible to move the fluids (whether liquid or air) without mechanical aid, by natural convection or *thermosiphoning.* As the fluid is heated, it tends to rise, and cooler fluid flows in to take its place. If the collector is tilted or vertical, this effect will move fluid across the collector plate and off without any external help. Some of the simplest systems work this way and, in the right application, they are very effective. Pumping, however, usually gives greater collection efficiencies and allows more versatile use of the collected heat.

This heat is directly used for warming the living spaces of a building in conventional ways, e.g., through radiators and hot air registers. When the building does not require heat, the warmed air or liquid from the collector can be moved to a heat storage container. In the case of air, the storage is often a pile of rocks, or some other heat-holding material; in the case of liquid, it is usually a large, well-insulated tank of water, which has considerable heat capacity. Heat is also stored in containers of chemicals called eutectic or *phase-changing salts.* These salts, which store large quantities of heat in a relatively small volume, melt when they are heated and release heat later as they cool and crystallize. When the building needs heat, the air or water from its heating system passes through the storage, is warmed, and is then fed through the conventional heaters to warm the space.

Because most systems cannot handle long periods of cold, sunless days, few of the solar projects completed to date provide all of the buildings' heat requirements by solar means. A system large enough to provide sufficient heat during these periods would be so over-sized for normal use that the added expense would seldom be justified. A solar heating system, therefore, almost always requires a full-sized, standard heating system as a back-up. The same is true for solar cooling systems.

In most cases, the conventional system and the solar system can be efficiently integrated. To do so easily may require some rather unconventional adaptations of the conventional system. For instance, solar systems are most efficient at low operating temperatures: the collectors gather more heat, and losses from the heat storage and transfer fluids are lower. A conventional hot water heating system, however, operates at relatively higher temperatures. The optimum system, then, will require an overall lower operating temperature and a slightly different approach to the whole heating system. For this reason, it is usually difficult to "plug in" a solar system to an existing conventional system, but there are several other ways of using solar energy in existing buildings, and some of these will be discussed later.

The heat from solar energy can be used to cool buildings, using the absorption cooling principle operative in gas-fired refrigerators. Presently available equipment, however, usually requires extremely high operating temperatures, far above those for efficient solar collection. A great deal of current research is being devoted to developing systems requiring lower operating temperatures and collectors which are more efficient at higher temperatures, but it will probably be several years before solar cooling systems which utilize solar collectors will be commercially viable. Other forms of solar cooling, both mechanical and natural, are also discussed in this book.

Solar energy units for heating domestic water are commercially available and are used by millions of people in various parts of the world. The Australian government requires electrically supplemented solar water heaters for all new housing in the northern part of the country. In Israel, solar water heaters are widely used, and simple, plastic, non-supplemented water heaters have been widely used in Japan. There is a thriving, though small, solar water heater industry in Florida and California. Because of the low price of competitive fuels and the difficulty of designing solar water heaters which can operate successfully during freezing weather, solar domestic water heating has not been widely used in northern climates. However, with rising fuel prices and the increased development of solar collectors, solar heating of domestic water in cold climates is being adopted.

Probably the best way of using the sun's energy for heating domestic water is as a preheater, heating municipal or well water to the temperature of the collector, then moving it to a conventional domestic water heater, where its temperature is boosted yet higher, if necessary. Efficiencies of the collector will be relatively high and there will not be the problem of having to rely completely on solar energy, since the back-up is an integral part of the solar system itself. This is discussed further in the section on water heating.

In addition to its use for heating and cooling buildings and heating domestic water, solar energy can be converted into electricity in five basic ways:

1. Using solar heated upper layers and the cold lower depths of the ocean to operate a low temperature difference (30 or 40°F) heat engine. This heat engine is then used to drive a generator, producing electricity or hydrogen fuel. According to Dr. William Heronemus at the University of Massachusetts, such power plants are competitive in price with fossil fuel plants.

2. Using the solar-driven wind to power a wind turbine (a windmill), which in turn drives a generator to produce electricity. Wind can also be used to provide mechanical power such as water pumping.

3. Using solar energy through photosynthesis to grow trees, plants, and other organisms such as algae which can be used as a combustible fuel in place of coal after suitable processing, such as drying, chipping or grinding. Plants grown with sunshine could be used to fuel furnaces or boilers, for example. Photosynthesis efficiencies are estimated to range from 0.3 to 3%, depending on the vegetation used. Because of the low conversion efficiencies, large land areas would be required to supply significant amounts of energy: about 3% of the U.S. land area would be needed to supply current energy needs.

In December 1972, the Staff Report of the Committee on Science and Astronautics of the U.S. House of Representatives pointed out that with today's plant-growing technology the available farm land held in surplus by current farm policies could be utilized to produce combustible plant fuel. As of December 1972, 470,000 square miles of land were being cultivated and 94,000 square miles were being held as surplus. The authors of the Report estimated that this surplus land could supply enough fuel to meet 10% of our projected energy needs for 1985.

Organic material can also be converted into methane, hydrogen, or oil by destructive distillation (*pyrolysis*), by high-pressure chemical processing, or by biochemical fermentation. Pyrolysis is a process in which organics are heated in the absence of air. In high-pressure chemical processing, organics are heated under pressure in the presence of water and a cover gas of carbon dioxide. Several biochemical fermentation processes which produce methane, reused in the treatment process as fuel, have been in use over the past twenty years in the U.S. for sanitary waste treatment. The purpose of most treatment plants, however, is to process waste, not to produce fuel. Hydrogen can also be produced through solar powered *electrolysis* processes (the separation of water into hydrogen and oxygen) and burned as a fuel or used in a fuel cell to produce electricity.

Waste from forestry operations and from municipal trash collections can be used directly as fuel for power and/or indirectly to produce other fuels such as methane or methanol. It was estimated that waste from current forestry operations could have provided 10–20% of U.S. energy needs for 1975 (MOR.).

4. *Directly converting solar energy to electricity using solar photovoltaic cells.* Silicon solar cells have been widely employed on space craft, where typical efficiencies are about 10%, and costs range up to one million dollars per kilowatt capacity. For terrestrial applications, such as isolated, off-shore oil derricks, slightly lower quality cells produce efficiencies of 4 to 5%. The maximum theoretical efficiency for simple photovoltaic conversion is about 35%, but efficiencies of 20% are believed to be achievable. Polycrystalline cells have a lower efficiency than single crystal cells now in use in space crafts, but they would be less expensive. Concentrators can be used to increase solar intensities onto a small area, greatly increasing the output of the solar cells. Peak kilowatt installations could be reduced to $2,000, and possibly lower.

Photovoltaic cells can be used to generate electricity on a large scale by covering square miles of land, but they can also be used on individual buildings, as is being done on what is probably the most technologically-oriented solar powered house ever built. The house, at the University of Delaware, converts solar radiation not only to heat but also to produce electricity through the use of photovoltaic cells.

5. *Using concentrating-type collectors to heat fluids which can be used to operate heat engines which in turn drive generators to produce electricity.* Low temperature collectors, such as the *flat plate* types described at the beginning of this chapter, are now adequate for heating buildings and water and with increased development will be adequate for powering solar absorption-type cooling equipment. However, they are insufficient for the high-efficiency production of electricity or for making artificial fuels by thermal processes. For this, high temperature collectors such as concentrating collectors are required. Solar energy is focused from a

An artist's conception of a solar power plant. A field of mirrors focuses light onto a boiler. A conventional steam turbine converts the steam to electricity. (Photo courtesy of Martin Marietta Corporation.)

relatively large area into a small collector from which it is carried to storage. Such concentrators are usually parabolic or cylindrical in shape. Temperatures up to 1000°F and more are attainable.

Seasonal efficiencies of most focusing collectors are often lower than for flat plate collectors because of these higher operating temperatures; in addition, because they rely only on direct rather than diffuse radiation, they need clear skies to operate. On the other hand, flat plate collectors are able to use solar radiation that is not nearly as bright as that necessary for concentrating collectors. Concentrating collectors are usually comparatively expensive.

Scientists have proposed schemes for using the sun's energy to generate electricity on a large scale by creating *farms* of many square miles of concentrating solar panels in areas where the skies are rarely cloudy, such as the Arizona desert. Various proposals call for thousands of square miles of desert area in which hundreds of individual plants would be placed. There are more than one hundred thousand square miles of desert in the United States, and as much as 10% of it has been suggested for such use.

According to Walter E. Morrow, Jr., at The Lincoln Laboratory of MIT, assuming 30% efficiency, the present electric power demand of the United States could be supplied by solar energy plants having a total area of about 2,000 square miles. "This is about 0.03% of the U.S. land area (which is) devoted to farming, and about 2% of the land area devoted to roads; and it is about equal to the roof area of all the buildings in the U.S." (MOR.).

CHAPTER I·B

Solar Energy
Is Being Used Now

During the past several years, in particular, since the now historic oil embargo of the 1973–74 heating season, solar energy has been recognized as an energy alternative which can be immediately implemented. In the space of two fiscal years, Federal spending has risen from $13 million per year to around $75 million in FY 75–76. Not to be outdone, and in fact, leading government by the hand, is the private sector, which is implementing existing solutions and researching new ones. In the same short two years, the number of solar buildings in this country has risen from a mere handful to several thousand, including those which are still in the planning stages. Government officials, environmentalists, housing developers, local school boards, and individual taxpayers are all scurrying to demonstrate their desperate infatuation with the warm rays of the sun.

Solar energy for space heating and cooling will probably continue to have broad acceptance when

- The application uses the energy during a large fraction of the year
- The climate is sunny
- Conventional fuels are costly
- The application requires moderate temperatures

Contrary to popular belief, solar heating is often more applicable in cold climates than in warm ones. The longer and colder the heating season, the greater the demand for the heat which is absorbed and stored by the solar system. Thus, according to this single criterion, solar water heating for domestic use, which uses solar energy year round, is an even better use for solar energy than space heating. In combination with a long, intense demand, that area of the country which has the most sun will fare best; and eventually, every area of the country will be affected by increased costs for conventional fuels.

The fourth point includes applications where hot water or hot air at temperatures below 200°F (and preferably lower) will serve an end use, such as space heating, hot water, refrigeration, and air conditioning. In 1968, these uses represented about 11% of the national energy consumption. More importantly, perhaps, they accounted for 76% of the energy used by commercial enterprises. About 28% of all energy for industrial purposes was for such heat, accounting for 11½% of the nation's consumption in 1968.

Using traditional economic criteria, solar energy is now often competitive with fossil fuels and with electricity, particularly in the heating of buildings and domestic water. Most types of buildings can use solar energy: schools and other public buildings, industrial and commercial buildings, mobile homes, existing and new homes.

Schools are particularly suited to the use of solar energy for several reasons: (1) their high heating and cooling demands result in a large demand on the collectors if they are properly sized; (2) schools are usually three stories or less in height, providing sufficient roof area for the collectors, and non-classroom walls and roofs such as those of gymnasiums can be used for additional collector area; (3) with the relatively large thermal mass of schools, wide weather fluctuations are not experienced as greatly as with lightweight buildings, thus placing a more uniform demand on the solar system and giving a better return on the initial investment; (4) although school construction funds are limited and difficult to obtain, they carry low interest rates, resulting in lower effective initial costs and lower overall life-cycle costs than are incurred by commercial and domestic structures; and, (5) where a school board may have difficulty in obtaining outside funds for operation and maintenance (which are reduced with the use of solar energy), it may have a comparatively easy time obtaining funds for higher initial costs, particularly for solar energy in this time of ecological and energy conservation awareness.

If classes are scheduled to begin later in the day, the solar energy coming directly through windows and the thermal inertia of the building can be used more effectively. This would allow classrooms on the east and south sides to be heated by solar radiation through the glass on sunny mornings before classes start. It would also make better use of solar energy which is collected during the day since solar collection cannot usually be started until an hour or two after sunrise. This would ease the start-up energy required of the installed HVAC system, which would in turn ease the load on the municipal utility system at a time of relatively large demand. Excess heat from the solar heating system of a school could be used by neighboring buildings.

In an effort to speed the promotion of solar energy utilization and to show Congress some immediate results, the National Science Foundation awarded four contracts in January 1974, to provide for experimental solar heating systems in a high school, two junior high schools, and an elementary school. This program, called "Solar Energy—School Heating Augmentation Experiments," is aimed "at advancing the systems technology for using solar energy for space heating and hot water needs of buildings, and to provide important information on the degree to which such systems can be made economically justifiable and socially acceptable."

The construction of the projects was completed in March 1974. At all of the schools, the solar energy installations supply heat only. The 5,000 square feet of solar energy collectors at North View Junior High School, Osseo, Minnesota, and the 2,500 square foot Fauquier installation, Warrenton, Virginia, are mounted on their own structures placed on the ground. The 4,500 square foot Grover Cleveland Junior High School array, Dorchester, Massachusetts, and the 5,700 square foot Timonium array, Baltimore, Maryland, are installed on the roofs of the schools. The primary purpose of these projects was to compile data on "the performance of collectors, the reliability of the system, better estimates of operating and maintenance costs, social impacts, acceptability and generalizability to a variety of schools and locations, and to new and existing construction." Contracts covered the cost of the equipment, their installation and operation, and the collecting of data.

A 2500 sq. ft. solar collector at Fauquier High School, Warrenton, Virginia, 1974. Solar design by InterTechnology Corporation. (Photo courtesy of InterTechnology Corporation.)

The rooftop solar collector for the Grover Cleveland Junior High School, Dorchester, Massachusetts, 1974. Solar design by General Electric. (Photo courtesy of General Electric.)

The Energy Research and Development Administration (ERDA) provided funding to a consortium of firms to refit yet another school, George A. Townes, located in Atlanta, Georgia. This project, completed in 1975, is to demonstrate the use of solar energy for cooling as well as heating.

Other institutions besides the Federal Government are building solar heated and cooled academic buildings. The Colorado Department of Public Works gave approval to the A.B.R. Partnership, Architects, of Denver, Colorado, to design the north campus of the Community College of Denver, a 325,000 square foot complex heated by solar energy. Solar heating and energy conservation features add an additional 10% to the cost of the project. In an unprecedented action, the Colorado State Legislature voted to fund the additional expense. The investment should be returned to the state in a period of 10 to 15 years, after which the state would realize operating savings averaging more than $60,000 per year.

Other public buildings besides schools are using solar energy. The General Services Administration of the Federal Government has taken the initiative on two demonstration projects, both of which are office buildings. The purposes of the environmental demonstration project in Saginaw, Michigan, are: to dramatize the firm commitment of the GSA to environmental concerns in the design, construction, and operation of federal buildings; to provide a laboratory for the installation of both recognized and innovative environmental features and equipment; and to inspire others in the building industry toward environmental improvement. The second GSA project is an energy conservation demonstration project in Manchester, New Hampshire. The purposes of this project are the same as those for the Saginaw project, except that the emphasis is on the conservation of energy rather than on broader environmental concerns.

The United States Postal Service (USPS), the National Aeronautics and Space Administration (NASA), and the National Bureau of Standards (NBS) are also promoting the use of solar energy at the federal level. The solar heated post office in Ridley Park, Pennsylvania, serves as the determiner for future USPS projects. If Ridley Park proves successful, USPS expects that most post office buildings constructed in the late seventies will utilize solar heating and cooling.

NASA announced plans in 1974 to build a 53,000 square foot engineering building at the Langley Research Center in Virginia which will be heated and cooled with solar energy. The plans include a 15,000 square foot collector.

The National Bureau of Standards, in Gaithersburg, Maryland, is taking an active role in promoting the use of solar energy. In 1974 and 1975, the Bureau developed a standard test method for rating both solar collectors and thermal storage units. The initial recommendations will be modified after experimental testing at NBS. Another NBS project is a four bedroom house which has been used since 1972 to study the response of a total structure to changing environmental conditions. The house is fully furnished and instrumented. Formerly, it was enclosed in a large environmental chamber where the temperature could be varied from −50° to 150°F. NBS was able to predict the energy demand within 10 percent (usually much closer) for various summer and winter operating conditions. In 1974, the Bureau designed a solar heating and cooling system for the house and moved it outside, where the testing continues.

The New York Botanical Garden is using solar heating and other advanced concepts of environmental engineering in its administration and research build-

Seven black solar panels dominate the roof of the new administration and research center for the Cary Arboretum of the New York Botanical Garden in Millbrook, New York, 1976. Building design by Malcolm B. Wells, architect. Solar and mechanical design by Dubin-Mindell-Bloome, engineers. (Photo courtesy of the New York Botanical Garden.)

ing in Millbrook, New York, for its Cary Arboretum. This building has been designed by Malcolm B. Wells, a Cherry Hill, New Jersey, architect-conservationist, in such a way that it will be, in his words, "gentle to the environment."

Some of the building's features include the recycling of air, water, and wastes, and the covering of the roof with earth. Earth is also banked against the north and west walls, and trees are located on the north sides to shelter the building from cold winter winds. The walls are of concrete, and insulation is installed on the outside of the walls to increase the building's heat storage capacity. Windows are double-glazed and have shutters on the inside which can be closed at night to reduce heat loss. The 8,000 to 10,000 square feet of collectors are in a sawtooth arrangement of seven parallel sloping panels, one behind the other. The roof area between the panels serves to reflect additional heat onto the panel behind it. Dubin-Mindell-Bloome, Associates of New York City, is the mechanical engineering firm.

In addition to other houses and the recent NSF-sponsored school projects, other retrofitting of buildings includes a large-scale industrial application by General Electric at their Valley Forge, Pennsylvania, plant. The 4,900 square feet of solar collectors provides a large percentage of the heating and domestic hot water needs for its 20,000 square foot kitchen cafeteria complex.

Refitting (retrofitting) existing buildings, as was done with the five NSF-funded school projects, is one of the most important applications of solar heating. Most of the millions of buildings in this country will still be in existence and will be using excessive amounts of energy well into the next century. And we will double our stock of buildings in 20 to 30 years. For many existing buildings, it will be easier to reduce fuel consumption by adding solar heat than by adding insulation in the walls and roofs. Solar collectors can be placed on their own structure on the ground or applied to the walls of some buildings; others can be placed on top of an existing sloped or flat roof.

Refitting existing buildings will be discussed in more detail in subsequent pages, including Chapter V·H, "Separated Collectors" and Part III, "Low Impact Technology Solutions." Two of the first solar systems, one by George Löf in Boulder, Colorado (1945), and the other by Raymond Bliss and Mary Donovan in Amado, Arizona (1954), were applied to existing houses.

One of the oldest operating solar houses is a building which was refitted for solar energy and which is located near the campus of the University of Florida in Gainesville. It was built in 1955 by the Mechanical Engineering Department to measure heat flow into and out of a home and was converted to solar heat in 1968. Dr. Erich Farber, one of the world's leading authorities on solar energy, is head of the department's solar energy group, which is involved in a wide range of solar energy endeavors, including solar-powered engines, solar-powered pumps, solar cookers, solar cooling equipment, solar pool heaters and solar water distillers. The sun heats domestic water, heats and cools the house, heats the swimming pool, and actuates a solar, liquid-waste recycling system (through water distillation). It also partially powers an electric conversion system for TV, lights, radio, small appliances, and a solar/electric car.

According to Richard G. Stein, a New York City architect, one and one-half billion housing units will be built throughout the world during the next two decades; an average of 1.5 million houses will be built each year in the United States until the early 1980's. Of course, among solar scientists, new homes are the favorite type of building to which to apply solar energy. By 1976, hundreds had been built and several thousand more were being planned or were under construction.

Probably the most technologically complex solar home in existence is at the University of Delaware. According to Dr. Karl Böer, the first director of the University's Institute of Energy Conversion, "this is the first house in which there has been a substantial effort to collect both electric and thermal energy." The thermal energy is used for heating and cooling the building and for heating domestic water. Electrical energy is converted through the use of cadmium sulfide solar cells. Electric energy from the solar cells is stored in lead-acid auto batteries and used for lights and other purposes where DC energy can be used. Eventually, equipment to change the DC to 115 volt AC could be added to this house. These photovoltaic cells are part of the same panels through which air is moved and heated.

The heat storage system was developed by Dr. Maria Telkes, who has been in the field of solar energy since the early MIT houses back in the 1940's. It is made up of plastic containers containing eutectic salts. These chemicals absorb large amounts of heat as they melt; when they return to the solid state, they give up their stored heat.

A solar powered and heated house such as this is limited in its practicality because of the present high cost of the solar cells. However, these cadmium sulfide cells, or similar photovoltaic cells, will eventually be cheap enough for widespread use.

Colorado State University is testing and evaluating three residential structures. The first one, built in 1974, uses a liquid solar system for heating, cooling, and domestic hot water. It is said to be the first residential solar system designed, constructed, and evaluated using state-of-the-art components and using lithium bromide absorption air conditioning equipment. Another of the units is also a liquid system, but of a different design. The third house uses an air system. The performance of all three is being tested and compared.

A 1,300 square foot house was built in 1975, in Guilford, Connecticut, by Professor Everett Barber, of Yale University, for his own family. The energy-conserving house uses a flat plate solar collector and system designed and built by Barber's firm, Sunworks, Inc. Barber estimates that solar energy will provide approximately 60 percent of his houses's annual space heating and domestic water heating requirements. Two windmills will be used to augment local utility company power by providing about 80 percent of the electric power requirements.

A house for which Professor Barber was the engineer was completed in August 1974, on Long Island Sound in Connecticut. Donald Watson, AIA, of Guilford, Connecticut, was the architect. This three bedroom, year-round residence obtains more than half of its space heating and close to all of its domestic water heating energy requirements from Barber's modular flat plate collector system. A zoning height limitation of 20 feet required that the roof be kept low. Therefore, the three south-facing collectors are arrayed in a sawtooth fashion,

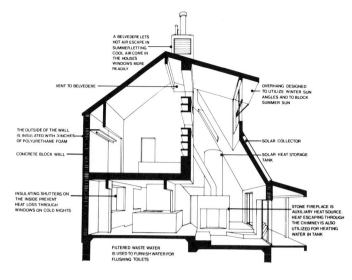

A BELVEDERE LETS HOT AIR ESCAPE IN SUMMER, LETTING COOL AIR COME IN THE HOUSE'S WINDOWS MORE READILY

VENT TO BELVEDERE

OVERHANG DESIGNED TO UTILIZE WINTER SUN ANGLES AND TO BLOCK SUMMER SUN

THE OUTSIDE OF THE WALL IS INSULATED WITH 3 INCHES OF POLYURETHANE FOAM

CONCRETE BLOCK WALL

SOLAR COLLECTOR

SOLAR HEAT STORAGE TANK

INSULATING SHUTTERS ON THE INSIDE PREVENT HEAT LOSS THROUGH WINDOWS ON COLD NIGHTS

STONE FIREPLACE IS AUXILIARY HEAT SOURCE. HEAT ESCAPING THROUGH THE CHIMNEY IS ALSO UTILIZED FOR HEATING WATER IN TANK

FILTERED WASTE WATER IS USED TO FURNISH WATER FOR FLUSHING TOILETS

Interior view of a solar house for Everett Barber in Guilford, Connecticut, 1975. Designed by Charles W. Moore Associates, architects. Solar engineering by Everett M. Barber, Jr. Solar equipment by Sunworks, Inc. (Drawing courtesy of Charles W. Moore Associates.)

which also provides clerestory lighting and excellent natural ventilation for the interior. Each module is two feet wide and about six feet high. In addition to the solar heating system, energy conserving features include the placement and sizing of windows for natural daylight and ventilation. The overhangs above the large window areas reduce summer heat gain but allow the sun's heat to penetrate the house during the winter. The house is very well-insulated.

In an effort to simplify solar systems, Total Environmental Action, Inc., building designers in Harrisville, New Hampshire, designed a solar house near Manchester, New Hampshire. The house combines sound architectural design with a solar system which is integrated with numerous windows into the south wall of the house. The only mechanical equipment for the solar heating system is the movable insulation which allows the collector to be exposed to the sun during the day but to be protected from losing heat at night.

The collector, which stores heat as well as absorbs it, is a 12-inch thick concrete wall. The concept was used widely by Felix Trombe and Jacques Michel in Odeillo, France (see Part III, "Low Impact Technology Solutions"). A transparent wall panel is located between the collector/storage concrete wall and the outside. When the sun is shining, the panel is empty, exposing the black concrete. When the sun is not shining, tiny polystyrene beads fill the wall, forming an insulating barrier between the warm concrete and the cold outdoors. Heat is transferred to the living space in the house by conduction and natural air circulation. This Beadwall® concept was developed by Zomeworks Corporation.

An example of a move towards complete independence from outside energy sources is the homesteading project of Robert and Eileen Reines located about an hour from Albuquerque, New Mexico. The Reineses have built a solar heated dome which they claim to be the world's first 100 percent solar heated and wind powered residence. They have designed the home and their life style so that very little energy is required. What energy they do use comes from the sun as heat and from the wind as electricity.

The solar heating system is built separately from the dome structure. Radiators are used for heating the dome, using water from the 3,000 gallon heat storage tank. The dome can be kept warm for ten days of cloudy, sub-freezing weather; Reines has been able to maintain temperatures inside the dome at between 65°F and 85°F. The wind-driven generators charge sixteen, high capacity storage batteries, heat domestic water electrically, and drive solid state power equipment. The stored power is the exclusive source of electrical energy for the house. Bulbs totaling less than 150 watts provide sufficient light for reading comfortably anywhere in the house.

Bob Reines estimates that a conventional structure of the same size would require about ten times the heat to maintain the same temperature and would use about five times as much electricity. He and his colleagues continue to work on developing energy-efficient refrigerators, lights, and cooking equipment, water use and waste disposal devices, and enclosed food production systems which could provide the homeowner with year-round fruits and vegetables.

In addition to single, solar powered structures, solar communities are also being designed to take advantage of possible cost reductions and higher system efficiencies. Studies by Sandia Laboratories of Albuquerque, New Mexico, show that a solar community in Albuquerque could be supplied with 60 percent of its energy requirements by the sun. Solar energy would be centrally collected and stored, then distributed to individual homes and businesses in the form of electricity, heating, air conditioning, or hot water. The projected size of the community is somewhere between 100 and 1000 units, including houses, apartments, and small businesses. The concept could also be applied to existing communities of comparable size. A pilot project to test the new concept is planned.

Grassy Brook Village, a 10-unit solar condominium project in Brookline, Vermont, 1976. Design by People/Space Co., architects. Solar and mechanical design by Dubin-Mindell-Bloome, engineers. Construction by Richard Blazej. (Drawing courtesy of Grassy Brook Village, Inc.)

One of the most exciting and farsighted projects for using solar energy is Grassy Brook Village in Brookline, Vermont. The first phase of the project includes 10 housing units; the second phase will include an additional 10. According to Richard Blazej, the developer, the units are "designed to embody principles of ecological integrity in their construction and operation ... (to) create a minimum of impact on the natural environment and provide a pattern of utilities and life-support systems that derive their energy from natural and non-polluting sources, and over which the owner/resident can be assured a high degree of financial and quality control." The development features the use of solar energy for house heating and hot water, the generation of electricity from wind power (in the second phase), the on-site handling of wastes in a system designed for experimental operation in several recycling and pollution-free modes, and any applied principles of energy conservation.

The three-bedroom houses, about 1200 square feet each, are being clustered in two groups of 10. Each cluster is integrated by its own central solar heating system and waste handling system as well as by common amenities such as walkways, decks, a central green or common, a central laundry facility, workshops, storerooms, carports, and a pond for recreation and fire protection purposes. Robert Shannon, a partner of People/Space Company of Boston, Massachusetts, is the architect.

The Grassy Brook Village solar heating system is unique in that a single collector and storage system will serve the entire cluster of 10 houses. The collector is on three separate support structures, each about 16 feet high and 100 feet long, and is separated from the houses to allow freedom of architectural design. The three structures are arranged in a sawtooth configuration in order to reduce the effect of wind loads and to permit easier servicing. The water heat storage system is located under some of the housing units. A heat pump, used in combination with the solar system, is supplemented by an oil-fired boiler during extended periods of cloudy or extremely cold weather. Each house also has a woodburning stove.

These examples of some recent solar powered projects are representative of the exciting work taking place in the field of solar heating and cooling. Although there are several thousand other projects planned and/or built, the pace of their acceptance by and incorporation into the construction industry seems excruciatingly slow, especially when compared with the millions of buildings constructed each year. The number of solar projects is insignificant in relation to what should be happening in this country in terms of what our national needs are and of what our national priorities should be.

CHAPTER I·C

Solar Energy: Obstacles And Outlook

Dr. Erich Farber at the University of Florida cites three primary reasons why the use of solar energy has not been more widely accepted. The first is ignorance on the part of the American public about what solar energy is able to do. The second is that the equipment for using solar energy has not been available, or, if people want it, they have not known how to find it. The third is that most homes using solar energy have been custom built and have a relatively high cost. Most buildings today are not built by those who will live in them, but by developers who construct the buildings for the lowest possible initial cost in order to sell them at a profit.

There are many other similar barriers to the utilization of solar energy. Clients want assurances that their new buildings will function properly and that the solar components will not fail. One of the purposes of prototype projects is to provide this reassurance. Also, people generally find it more difficult to accept solar energy merely as a means of reducing fuel consumption than as a complete alternative to the conventional heating system, but providing 100 percent of the heating with solar energy usually results in far too large and costly a system to be practical (50 to 75% is a more realistic goal). Systems providing 100% of the heat have to be large enough to store heat for cloudy spells of five, six, even ten days and for even longer periods occasionally.

A solar system usually has a high initial cost compared to that of a conventional system, and it is extremely difficult to finance 5 to 40 years of heating bills at one time, especially with traditional economics. The greatly increased cost of fuel, however, is changing the economics. Dr. Farber estimates that his system in a totally solar heated and cooled house would cost perhaps $5,000 more than a conventional one. Harold Hay, on the other hand, with his flat waterbed-type collector (see Part III, "Low Impact Technology Solutions") estimates that the new "equipment" on his house would cost no more than the furnace and air conditioning system it replaces.

Financing an increased initial cost is particularly difficult when building costs are high, when interest rates are high, and when mortgage money is difficult to obtain. Unquestionably, one of the most significant reasons why people who have decided to use solar energy change their minds is not that they fear the system might not pay for itself but that they just cannot obtain the money required for the higher initial cost. Public utilities and energy companies should prepare themselves to provide the consumer with solar components on a lease basis as an alternative to building more power plants or to discovering and providing more oil, gas, or coal. Some utility companies have established a precedent for this by leasing conventional hot water heaters to homeowners as an inducement for using their product (gas, oil, or electricity). Solar components would conserve energy and thus reduce the burden on the utilities of meeting rising energy demands through construction of more and expensive power plants.

Unfortunately, designs using alternate forms of energy appear to be hampered by their seeming inability to reduce peak demand on the utilities. Most solar energy systems for the space conditioning of buildings require a full-sized back-up system for long periods of cloudy weather. If gas or electricity is a source of energy for the back-up system, not only does the building owner have to provide both a solar energy system and a back-up system, but the utility company has to build and maintain full-sized facilities to provide for demand by the back-up system during peak load conditions. Consider, for example, the diurnal load curve for residential consumption. This curve shows an evening peak which would likely be dampened in magnitude and/or duration by solar energy heating/cooling, depending on sunshine conditions and thermal energy storage capacity. This effect could save the utility considerable long-run costs associated with peaking capacity. These savings, incorporated into the rate structure, may provide the incentives for the design and purchase of peak-reducing solar energy facilities (FEL.).

For the above reasons concerning the nature of solar energy, the utilities will be primary motivators or inhibitors of adoption of this alternate energy source. The utilities, together with the Federal Energy Administration, are already experimenting with direct and indirect methods of load management, especially in regard to redesigning rate structures to reflect the difference between peak and off-peak costs of electricity supply. Solar building design should be integrated into such programs.

Many analyses have been done comparing the cost of solar energy with that of fossil fuels. Dr. George Löf, Colorado State University, and Dr. Richard Tybout, Ohio State University, have carried out some of the most extensive calculations on the costs of solar heating and cooling of buildings compared with those of oil, gas, or electricity. The results are optimistic (see Figure I·C·1). In the seven U.S. cities studied, solar heating was cheaper than electricity and, in some instances, cheaper than gas.

Financial institutions must re-evaluate their lending procedures to take into account *life-cycle costing*, which looks at the cost versus the benefits and compares higher initial costs to lower owning, operating, and maintenance costs. Although

Location	Optimized solar heating cost in 25,000 BTU/degree-day house, capital charges @ 6%, 20 years		Electric heating, usage 30,000 kwh/year	Fuel heating, fuel cost only	
	Collector @ $2/ft.^{2}	Collector @ $4/ft.^{2}		Gas	Oil
Santa Maria	1.10	1.59	4.28[1]	1.52	1.91
Albuquerque	1.60	2.32	4.63	0.95	2.44
Phoenix	2.05	3.09	5.07	0.85	1.89
Omaha	2.45	2.98	3.25[3]	1.12	1.56
Boston	2.50	3.02	5.25	1.85	2.08
Charleston	2.55	3.56	4.22	1.03	1.83
Seattle-Tacoma	2.60	3.82	2.29[2,3]	1.96	2.36
Miami	4.05	4.64	4.87	3.01	2.04

Notes: [1] Electric power costs are for Santa Barbara. Electric power data for Santa Maria were not available.
    [2] Electric power costs are for Seattle.
    [3] Publicly owned utility.

Solar heat costs are from optimal design systems yielding least cost heat.

Electric power heat costs are from U.S. Federal Power Commission, All Electric Homes, Table 2 (1970). Conventional heat fuel costs are derived from prices per million BTU reported in P. Balestra, The Demand for Natural Gas in the United States, Tables 1.2 and 1.3 (North Holland Publishing Co., 1967). The 1962 costs were updated to 1970 by use of national price indexes on gas (121.1 in 1970 versus 112.8 in 1962) and on fuel oil (119.22 in 1970 versus 101.2 in 1962) as adjustment factors on each fuel price in each state. Bureau of Labor Statistics fuel prices indexes obtained from Gas Facts. Fuel prices were converted to fuel costs by dividing by the following national average heat (combustion) efficiencies: gas, 75%; oil, 75%. Heat efficiencies are from American Society of Heating, Refrigerating and Air Conditioning Engineers, Guide and Data Book 692-694 (1963 ed.).

All solar heat costs based on amortizing entire solar system capital costs in 20 years at 6 percent interest. Capital investment based on current prices of solar water heaters at $4 per sq. ft. plus current costs of other components, and on anticipated near-term solar collector price of $2 per sq. ft.

Figure I·C·1. Costs of space heating (1970 prices) in dollars per million Btu useful delivery. Source: (TYB.)

the use of solar energy usually costs more initially, it results in lower energy costs. Lenders should allow higher annual mortgage payments if annual energy costs are lower, amortizing the initial cost over the life of the solar energy system. With rising fuel costs and with the widespread use of (expensive) electricity for heating, solar energy is economically competitive on a life-cycle cost basis in most parts of the country.

Such economic analyses assume that past and present trends will continue, but economic analysis in the traditional sense is losing its force as the actual availability of fuel becomes the real issue. Some people have decided that the shortage of fuel is reason enough for using solar energy and that the higher cost of the solar energy system is less important.

The attitude throughout this country regarding alternative sources of energy is changing dramatically. In addition to the concern over the unavailability of fossil fuel and electricity, a subtle but powerful move toward self-sufficiency is abroad in the land. Just as individuals are moving to free themselves from the grip of the energy giants, so the nation, with Project Independence, is striving to free itself from a similar but larger-scale energy blackmail.

Traditionally, developers have been interested in trying to keep costs as low as possible, a goal frequently incompatible with some solar energy systems. However, developers are also often having great difficulty in obtaining gas and oil for new homes. If they cannot find an alternative source of energy, they are going to be unable to develop new projects.

Some people may have money now for higher initial investment in a house but would like to be reassured that their fixed retirement income will not be eaten up by increased fuel costs. Solar energy provides a solution to this problem: although the initial cost is higher, the low yearly cost is relatively stable and will not greatly affect an elderly person's fixed income as fuel prices rise.

Since property taxes are based on value, however, higher initial cost results in higher property taxes. For buildings which use solar energy, taxes should be reduced so as to encourage its use. Indiana took the lead in this by offering a real estate tax incentive on solar energy components for both new and existing buildings. The real estate tax is based on the property value, which is determined by subtracting the value of the solar heating system from the total valuation, or by subtracting $2,000, whichever total gives the greater taxable value. Dozens of other states, such as Arizona, California, and New York, have followed this lead.

Other possible incentives include the availability of low interest loans, probably government subsidized, to building owners and to manufacturers wishing to develop products. HUD is developing interim solar energy design criteria for FHA financed homes and may insure home improvement loans of up to $5,000, or 10% of the value of the home, toward the installation of solar energy systems in existing homes. Tax write-offs based on a percentage of the installation cost or energy savings of the system would be a direct subsidy, a form similar to oil depletion allowances.

Although the use of solar energy can add to the initial cost of a building, the extra work stimulated by its development can help a local economy. For example, in New England most of the heating oil and gas is brought in from overseas or from other sections of the country; the annual outward cash flow for this energy is in the billions of dollars every year, and increasing annually. The manufacture of solar panels, however, generally uses American-made components such as glass, and they are locally assembled, diverting money from foreign markets to the local economy. The installation also requires local labor, additionally boosting the local economy.

A large number of solar collectors are available from manufacturers, but on-site construction of collectors remains a viable alternative, depending on the labor costs involved. If a water heating system is used and if plumbing costs are high, on-site construction may be impractical. Air heating systems, as well as those which are integrated into the fabric of the building, will more often be built on the site.

There are at least 100 manufacturers who are seriously involved in producing equipment and several hundred more who are at least toying with the possibility, hesitantly putting money into their programs and waiting for a market to develop. The Solar Energy Industries Association (SEIA), comprised primarily of manufacturers, was organized to "stimulate prompt, orderly, widespread, and open growth of economic utilization of solar energy." The Association was formed in 1973 in conjunction with the Washington, D.C.-based Solar Energy Research and Information Center, a "specialty service organization devoted exclusively to assisting persons, companies, governments, associations, and other organizations" in promoting the use of solar energy.

One of the main difficulties for manufacturers is marketing economical solar collectors and storage systems that will perform well for many years. In most instances, the manufacturer has to design a collector which will not conflict with one already patented. The necessary research and testing is expensive and arduous. It requires a sophisticated understanding of how solar energy works and of the pitfalls that have been encountered in the past. The manufacturer also has to find (and to help develop) a market which is scrambled in tiny pieces throughout the country. He is reluctant to tool up an assembly line to achieve economies from mass production before the market has developed, but the market cannot easily develop until good collectors are available for sale at reasonable prices.

A barrier to developing the market for solar collectors is the shortage of contractors and architectural engineering designers who are equipped to handle the present complexity of solar energy design. Unfortunately, the powers that are allocating money to promote solar energy (e.g., the National Science Foundation and ERDA) are expending little of their budgets on the training of contractors and designers. Many solar experts and enthusiasts, however, are affiliated with universities which support the proliferation of public and private seminars, workshops, and education courses.

Present designers of solar systems and buildings delay the selection of components until the last possible moment as new technology develops and as mass production brings costs down. The extra time and money required by architects and engineers to design systems using solar energy deters many whose offices may already be struggling financially.

In addition to the obstacles mentioned so far, there are several other potential problem areas, many centering on the construction industry which has a record of excruciatingly slow adaptation to change, particularly if the change means higher construction costs. There are about 30,000 building code jurisdictions, often with mutually incompatible requirements. Also, the building industry is highly fragmented. There are thousands of builders, and ninety percent of all construction is done by companies which produce fewer than 100 units per year. Even the giants in the field each produce less than one half of one percent of all housing units. Not only is profit margin small, but innovation in an initial cost oriented industry is a risk which few builders will take.

Labor is the greatest single factor in the overall cost of a solar energy system. Since the installation of sophisticated solar energy systems is likely to involve plumbers, who are among the most highly paid tradesmen, these systems should be designed to minimize on-site plumbing labor. This financial difficulty can be ameliorated by using non-liquid systems.

The solar energy designer must be familiar with building codes which may restrict the use of solar energy. Health codes may apply when using ethylene glycol (mixed with water to prevent freezing) as a heat transfer medium. If the domestic hot water supply comes in contact with the heat exchangers containing the glycol, precautions must be taken to insure that pinhole leaks in the exchangers do not contaminate and poison drinking water.

Solar easements may soon be incorporated into deeds when land is plotted.

Architectural covenants allowing for unobstructed solar energy *vision* may also be included. Street layouts and vegetation locations will influence design and planning. A building height limitation imposed by a building code on a home in Connecticut (discussed earlier) resulted in an unfamiliar but pleasant building design.

A user of natural forms of energy must be assured that the energy will not be removed from him by the actions of other people. When the energy source is the sun, this will most commonly be done through the shading effect of one building on another. It is unlikely, in situations where buildings are now shaded partially during the day by other buildings, that such a building will be able to regain full *sun rights* through judicial action. As more buildings begin to rely on solar radiation which hits them as a means of obtaining energy for their operation, however, judicial steps will have to be taken to guarantee that other people's construction or vegetation will not reduce the total solar energy which strikes the building. California is leading the country in attempting to provide sun rights through legislative action. Until such legislation is passed, however, the use of solar energy may be subjected to rude shadows from new neighboring buildings or vegetation.

The potential for vandalism of the transparent cover plates of solar collectors (which are sometimes made of glass) may be one of the most significant disruptive factors in the use of solar energy in both urban and non-urban areas. However, in the 20 or 30 year history of the use of solar energy, in the approximately 25 solar energy projects which had been completed prior to 1965, vandalism has not been a problem. The situation is similar to that for all-glass buildings, which have experienced relatively minor difficulty.

The solution to the potential problem of vandalism is partially an architectural one. The use of vertical wall collectors using glass as a cover plate is not as likely a solution in urban areas as is the use of collectors which are part of or on top of the roof. Collectors which are on top of the roof and which are arranged in a sawtooth configuration using setbacks from the edge of the roof can reduce the threat of vandalism considerably. Portions of buildings which are in less direct contact with people than are other parts of buildings, e.g., the walls of the top floors, should be used for collectors.

In addition, cover plates which are more resistant to impact and breakage than ordinary glass already exist, and more are being developed. Some of these alternatives include tempered glass; Lexan, which can be vandalized in other ways, however, and which can deform at high collector temperatures; plexiglass, which scratches and which is also sensitive to high collector temperatures; rigid fiberglass-reinforced polyester sheets, which are also sensitive to high temperatures; and plastic vinyls, most of which are sensitive to ultraviolet radiation and which therefore deteriorate with time. (These alternatives are covered more completely in the last two Parts.) The cost of some of these cover plates is as much as one dollar more per square foot of collector.

Another design difficulty is the reflection of sun off large expanses of glass-covered collectors and into the eyes of pedestrians, drivers, and people in other buildings. Glass buildings have caused similar reflection problems. Vertical wall collectors pose the same problem as do walls of such buildings. However, except in rare cases, the expanse of solar collector will not nearly approach the extremely large expanses of some large glass buildings. The most severe reflection off vertical walls occurs when the sun is low in the sky at sunrise or sunset, at which time the intensity is less because of the greater length of atmosphere through which the sun has to travel in order to reach the reflective surface. The glare is not as bad as that experienced when driving directly into a setting or rising sun.

Tilted collectors pose slightly different problems. Reflections off tilted collectors will occur when the sun is relatively high in the sky, at which time the intensity of the sun is at its peak; the discomfort is similar to that of facing into the sun. This should not be a serious hazard, since someone who is moving with respect to the collector would probably pass through the reflection area fairly quickly. However, the hazard does exist and can be dangerous in some instances. Discomfort can also occur when the sun is reflected off the collector and onto nearby or adjacent buildings, directly into areas which had not formerly experienced direct sunlight. It is unlikely that such a reflection would remain very long in the same spot because of the relatively fast movement of the sun across the sky. Also, tilted collectors, because of the architectural constraints of roofs, usually do not have large surface areas. Where large expanses of glass for collectors are required in buildings, a designer can choose to use several smaller collectors, possibly in a sawtooth type of configuration.

Non-reflective cover plates for collectors exist and are being further developed. However, their use can increase the cost of the collector as much as 50 cents per square foot and affect its seasonal efficiency not at all or as much as 3% (e.g., from 40% up to 43%).

As if the socio-economic issues impeding the use of solar energy were not severe enough, government and political indecision, derision, and turmoil may thwart its potential implementation. In fact, when President Nixon outlined his concept of Project Independence, he did not mention solar energy. Although this omission may not seem incredible, it is extremely disconcerting, and, in fact, bodes ill for our country.

In 1974, the National Science Foundation, the National Aeronautics and Space Administration, the Department of Housing and Urban Development, and the Atomic Energy Commission were all scrambling for position to obtain authority and funding in regard to solar energy. Until then, NSF had dominated government participation in the promotion of its use. Through a five-phase project sequence, NSF planned to develop advanced equipment and devices from conceptual design, through preliminary and detailed design, to system construction and evaluation in the shortest possible time in each of the following areas: heating and cooling of buildings; wind energy conversion; bio-conversion to fuels; solar thermal conversions; photovoltaic conversion; and ocean thermal conversion. By this Proof-of-Concept sequence, it was anticipated that the equipment in the first three areas listed would be "ready for commercial application before 1980," and indeed, they already were ready at that time (1974). Devices in the remaining areas are expected to require more time but to become commercial in the following decade.

In the Proof-of-Concept program for solar heating and cooling of buildings, more than one-half million dollars was awarded to each of three teams: General Electric with the University of Pennsylvania; Westinghouse with Colorado State University and Carnegie-Mellon; and TRW Corporation with Arizona State University. Phase Zero was a "multidisciplinary" examination of the potential of solar heating and cooling in buildings of all kinds; Phases I and II involve solar energy "hardware" development; and Phases III and IV include the design, construction, and testing of systems and buildings.

The focus for Phase Zero was: "1) to identify the regions of the nation where population and industrial densities cause the heaviest consumption of energy for space heating and cooling; 2) to select those building types such as single family residences, multi-family buildings, commercial buildings, and so forth that consume the most oil, gas and electricity; and 3) to specify the functional performance and operational requirements for the required solar system."

Unfortunately, the Proof-of-Concept program was more a cautious delaying tactic than an all-out effort to promote solar energy in as short a time as possible. The companies to which large sums were awarded are involved in the construction industry primarily because of their size, not because of their influence. The difficulty of stimulating the construction industry to accept change is well-documented. Seventeen trade unions, 300,000 contractors (including specialty contractors), and approximately 160,000 real estate people must be placated and appeased in order for real change to occur. Yet NSF opted to give much of its money to three large companies to evaluate the market, instead of distributing it to many small organizations or using it to finance construction, to finance tax incentives for potential manufacturers or home builders, or to promote solar energy in an educational and expansive way. Instead of trying to implement Model T's on a large scale, NSF delayed in favor of slowly developing the Cadillac.

In order to adopt and expand solar energy use, several factors must exist: evidence that a change is technologically and economically practical; a manufacturing industry to produce components; a service industry to maintain the components; designers to incorporate them into the design of buildings; and contractors and clients and users willing to adopt this new change. Except for the development of solar cooling equipment and of components which are economically competitive with fossil fuel systems, technology exists for using solar energy.

The Energy Research and Development Administration (ERDA), established in 1974, assumed charge of most NSF functions and personnel relating to the development of solar heating and cooling, as well as the small amount of solar energy research being done by the AEC. Energy programs under the Environmental Protection Agency (EPA) and the Department of Interior also transferred to ERDA. In addition, the Administration is responsible for non-nuclear research and development programs passed later by Congress. In order to promote uniform distribution of attention and money among the many forms of energy sources, the bill establishing ERDA reads that "the Congress intends that no energy technology be given an unwarranted priority." A Senate amendment requires that the Administrator and Deputy Administrator of ERDA be "especially qualified to manage a full range of energy R & D programs, but not limited to a specific energy technology."

In other legislation, the Solar Heating and Cooling Demonstration Act of 1974 was designed to "provide for the early commercial demonstration of the technology of solar heating by the National Aeronautics and Space Administration and the Department of Housing and Urban Development, in cooperation with the National Bureau of Standards, the National Science Foundation, the General Services Administration, and other Federal agencies, and for the early development and commercial demonstration of technology for combining solar heating and cooling."

The Solar Heating and Cooling Demonstration Act of 1974 authorized (or implied) the sponsorship of 4,000 demonstration solar homes. Half of the homes are to be built on military and other federal land and the other half on private land. Those in the private sector include homes, apartments, and townhouses. The bill authorized the National Aeronautics and Space Administration, the National Science Foundation, the National Bureau of Standards, the Department of Housing and Urban Development, and the Department of Defense to interact in bringing the substance of the Act to fruition. The initial work is on the actual development and testing of heating units, estimated to require three years, with cooling units available in five years.

The importance of government incentive and development programs cannot be underestimated. All areas of industry, including the energy industry, are being subsidized and supported by government programs. Solar energy must overcome many difficulties, apart from its competition with established and government-supported energy supplies such as gas, coal, oil, and nuclear. These difficulties and the global importance of solar energy as an alternative to present commercial forms of energy are more than adequate justifications for a concerted action on the part of the American people to demand equal and better-than-equal treatment for the promotion of solar energy use.

The effort in Congress, however, hardly looks like a set of emergency measures, despite the "energy crisis." Timid lawmakers must change their priorities and produce efforts similar to the cold war fallout shelter program of the early sixties. Massive financial (and other) support must be given by the government to almost every possible avenue of promoting energy conservation and the use of renewable, non-polluting energy sources such as the sun (including heating and cooling, wind, photosynthesis, ocean thermal gradients, and solar electricity). Every citizen must also participate in the effort and not rely on government to do all the work.

REFERENCES FOR PART I

(FEL.) Feldman, Stephen L. and Anderson, Bruce. "Financial Incentives for the Adoption of Solar Energy Design: Peak-Load Pricing of Back-up Systems." *Solar Energy* 17 (1975):339-343.

(MOR.) Morrow, Walter E., Jr. "Solar Energy: Its Time is Near." *Technology Review,* December 1973, pp. 31-43.

(TYB.) Tybout, Richard A. and Löf, G. O. G. "Solar House Heating." *Natural Resources Journal* 10 (April 1970):268-326.

PART II

Let The Sun Shine In: Designing With The Sun

The possibility of harnessing solar energy can side-track us into using it in ways which overshadow and neglect some of the reasons which led us to its use. Energy conservation, for example, is less exciting than inventing ingenious means of capturing the sun's heat. There is nothing new and exciting about letting sunlight in to heat a building during the winter and figuring out a way to keep it out during the summer; we'd rather design an elaborate, highly technical machine to do this simple task.

Nature, however, creates simple but sophisticated designs to compensate for the undependability and irregularity of solar radiation and temperature. Witness the flowers which open and close with the rising and setting sun and the animals which find shelter to shield themselves from intense heat, as in the desert, but bury themselves in the warmth of the earth during the winter.

Over the course of history, man has taken his clue from nature in the design of shelters and clothing. But as we learned to overcome many of the difficulties of survival and to make ourselves seemingly impervious to the idiosyncracies of nature, we also created a gap which separates us from a basic understanding and appreciation of natural phenomena. More and more we rely on technology and on our own abilities to solve problems, not only without nature's assistance, but often in direct conflict with nature's laws.

Part I included a brief introduction to some of the concepts of solar energy. One of the most important of these involves those properties of glass which allow short-wave radiation to penetrate but prevent long-wave heat radiation from escaping. This principle, or "greenhouse effect," also operates in our atmosphere. In a sense, the earth is a giant solar collector. The sun's rays (over 400 Btu/hr/ft²) hit the atmosphere in their trip from the sun to the earth. Some of the heat is reflected by the atmosphere and some is stopped by the atmospheric particles, which are heated up in the process. The rest of it is transmitted to the earth itself, at which point it is reflected back into the atmosphere or absorbed, to be reradiated to the atmosphere in the form of heat. This reradiation process may be instantaneous, e.g., when it hits metal; or it may be delayed millions of years, e.g., when fossil fuels are formed. Such stored solar energy is released into the atmosphere in the form of heat when we burn these fuels.

The atmosphere reradiates some of its heat back into space. Were it not for the enormous quantities of heat which we are releasing into the atmosphere through nuclear fission and fusion and the burning of fossil fuels, the earth would maintain a temperature equilibrium with space. That is, without this release of energy by man, the earth would reradiate heat to space at approximately the same rate at which it receives it from the sun. If this were not the case, the average temperature of the earth would continue to change, becoming either higher or lower, until this balance is attained.

Because of this principle, many question the desirability of capturing the sun's energy for use at a later time to generate electric power or to heat buildings. Fortunately, however, the effect of this practice on the earth's temperature would be negligible. Within a very short period of time, usually two or three days at the most, the solar energy being used (e.g., to heat a building) is converted to heat which is eventually released to the atmosphere (e.g., as heat loss through the walls of a building).

The earth experiences wide fluctuations in temperature at particular locations, but its large heat storage capacity (and the atmospheric envelope) prevents it from cooling off too much at night and from getting too hot during the day. Because of its large heat storage capacity, the earth takes a long time to cool off after the sun goes down and a long time to warm up after the sun rises. This accounts for afternoon temperatures being higher than morning temperatures, in spite of what are usually similar amounts of solar radiation at both times. The same principle accounts for the time lag between the earth's and the sun's seasons. Midsummer for the sun is the summer solstice, around June 21 in northern latitudes, but the warmest weather usually occurs in July and August.

Building design should be based on similar principles. It should not "notice" extreme weather variations from one hour to the next nor from cold night hours to warm daytime hours. If possible, it should not even "notice" the wider extremes of summer and winter.

There are countless examples of indigenous architecture based on these criteria. The example with which we are most familiar is the heavy adobe home of the Pueblo Indians in the southwestern United States. During the day, the thick walls of hardened clay store the heat of the sun, preventing it from reaching the interior of the home. At night, the stored heat warms the interior of the space while the temperatures of the desert night plummet. The coolness of the night air is then stored in the walls, keeping the home cool during the day. Buildings which are made of heavy material such as stone and concrete will perform in a similar way in many climates.

In addition to eliminating the effects of daily extremes, a design should allow a building to modulate itself in such a way that it does not notice the extremes of the seasons. Caves, for example, have fairly constant temperatures and humidities all year round. Buildings which are covered with earth or have earth piled against the outside walls and those which are molded into the side of a hill will be less influenced by seasonal temperature variations.

A building should be designed to respond to the outdoors in another way. On sunny winter days, the building should be able to open up, in a sense, to let the sun shine in, and then to button itself up tightly, like a cocoon, to keep that heat from escaping. During summer days, it should close itself up to keep out the heat, but at night, it should open itself up to receive the cool night air (see Figure II·1).

Figure II·1.
Diagrams of how buildings should perform as solar heat collectors, storage devices, and traps.

SUNNY WINTER DAY
Building opens up to absorb heat.

WINTER NIGHT
Building closes itself up, like a cocoon.

If a building is designed properly, it will function as a solar collector, collecting heat when the sun is shining and storing it for later use. It ceases to operate when there is enough heat in storage and when the sun is not shining. During the summer, for instance, the collector will not operate if the building does not need heat. But it might be able to operate in reverse, circulating the heat storage medium through the building as if it were opening itself up to the night.

The importance of designing a building to complement and to interact with the climate and the powerful influence which this can have on the success of a building as a life-support system cannot be over-emphasized. It is extremely important that a discussion of the utilization of solar energy begin with an appreciation of this premise. This forces us to deal first with the fundamental methods of interfacing with the sun's energy directly, without the use of complicated and highly technical machinery and technology. Among solar energy enthusiasts and scientists, these methods (and materials) are described as *low impact technology* and *passive*. This means that there are few moving parts and that controls or machines of high technological origin (such as those used in the space program, or those that must be mass produced) are probably not used. If they are used, they complement other components in a sophisticated but simple system. Designers of these systems recognize that the human element of control is often more reliable than control by machines; if not, it is at least more flexible in its ability to provide for human need. In addition, these designers carefully weigh the energy and resource consumption in the manufacture and use of these systems and materials against their environmental benefits.

One of the clearest illustrations of human versus machine control is the use of operable or inoperable windows in buildings. A larger and larger percentage of new buildings use windows which cannot be opened by the people inside; such windows are termed "inoperable." The theory is that the indoor climate of the building can be controlled more easily by machines than by the people living or working there. The machines are designed to provide conditions which the average person would find most comfortable. But there are few "average" people: many are uncomfortable under "average" conditions, and even larger numbers are uncomfortable when these machines are poorly designed or fail. When air conditioning systems break down (usually during the hottest weather) in buildings which have inoperable windows, the people inside are unable to let in cool outdoor air until the machinery is repaired.

The best way of using sun for heating, then, is to design and use a building as a natural solar collector, trying to avoid a reliance on high technology. A building must satisfy three basic requirements to achieve this:

1. *The building must be a solar collector.*

   It must let the sun in when it needs heat, and it must keep it out when it doesn't. It must also let coolness in when it needs it. This is done primarily by orienting and designing the building to let the sun penetrate through the walls and windows during the winter and by keeping it out during the summer with shading devices such as trees, awnings, venetian blinds, and a myriad of other methods.

2. *The building must be a solar storehouse.*

   It must store the heat for cool (and cold) times when the sun is not shining, and the cool for warm (and hot) periods when the sun is shining. Buildings which are built of heavy materials such as stone and concrete do this most effectively.

3. *The building must be a good heat trap.*

   It must make good use of the heat (or cool) and let it escape only very slowly. This is done primarily by reducing the heat loss of the building through the use of insulation, reduction of air infiltration, and storm windows.

Each of these areas will be discussed in the next three chapters. Since the main thrust of this book is heating, there will be only occasional references to the energies involved in cooling. Heat theory is explained in some detail in the Resource Section and will not be dealt with here, except for a few fundamentals. The physiological factors which affect human comfort are also discussed in the Resource Section. These are:

- production and regulation of heat in the human body;
- heat and moisture losses from the human body;
- the effects of cold and hot surfaces in the space;
- the stratification of air; and
- effective temperature (the combination of the effects of air, temperature, moisture content, and air movement).

There are countless other factors which affect energy consumption in a building in addition to the requirements for heating and cooling: the ventilation and movement of air within a building; the purification of that air (especially in urban buildings); the control of water content (humidity) of that air; the use of energy for lighting, for pumps, fans, controls, and other mechanical equipment; and the layout of the building's interior (e.g., putting storage areas and stairwells on the cold north ends of buildings).

As designers try to insure human comfort, the users of buildings should be encouraged to participate in the operation of buildings. Through the combined efforts of designers and users, the consumption of fossil fuels will be reduced, thereby lowering fuel costs and reducing pollution. Simple participatory means of reducing consumption might include: turning down the thermostat whenever possible; drawing curtains, pulling shades, closing venetian blinds, closing shutters; putting on storm windows; weatherstripping windows and doors; keeping the furnace running cleanly and efficiently; and putting up temporary windbreaks during winter.

It is hoped that such owner/user efforts to reduce energy consumption will accompany attempts to increase the effective utilization of solar energy.

CHAPTER II·A

The Building As A Solar Collector

Total heating demand can be lessened by designing buildings which can increase their heat gain from insolation (solar irradiation) and decrease it during hot weather. Customarily, solar gains have not been included in the computation of seasonal heating supply and demand. When engineers size a furnace, they usually provide for the coldest conditions when there is no sun. This may be logical, but building designers should be concerned with reducing total *seasonal* energy consumption for heating. Unfortunately, most research on solar gain has focused on reducing energy required for cooling and refrigeration in hot weather rather than on reducing heating energy needs in cold weather. The data applicable to heating is difficult to understand and even harder to use in the design process. The construction of useful design tools from this data has begun and will be continued here, but extensive work in this area would be of great benefit in reducing our energy needs.

Probably the best way of using the sun's energy for heating is to let it penetrate through the roof, walls, and windows of a building. Insulated roofs and walls do not allow this to occur nearly as readily as do windows. However, if roofs and walls are uninsulated, as is the case with many metal-roofed shanties throughout the world, they will not let the sun through directly but will be warmed and will conduct and radiate considerable amounts of heat into the building.

The color of the roofs and walls will affect the amount of heat which enters the building, since dark colors usually absorb more sunlight than light colors do. Color is particularly important when little or no insulation is used, but has a decreasing effect as insulation increases. In warm and hot climates, exterior surfaces which face the sun should be light in color; in cool and cold climates, such surfaces should be dark. Heat gain through opaque surfaces such as roofs and walls is discussed in more detail in the Resource Section.

Since solar radiation strikes differently oriented surfaces with varying intensity, a building might benefit if its walls and roof were oriented in such a way so as to receive this heat in the winter and shed it in the summer. Henry Niccolis

Wright studied this possibility in "Solar Radiation as Related to Summer Cooling and Winter Radiation in Residence" (WRI.). Some of his conclusions concerning New York City may be summarized as follows:

- The maximum heat value of solar radiation at ground level is 350 Btu/hr/ft².
- The maximum heat value of solar radiation is the same throughout the year in spite of the fact that the sun is lower in the sky in winter than in summer. This similarity is probably due to the lower humidity in winter (less atmospheric absorption). Also, the earth is closer to the sun during the winter.
- "The effective solar radiation on a wall facing south is almost five times as great in the winter as in the summer."
- "The effective radiation on a wall facing west-northwest is six times as great in the summer as in the winter."
- "The greatest effective solar radiation on vertical walls occurs in the winter."
- Therefore, "houses placed broadside to the south-southwest, with most of the important rooms and large windows located on that side—and with a minimum of window area on the west-northwest end—will be a great deal easier to cool in the summer, and more pleasant to live in and easier to heat in the winter."

Figure II·A·1 (WRI.) is self-explanatory. The solar radiation on a house with the worst possible orientation is compared with that on one with the best orientation. However, Victor Olgyay, in *Design with Climate*, says that Wright used "exaggerated values in his radiation calculations," although his principals investigation were probably correct. Most other theorists on the subject of orientation suggest that the principal facade of a building should be oriented within 30° of due south (between south-southeast and south-southwest), with due south being preferred. Of course, it is more important for windows to be oriented south than for walls, since windows allow a significantly greater heat gain than walls do. This is discussed further in the following pages.

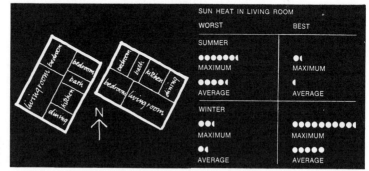

Figure II·A·1.
Effect of house orientation and design on potential solar heat gain in New York City. Source: (WRI.)

Olgyay cautions against conclusions which generalize for all locations. He promotes the use of *sol-air temperatures* (see Resource Section) at particular locations as a means of determining optimal orientation (*Design With Climate*, pp. 54–62, explains Olgyay's excellent theories in detail). Due south may not be optimal for all locations but will almost always be better than 30° east or west of it.

A building can benefit from its orientation; for similar reasons, it also benefits from different ratios of its length to its width to its height. The optimum shape loses the minimum amount of outward moving heat and gains the maximum amount of solar heat in the winter, and retains the minimum amount of solar heat in the summer. Olgyay (OLG.) has shown that:

- In the upper latitudes (40°N+), south sides of buildings receive nearly twice as much radiation in winter as in summer. East and west receive 2½ times more in summer than in winter.
- Lower latitudes (35°N−) gain even more on south sides in the winter than in the summer. East and west walls can gain two or three times more heat than those on the south in the summer.
- Well-insulated buildings and those with shading devices on the south side show even greater variances, but those with windows which are small or fully shaded show less.
- "The square house is not the 'optimum' form in any location."
- All shapes elongated on the north-south axis work with less efficiency than the square one in both winter and summer.
- The optimum form in every case is elongated along the east-west direction.

Besides that of saving energy, there are other considerations in building shape, some of which also affect total energy and resource savings and environmental well-being. For instance, the orientation or size of the site may not accommodate the optimum shape; the needs and purposes of the building may require other shapes; or, if natural lighting is desired, more perimeter exterior surface areas may be needed for the placement of windows.

The determination of relative solar heat gains for buildings with various facade orientations can aid in deciding on the shape and orientation of a building and the location of the windows. Building orientations can be generally classified according to the four plan diagrams (see Figure II·A·2). Within these four orientations, there are three basic variations in the facade and area configurations of buildings:

1. The building has facades approximately equal in area and is represented in plan by a square.
2. The building has facades of greatly differing areas (a ratio of 1½:1 or greater); the long axis of the rectangle is in the direction of the compass orientation.
3. The building has facades of greatly differing areas (a ratio of 1½:1 or greater); the short axis of the rectangle is in the direction of the compass orientation.

Figure 11·A·3 shows the combinations of orientation, shape, and floor and wall area. The values are relative and are based on actual Solar Heat Gain Factors taken from the ASHRAE *Handbook of Fundamentals* and from information provided by Koolshade Corporation. Both sources contain data for numerous latitudes.

In the "total" column of Figure II·A·3, a building with its long axis oriented in the east-west direction has greater potential for solar heat gain in January (3,210 Btu/day per unit area W^2) than one which is oriented north-south (2,668) or one which is square. In fact, east-west is the best of all other variations shown. The poorest shape and orientation is the square with facades facing northeast, northwest, southeast and southwest.

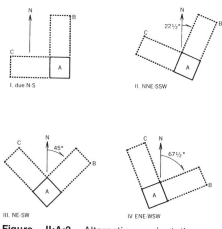

Figure II·A·2. Alternative orientations and configurations for determining potential relative solar heat gain. (From "A Feasibility Study for the Utilization of Solar Energy in New and Existing Schools in New York City" by Total Environmental Action, Inc. 1974).

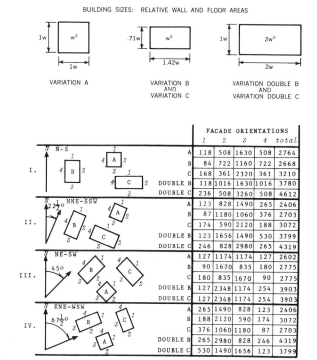

		FACADE ORIENTATIONS				
		1	2	3	4	total
I. N-S	A	118	508	1630	508	2764
	B	84	722	1160	722	2668
	C	168	361	2320	361	3210
	DOUBLE B	118	1016	1630	1016	3780
	DOUBLE C	236	508	3260	508	4612
II. NNE-SSW	A	123	828	1490	265	2406
	B	87	1180	1060	376	2703
	C	174	590	2120	188	3072
	DOUBLE B	123	1656	1490	530	3799
	DOUBLE C	246	828	2980	265	4319
III. NE-SW	A	127	1174	1174	127	2602
	B	90	1670	835	180	2775
	C	180	835	1670	90	2775
	DOUBLE B	127	2348	1174	254	3903
	DOUBLE C	127	2348	1174	254	3903
IV. ENE-WSW	A	265	1490	828	123	2406
	B	188	2120	590	174	3072
	C	376	1060	1180	87	2703
	DOUBLE B	265	2980	828	246	4319
	DOUBLE C	530	1490	1656	123	3799

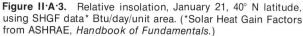

Figure II·A·3. Relative insolation, January 21, 40° N latitude, using SHGF data* Btu/day/unit area. (*Solar Heat Gain Factors from ASHRAE, *Handbook of Fundamentals.*)

By doubling the ground floor area, the optimal gain of 3,210 Btu increases by one-third to 4,612 Btu because the perimeter increases by only one-third. If the floor area were doubled by adding a second floor, the perimeter area would double, as would the solar gain (6,420 Btu).

The analysis does not take into account the color of the walls, the solar impact on the roof, the variations in window location and size, the effect of heat loss, or the implications of windows for natural lighting. A detailed analysis would also take into account the actual weather conditions. The simplified analysis, however, does serve to produce relative values useful in making preliminary decisions.

Although the color, orientation, and shape of the building are important factors in solar heat gain, the most significant component is its windows. Openings in shelters were once without the benefit of glass. They were used for the passage of people and their accompanying possessions, for the passage of air providing natural ventilation, and for the passage of natural light into the interiors. But they also had drawbacks: animals and bugs had free access; the inside temperature was difficult to regulate; and air movement, humidity, and air cleanliness could not be controlled.

Although glass probably existed as far back as 2300 B.C., it was not used in windows until the time of Christ. But only in the last 75 years has it become economically and technologically possible to produce and use panes of glass larger than 8 or 12 in. on a side. As technology and economics improve, glass is increasingly used to replace the traditional solid (masonry or wood) exterior wall. The design problems accompanying this substitution have often been ignored or under-rated.

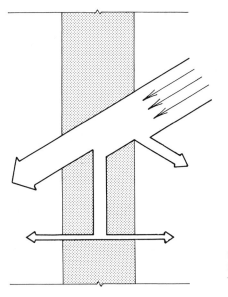

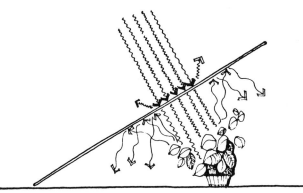

Figure II·A·5.   Solar trap.

Figure II·A·4.
Dynamics of solar heat gain. Section through glass. 183 Btu/ft²/hr on a typical, clear afternoon in mid-summer at 40°N latitude facing southwest.

Besides reducing the electrical energy required for lighting, glass exposed to sunlight admits heat (see Figure II·A·4). Experimental houses which have glass forming the major parts of south-facing walls are designed to benefit from this fact. Heat savings in a house near Chicago, designed by G. F. Keck and sponsored by the Illinois Institute of Technology, may have been as high as 18 percent; the house admitted so much solar heat that it became overheated even on clear winter days (TEL.). Other "solar houses" have reportedly saved up to 30% in fuel bills. The greenhouse effect is primarily responsible for this phenomenon. Glass readily transmits the short-wave light radiation as shown in Figure II·A·5, but does not readily transmit in the other direction the long-wave thermal radiation resulting when the light energy changes to heat energy as it hits an interior surface.

The designer should be able to determine by looking at a chart (or several charts) how many useful Btu of solar energy will enter through a wall or window. The ASHRAE *Handbook of Fundamentals* contains extensive tables on solar heat gain for latitudes 24, 32, 40, 48, and 50°N on the twenty-first day of each month. But a designer would find it difficult to use these tables to estimate solar heat gain for heating (the tables are designed for use in sizing cooling equipment). The basic principles of "Solar Heat Gain Through Windows" are discussed in the Resource Section by that name.

Extensive work on the solar house concept was done by F. W. Hutchinson at Purdue University. In 1945, under a grant from Libbey-Owens-Ford Glass Company, two nearly identical houses were built side by side. The only difference was that one of the houses had considerably more south-facing glass (two ¼ in. thick clear glass panes separated by a ½ in. air space, HUT. 1). Based on the performance of these two houses, Hutchinson reported in May 1947, that "the available solar gain for double windows in south walls in most cities in the U.S.A. is more than sufficient to offset the excess transmission loss through the glass" (HUT. 2).

The design of south-facing windows requires that the thermal capacity of the inside of the building be great enough to absorb and store the excess heat so that the interior space does not require venting. The better the insulative value of the walls and windows, the less heat will be lost through heat transmission and the greater the heat capacity will have to be. Figure II·A·6 shows that the inside temperature in the *unheated* solar house was 80° on January 15 while outside it was below freezing.

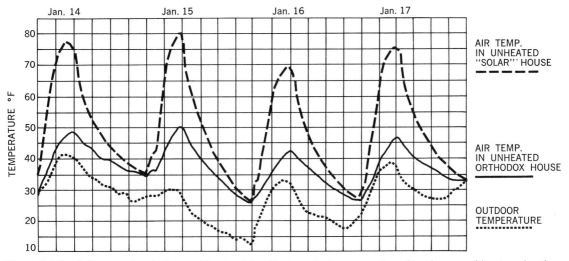

Figure II·A·6. Influence of periodic sunshine in winter: Test results from operation of two houses without any heating. Source: (HUT.2)

It must be pointed out that large glass areas do require a larger initial cost of a heating system because of the extra heat loss through the glass which would otherwise be solid wall. Also, for a given latitude, solar intensity does not vary (although cloudiness does), but heat loss does vary according to the outside temperature. It follows then that the use of glass in mild climates has greater potential for reducing seasonal heating demand that it does in cold climates at the same latitude.

Earlier in this section it was indicated that the quantity of solar energy which gets through a south-facing window on an average sunny day in the winter is greater than that which is received through that same window on an average sunny day in summer. There are a number of reasons for this:

1. Although there are more daylight hours during the summer than during the winter, there are more hours of possible sunshine on a south-facing window in winter than in summer. For example, at latitude 35°N, 14 hours of sunshine are possible on June 21, but the sun remains north of east until after 8:30 A.M. and goes to north of west before 3:30 P.M., so that direct

sunshine occurs for only seven hours on the south-facing wall. On December 21, however, the sun is on the south wall for the full 10 hours that it is above the horizon.

2. The intensity of insolation on a plane normal to the sun's rays is approximately the same in summer as in winter. The extra distance that the rays must travel through the atmosphere is offset by the sun's being closer to the earth during the winter than during the summer.

3. Since the sun is closer to the horizon during the winter, the rays strike the windows at more nearly right angles than they do in the summer when the sun is at a higher altitude. At 35°N, 150 units of energy may strike a square foot of window during an average winter hour; during the summer this number would be 100 units.

4. Winter sky radiation (due to the scattering effect of the atmosphere) is twice the amount of summer sky radiation.

5. The more nearly the sun's rays hit the window at right angles, the greater the transmittance (shown earlier). They are more nearly so in winter than in summer.

6. With proper shading, the window can be shielded from most of the direct summer radiation.

Hutchinson's conclusions are that more than twice as much solar radiation is transmitted through south-facing windows in winter than in summer. If the windows are shaded in summer, the difference is even greater.

An extremely useful graph, prepared by Hutchinson, is reprinted in the Resource Section "Solar Heat Gain Through Windows." It can be used as a design tool to approximate solar heat gain through south-facing windows for the seven-month season. One conclusion which can be drawn from these statistics is that the effects of window type and latitude are relatively small compared with the effects of normal outside air temperature and with the fraction of possible sunshine. These two values can be found in Figure II·A·7. The first column gives the fraction, F, of the average number of sunshine hours during the heating season (October 1 to May 1) to the maximum possible sunshine hours. The last two columns show transmission losses through single and double windows; such losses are useful in sizing the heating plant for a building but are less helpful in determining operating cost.

The fourth and fifth columns show the net gain of energy (a negative number represents a loss) resulting from the use of one square foot of single or double glass. All 48 cities show net energy gains through the double glass. (The losses for single glass in some cities should be compared with corresponding heat loss through walls.) The approximate seasonal heat gain is the product of the value in column four or five, times the window area, times the number of hours in the heating season. Of course, there will be many days when all of this heat cannot be used. Often, too, other factors such as pulled shades or closed curtains will reduce solar gain. Using thermal barriers (e.g., insulating shutters) to cover windows at night greatly reduces heat loss and increases the overall net heat gain. The analysis does not take into account the solar gain of south walls. Hutchinson's work showed that this factor can be significant unless the wall is extremely well-insulated.

City	Fraction, F, of maximum possible sunshine.	Normal temperature during seven month heating season.	Design outside winter alt temperature.	Net energy gain, Btu hr sqft due to use of glass.		Glass transmission losses, Btu/hr/ft²	
				Single glass	Double glass	Single	Double
1. Albany, N.Y.	.463	35.2	−24	−12.8	5.6	106.2	56.0
2. Albuquerque, N.M.	.770	47.0	−10	18.05	30.2	90.4	48.0
3. Atlanta, Ga.	.522	51.5	−8	9.0	18.8	88.1	46.8
4. Baltimore, Md.	.553	43.8	−7	2.0	15.9	87.0	46.2
5. Birmingham, Ala.	.510	53.8	−10	10.9	19.5	90.4	48.0
6. Bismarck, N.D.	.546	24.6	−45	−20.1	4.0	129.9	69.0
7. Boise, Id.	.540	45.2	−28	22.9	16.0	110.7	58.8
8. Boston, Mass.	.540	38.1	−18	5.2	11.7	99.44	52.8
9. Burlington, Va.	.419	31.5	−29	−19.5	.9	111.9	59.4
10. Chattanooga, Tenn.	.503	49.8	−10	5.9	16.7	90.4	48.0
11. Cheyenne, Wyo.	.666	41.3	−38	5.7	20.3	122.0	64.8
12. Cleveland, Ohio	.408	37.2	−17	−13.7	3.7	98.3	52.2
13. Columbia, S.C.	.511	54.0	−2	11.2	19.6	81.4	43.2
14. Concord, N.H.	.515	33.3	−35	−12.0	7.4	118.6	63.0
15. Dallas, Texas	.470	52.5	−3	7.1	16.4	82.5	43.8
16. Davenport, Iowa	.539	40.0	−27	−3.1	12.8	109.6	58.2
17. Denver, Colo.	.705	38.9	−29	5.2	21.7	111.9	59.4
18. Detroit, Mich.	.429	35.8	−24	14.1	44.0	106.2	56.4
19. Eugene, Ore.	.439	50.2	−4	2.7	13.2	83.6	44.4
20. Harrisburg, Pa.	.495	43.6	−14	−1.5	12.5	94.9	50.4
21. Hartford, Conn.	.532	42.8	−18	−.3	14.1	99.4	52.8
22. Helena, Mont.	.521	40.7	−42	−3.3	12.2	126.6	67.2
23. Huron, S.D.	.579	28.2	−43	−14.1	8.0	127.7	67.8
24. Indianapolis, Ind.	.507	40.3	−25	−4.6	11.2	107.3	57.0
25. Jacksonville, Fla.	.400	62.0	−10	13.9	18.1	67.8	36.0
26. Joliet, Ill.	.530	40.8	−25	2.9	12.8	107.3	57.0
27. Lincoln, Neb.	.614	37.0	−29	−2.2	15.3	67.8	69.4
28. Little Rock, Ark.	.513	51.6	−12	8.5	18.3	92.7	49.2
29. Louisville, Ky.	.514	45.3	−20	1.5	14.6	101.7	54.0
30. Madison, Wis.	.504	37.8	−29	−7.6	9.5	111.0	59.4
31. Minneapolis, Minn.	.527	29.4	−34	−15.74	5.8	117.5	62.4
32. Newark, N.J.	.550	43.4	−13	1.4	15.5	93.8	49.8
33. New Orleans, La.	.370	61.6	7	11.7	16.1	71.2	37.8
34. Phoenix, Ariz.	.590	59.5	16	21.9	27.5	61.0	32.4
35. Portland, Me.	.525	33.8	21	−7.2	12.0	55.4	29.4
36. Providence, R.I.	.542	37.2	−17	−6.1	11.3	98.3	52.2
37. Raleigh, N.C.	.570	50.0	−2	−10.0	20.6	81.4	43.2
38. Reno, Nev.	.637	45.4	−19	8.6	21.7	100.6	53.4
39. Richmond, Va.	.594	47.0	−3	8.0	20.2	82.5	3.8
40. St. Louis, Mo.	.567	43.6	−22	2.6	16.6	104.0	55.2
41. Salt Lake City, Utah	.592	40.0	−20	0.0	15.9	101.7	54.0
42. San Francisco, Cal.	.615	54.2	27	17.3	25.7	48.8	25.8
43. Seattle, Wash.	.340	46.3	3	−7.3	5.2	75.7	40.2
44. Topeka, Kan.	.613	42.3	−25	3.8	18.4	107.3	57.0
45. Tulsa, Okla.	.560	48.2	−16	7.4	19.0	97.2	51.6
46. Vicksburg, Miss.	.447	56.8	−1	−10.7	17.7	80.2	42.6
47. Wheeling, W.Va.	.408	46.1	−18	3.7	9.0	99.4	52.8
48. Wilmington, Del.	.558	45.0	−15	3.7	16.9	96.1	51.0

Figure II·A·7. Solar benefit values for 48 U.S. cities. Source: (HUT. 2)

Heat gain (and heat loss, discussed in Chapter II·C, "The Building as a Heat Trap") through windows varies with the type of window frame. Figures II·A·8 and II·A·9 are taken from work done at the National Bureau of Standards.

They show the relative difference in heat loss and heat gain for aluminum and wood windows for the four exposures: north, east, south, and west. The reduction in summer heat gain and winter heat loss for windows framed in wood instead of aluminum is significant. Thus wood should be considered for both new buildings and for the replacement of windows in existing buildings, even more so because of the large quantity of energy required to refine aluminum.

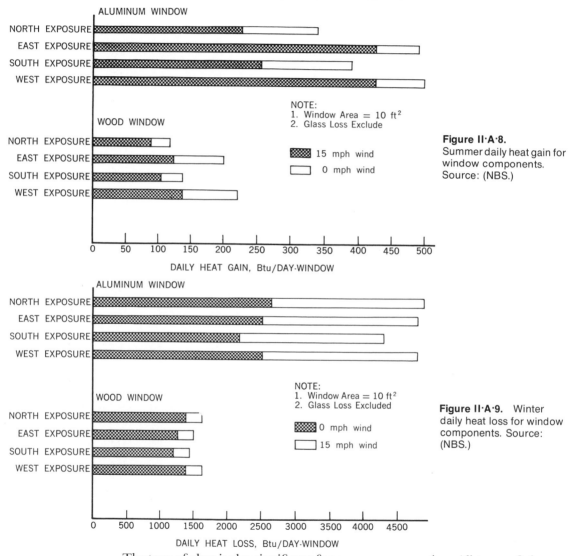

Figure II·A·8. Summer daily heat gain for window components. Source: (NBS.)

Figure II·A·9. Winter daily heat loss for window components. Source: (NBS.)

The type of glass is also significant for energy conservation. All types of glass—clear, heat-absorbing, or reflecting—lose about the same amount of heat through conduction. However, there is a great variation in the amount of solar heat which is transmitted through these three types of glass. Note the diagrams and the summary table of "Relative Net Heat Gain" for various configurations of single and double glazing (Figures II·A·10 through II·A·18). The heat gains are approximate and not absolute for the sunny day conditions shown. To reduce summer heat gain, reflecting glass should face the outside and clear glass the inside of two-pane windows subject to solar heat gain. This also reduces heat loss during the winter. Unfortunately, however, solar heat gain would be greatly reduced in the winter when the preferred position for these two layers of glass is just the opposite of what it is during the summer, i.e., clear glass outside and reflecting glass inside (see Figure II·A·16).

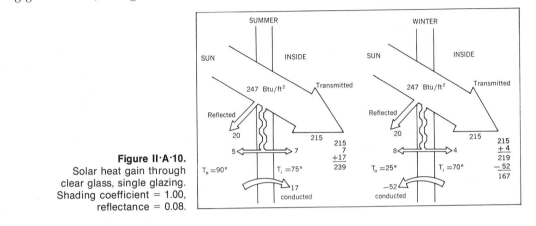

Figure II·A·10.
Solar heat gain through clear glass, single glazing. Shading coefficient = 1.00, reflectance = 0.08.

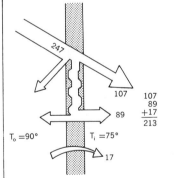
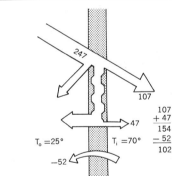

Figure II·A·11.
Solar heat gain through heat-absorbing glass, single glazing. Shading coefficient = 0.05, reflectance = 4.

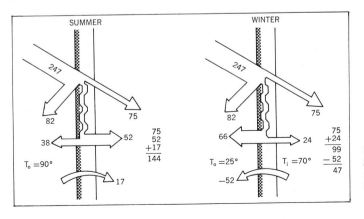

Figure II·A·12.
Solar heat gain through reflecting glass, single glazing. Shading coefficient = 0.35m reflectance = 33.

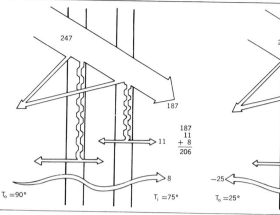
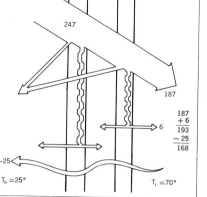

Figure II·A·13.
Solar heat gain through clear glass, double glazing. Shading coefficient = 1.00, reflectance = 0.08.

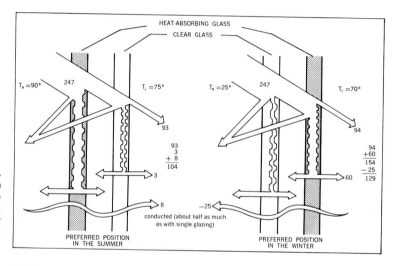

Figure II·A·14.
Solar heat gain through heat-absorbing glass, double glazing. Heat-absorbing glass: shading coefficient = 0.50, clear glass: shading coefficient = 1.00; reflectance = 0.08.

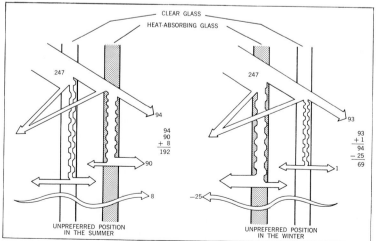

Figure II·A·15.
Solar heat gain through heat-absorbing glass, double glazing.

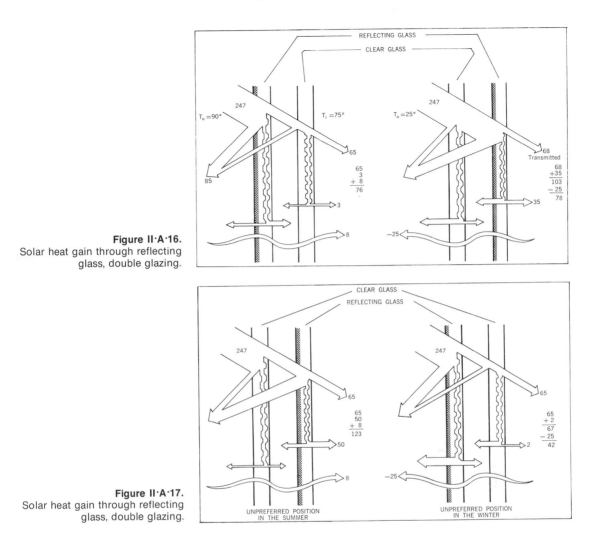

Figure II·A·16.
Solar heat gain through reflecting glass, double glazing.

Figure II·A·17.
Solar heat gain through reflecting glass, double glazing.

If heat gain is to be maximized and heat loss minimized during the winter, two layers of clear glass must be used. Although this configuration may transmit considerable heat during the summer, shading devices such as trees, awnings, and venetian blinds can reduce the need for reflecting or heat-absorbing glass. These expensive glasses might be considered for east- or west-facing windows because they are usually more difficult to shade than south-facing ones. (North-facing glass does not require shading except in the southern part of the country where the sun may strike north-facing glass to a significant extent at sunrise and sunset during the summer.) For example, in New York City there is little solar gain during the winter and a good deal of solar gain during the spring and fall through east- and west-facing windows; heat-absorbing or reflecting glass may provide comfort when clear glass would otherwise result in overheating.

GLASS TYPE	SUMMER	WINTER
Single Glazing		
Clear glass	239	167
Heat-absorbing glass (shading		
coefficient = 0.50)	213	154
Reflective glass (shading		
coefficient = 0.35)	144	47
Double Glazing		
Clear glass, both panes	206	168
Heat-absorbing glass outside,		
clear glass inside	104	69
Clear glass outside, heat-		
absorbing glass inside	184	129
Reflective glass outside, clear		
glass inside	76	42
Clear glass outside, reflective		
glass inside	123	103

Figure II·A·18. Summary of Figures II·A·10 to II·A·17. Relative net heat gain through various types and combinations of glass.

Just as it is important to let the sun in at the proper time of the year, it is also important to keep it out at other times. In many climatic regions, keeping the sun out during critical warm weather is actually more important to human comfort than letting it in during cold weather. The use of different types of glass for different sun orientations is one of the methods of obtaining shading. In many parts of the country where the reduction of heat gain is critical, heat-absorbing and reflecting glass can be beneficial, especially for east and west facades. The important factors to consider in the use of such glass have already been discussed, but they bear repeating.

1. Such glass reduces solar heat gain, but this can be as much of a disadvantage in the winter as it is an advantage in the summer.

2. Except for glare control, reflecting and heat-absorbing glass are almost always unnecessary on north, north-northeast, and north-northwest orientations. There is little solar heat gain on these facades except in the latitudes south of 30°N where they might be considered.

3. In almost all latitudes, except those north of 40°N, heat-absorbing and reflecting glass should usually not be considered for south-facing windows (except as a means of excluding winter solar heat, if necessary, e.g., in large office buildings). The heat gain through south-facing glass is relatively small in the summer (see Figure II·A·19).

4. More sensible solutions than heat-absorbing or reflecting glass for south, southeast, and southwest orientations are the use of vegetation and operable shading devices. Shading devices on the exterior of the building are most effective; those between two layers of glass (such as venetian blinds) are next most effective; and interior devices such as blinds, shades, and draperies are least effective since they stop the sun's rays after they have penetrated the building instead of before (see Figure II·A·20).

5. Advertisements for heat-absorbing and reflecting glass suggest that these products will reduce both the initial cost of air-conditioning equipment and the cost of its operation, especially as it affects energy consumption. The savings, however, are usually obtained by comparing costs with those for all-glass buildings rather than with those for buildings already designed to conserve energy. No mention is made that substantial savings could be achieved by switching to opaque, well-insulated walls with reduced glass areas (such as north, east and west) and to a well-designed building which allows the sun to penetrate through the south glass during the winter but which shades itself during the summer.

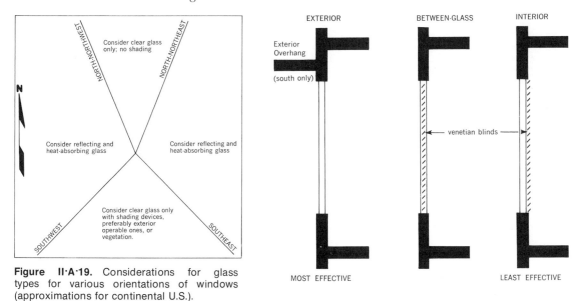

Figure II·A·19. Considerations for glass types for various orientations of windows (approximations for continental U.S.).

Figure II·A·20. Possible locations for shading devices.

All four (or more) facades of buildings need not, and in fact should not, be of identical appearance. This is particularly true of buildings which use large areas of glass. Although there may be economic, social, and personal reasons for building glass boxes, it should be clear that glass has been misused as a design element.

South-facing glass is easily shaded with exterior overhangs above the glass. The primary difficulty with fixed overhangs is that the amount of shading (resulting from the shadow of the overhangs on the glass) follows the seasons of the sun rather than those of the earth. The middle of the summer for the sun is June 21; for the earth it is the end of July or the middle of August. A fixed overhang designed for optimal shading on August 10 causes the same shadow on May 1. The overhang designed for optimal shading on September 21, the fall equinox when the weather is still somewhat warm and solar heat gain is relatively unwelcome, causes the same shading situation on March 21, the spring equinox when the weather is cooler and solar heat gain is relatively welcome. The use of vegetation, which more closely follows the sun's seasons, results in more optimal shading year round: on March 21, there are few or no leaves, and the sun's heat is admitted; on September 21, the leaves are still full, providing necessary shading (see Figure II·A·21). The determination of the proper shading angle is discussed in the Resource Section "Solar Angles and Shading."

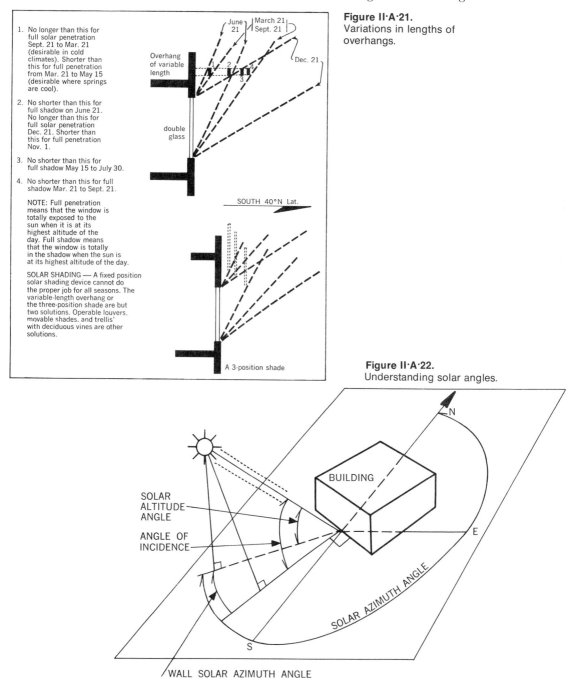

1. No longer than this for full solar penetration Sept. 21 to Mar. 21 (desirable in cold climates). Shorter than this for full penetration from Mar. 21 to May 15 (desirable where springs are cool).

2. No shorter than this for full shadow on June 21. No longer than this for full solar penetration Dec. 21. Shorter than this for full penetration Nov. 1.

3. No shorter than this for full shadow May 15 to July 30.

4. No shorter than this for full shadow Mar. 21 to Sept. 21.

NOTE: Full penetration means that the window is totally exposed to the sun when it is at its highest altitude of the day. Full shadow means that the window is totally in the shadow when the sun is at its highest altitude of the day.

SOLAR SHADING — A fixed position solar shading device cannot do the proper job for all seasons. The variable-length overhang or the three-position shade are but two solutions. Operable louvers, movable shades, and trellis' with deciduous vines are other solutions.

June 21   March 21 / Sept. 21

Overhang of variable length

Dec. 21

double glass

SOUTH 40°N Lat.

A 3-position shade

Figure II·A·21.
Variations in lengths of overhangs.

Figure II·A·22.
Understanding solar angles.

SOLAR ALTITUDE ANGLE

ANGLE OF INCIDENCE

BUILDING

N

E

S

SOLAR AZIMUTH ANGLE

WALL SOLAR AZIMUTH ANGLE

Operable shading devices are even more adaptable (see Figure II·A·23). But devices which are attached to the outsides of buildings are difficult to maintain and are subject to deterioration. Efforts to make them more durable have not usually been successful. In addition, they must be designed so that people can operate them easily; people must be thus encouraged to participate in providing themselves a comfortable environment (such encouragement extends to the operation of windows, temperature controls, lighting levels, and other areas of the physical environment).

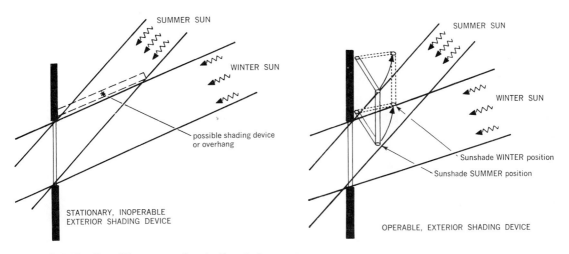

Figure II·A·23. Two different exterior shading devices.

Operable shading located between two layers of glass is not as effective as are exterior devices, but it is more effective than interior ones. A between-glass device such as venetian blinds is often expensive, and it is usually difficult to repair and clean. Interior shading devices such as roller shades, draperies, and venetian blinds are the least effective shading devices but are probably the most versatile for operation by the people in the building (see Figure II·A·20).

Skylid ® is an operable shading device which is actuated by the sun. Developed by Zomeworks Corporation in Albuquerque, New Mexico, the louvered device is located inside of the building (for protection from the weather). The louvers pivot simultaneously: when open, they admit the sun's rays; when closed, they keep the sun's rays out and insulate the window to keep the heat (or cool) in. Two cannisters connected by a small tube are mounted on one of the louvers. Freon flows between the cannisters, expanding and contracting according to its temperature, which is determined primarily by the solar heat which strikes the outward-facing cannister. When the sun heats the Freon, it flows from the outward-facing cannister to the other, counter-balancing the louvers and causing them to close. A manual lever is provided to override the automatic operation. During the winter, the opposite system can be put into effect: the louvers are triggered by the sun to open during sunny hours and to close at night to keep the heat in (see Figures II·A·24 through II·A·27).

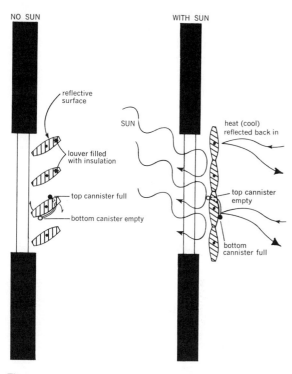

NO SUN

WITH SUN

SUN

reflective surface

louver filled with insulation

top cannister full

bottom canister empty

heat (cool) reflected back in

top cannister empty

bottom cannister full

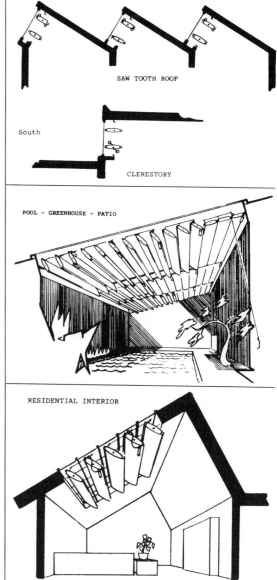

SAW TOOTH ROOF

South

CLERESTORY

POOL - GREENHOUSE - PATIO

RESIDENTIAL INTERIOR

Figure II·A·24. Skylid: summer operation. During the winter the skylid opens to allow sun to penetrate the building and closes at night to prevent the heat from escaping.

Figures II·A·25, II·A·26 and II·A·27. Some other applications of Skylid. (Drawings courtesy of Zomeworks Corporation.)

East- and west-facing glass is extremely difficult to shade because the sun is low in the sky both in summer and winter (see Figure II·A·28). Overhangs do not prevent the penetration of the sun during the summer much more than they do during the winter. Vertical louvers or extensions are probably the best means of shading such glass, but consideration may be given to reflecting and heat-absorbing glass.

One method of shading east- and west-facing glass is to orient it to face north or to face south. By facing the glass north, only indirect light will be admitted, a

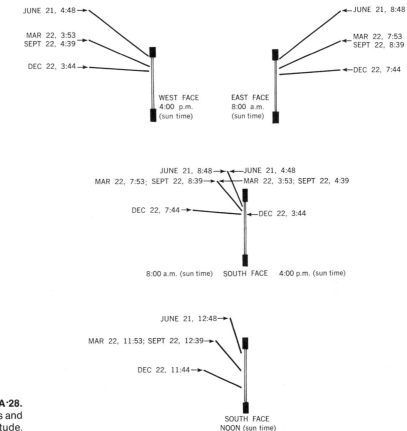

Figure II·A·28.
Sun Angles for various dates and
times, 40°N. latitude.

favorable lighting effect for many human tasks. By facing the glass south, however, solar heat is admitted during the winter. In Figure II·A·29, the method of orienting the glass toward the south also provides for full shading during the summer. Figure II·A·30 shows other configurations for south windows.

The notion of *shading coefficient* is important in comparing the relative effectiveness of various shading devices. A single layer of clear, double-strength glass has a shading coefficient of 1.00. The shading coefficient for any other glazing system in combination with shading devices is the ratio of the solar heat gain through that system to the solar heat gain through the double-strength glass. Thus, solar heat gain through glazing systems is the product of its shading coefficient times the Solar Heat Gain Factors listed in the ASHRAE *Handbook of Fundamentals* for clear, double-strength glass. Figure II·A·31 shows some typical shading coefficients for various shading conditions.

Large buildings with small exterior surface areas compared to relatively large amounts of useful floor area often generate tremendous amounts of internal heat from the activities of people, the burning of lights, and the operation of equipment. It is not uncommon for such buildings to require air-conditioning year round, even in the dead of winter and particularly during the summer. The

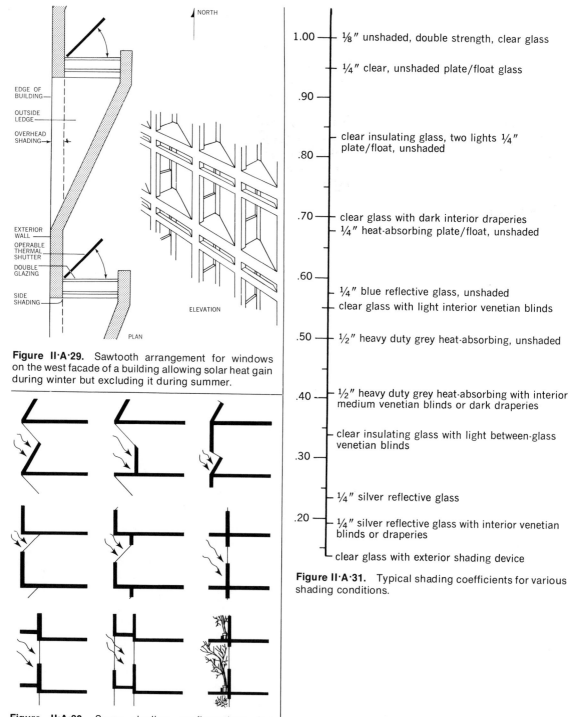

Figure II·A·29. Sawtooth arrangement for windows on the west facade of a building allowing solar heat gain during winter but excluding it during summer.

Figure II·A·30. Some shading configurations for south windows.

Figure II·A·31. Typical shading coefficients for various shading conditions.

Shading coefficient scale:

1.00 — ⅛″ unshaded, double strength, clear glass

— ¼″ clear, unshaded plate/float glass

.90 —

— clear insulating glass, two lights ¼″ plate/float, unshaded
.80 —

.70 — clear glass with dark interior draperies
— ¼″ heat-absorbing plate/float, unshaded

.60 —

— ¼″ blue reflective glass, unshaded
— clear glass with light interior venetian blinds

.50 — ½″ heavy duty grey heat-absorbing, unshaded

.40 — ½″ heavy duty grey heat-absorbing with interior medium venetian blinds or dark draperies

— clear insulating glass with light between-glass venetian blinds
.30 —

— ¼″ silver reflective glass

.20 — ¼″ silver reflective glass with interior venetian blinds or draperies

— clear glass with exterior shading device

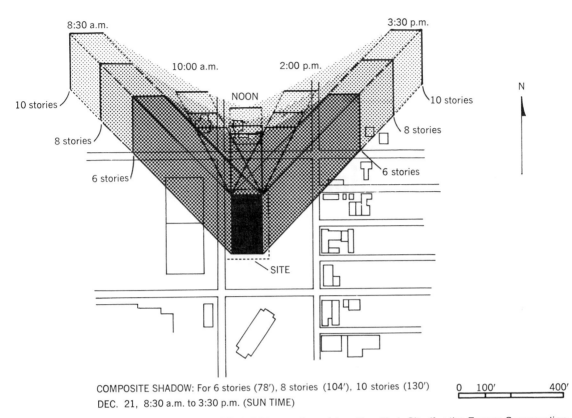

COMPOSITE SHADOW: For 6 stories (78′), 8 stories (104′), 10 stories (130′)   0   100'   400'
DEC. 21, 8:30 a.m. to 3:30 p.m. (SUN TIME)

Figure II·A·32. Shadow studies by Dubin-Mindell-Bloome Associates, New York City (for the Energy Conservation Demonstration Federal Offiice Building for the General Services Administration). Building located in Manchester, New Hampshire.

use of glass which is shaded 12 months a year, instead of only during the summer, will be most successful in these solutions. Of course, every effort should be made to reduce the amount of heat which is produced by the people, lights and machines. The dependence on artificial lighting should be reduced by using more natural lighting (through windows), by using lower lighting levels, or by placing lighting fixtures directly where light is needed ("task lighting").

Designers should also consider the shading effects of buildings on the surrounding environment, i.e., whether the shading occurs on buildings which directly or indirectly use the sun's heat or light, or on wild vegetation or gardens which need sun in order to grow (see Figure II·A·32).

CHAPTER II·B

The Building As A Solar Storehouse

A vital component of a system which uses solar energy for heating is a container for storing heat. When a building is a solar collector, it needs a method of soaking up or storing heat to prevent overheating when the sun shines and to retain (store) some of the heat for use when the sun does not shine.

Probably the most efficient storage container is the material of which the building is built—the walls, floors, roofs, and partitions. All materials (objects, things) absorb and store heat as they are warmed. When temperatures around them become cooler, the stored heat is released to the cooler surroundings and the materials themselves cool.

For a building, this phenomenon is very significant. Solar energy penetrates through the walls, roof, and windows of the building during the day. The short-wave light is stopped by the walls, floors, and furniture of the building after it penetrates through the glass. As it is stopped, it turns into heat, much of which is absorbed (see Figure II·B·1). If the objects and materials inside of the building are already warm—filled to capacity with heat—they release their heat to cooler objects and materials in the building. The air in the building is one of the "materials" which is likely to heat up the soonest, and it helps to distribute the excess solar heat gain to the rest of the materials.

However, if the building materials have already heated up to the temperature of the air or cannot absorb the heat fast enough, the air continues to warm and overheat, causing possible discomfort to the occupants. The materials in the building in turn continue to rise in temperature, storing more heat. The greater the heat storage capacity of the objects and materials in the building, the longer it will take for the air to reach temperatures of discomfort.

When the sun goes down, and if it is cold outside, the building begins losing heat. The many ways of reducing the heat loss of buildings will be discussed in

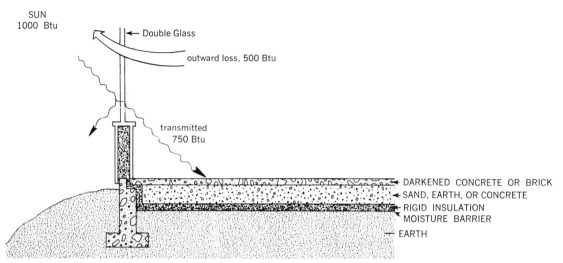

Figure II·B·1. Solar heat gain through windows.

the next chapter. However, even if the building loses very little heat, that heat must be replaced to maintain comfortable temperatures. For buildings which have not stored any solar heat during the day, the heat will have to be provided by other means, e.g., wood stoves, furnaces, space heaters, lights, machines, and people. However, if a building has a sufficient amount of material inside of it and if the sun is able to penetrate during the day to warm it, it will be heated by the sun even after the sun has set. The warmed materials will lose their stored heat to the inside air as it cools off. Depending on the quantity of heat storage material, the amount of sun which penetrates the building and is stored by the materials, and the heat loss of the building (which depends on a number of parameters such as the amount of insulation, the inside temperature, and the outside temperature), a building can remain comfortable for many hours, and possibly even days, without requiring extra heat from other sources.

The effects of varying outdoor temperatures on the indoor temperatures of different types of buildings are shown in Figures II·B·2, II·B·3, and II·B·4. The effect a sharp drop in outdoor temperature has on the indoor temperature of various types of buildings is shown in Figure II·B·2. Note that a lightweight building, such as a wood-framed one, drops relatively quickly in temperature, even if it is well-insulated. A heavy, massive structure, built of concrete, brick, or stone for example, and well-insulated, maintains its temperature over a longer period of time. To reduce heat loss, the insulation in such a building should be located on the outside of the thermal mass (between the mass and the outside air). The heavy materials which can store much heat are poor insulators and, to make use of their heat storage capacity, must be located within the confines of the thermal barrier (insulation) which separates the inside of the building from the outside.

The third indoor air temperature shown in the figure is for a building which not only has a great deal of heat storage capacity inside of it, but is also set into the side of a hill or is covered with earth. Rigid board insulation, such as a

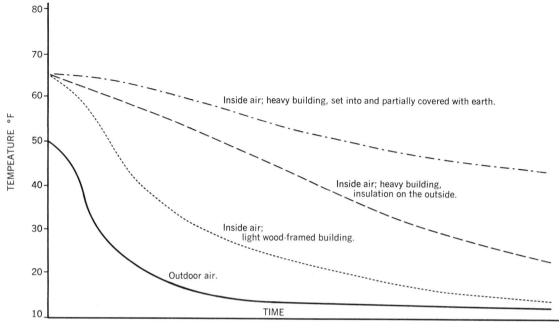

Figure II·B·2.   Effect of a drop in outdoor temperature on indoor temperature for various construction types assuming no source of heat energy within the building.

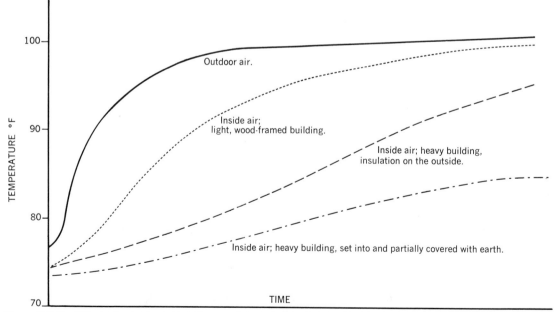

Figure II·B·3.   Effect of a rise in outdoor temperature or indoor temperature for various construction types assuming no source of cooling energy within the building.

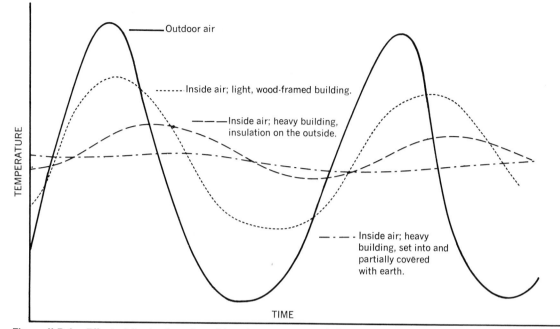

Figure II·B·4. Effect of fluctuations in outdoor temperature on indoor temperature for various construction types assuming no source of heating or cooling energy within the building.

polystyrene or a urethane, is placed between the concrete or stone walls and the earth. One or more walls could be exposed to the outside air, but the temperature, as shown in the figure, drops very slowly and levels off at a temperature close to that of the earth.

During the summer, the opposite conditions are in force. If the building is shaded so that little direct solar energy penetrates, the heat gain will be limited primarily to conduction through the walls, roof, and windows, a reverse heat loss, in effect. At night, when outside air is cooler than it is during the day, ventilation of that air into the building, either naturally through openings such as vents or windows or mechanically with fans, will cool the air and therefore all of the materials and objects inside of the building. Since they will be cool at the beginning of the warm day, they will absorb and store more heat before they become warm, cooling the air as they absorb heat from it. Thus, if they are cool in the morning, it will take a period of time before they will have stored up enough heat so that air conditioning will be required.

The effect of a sharp rise in outdoor temperature is shown in Figure II·B·3, for the same three buildings. Again, the lightweight building responds quickly to the change in outdoor temperature, and in spite of being well-insulated, it rises relatively quickly in temperature. The heavy buildings, on the other hand, absorb the heat and delay the temperature rise. The building set into or covered with earth delays most in its response to the outdoor air changes and, if properly designed, will never become too warm.

Figure II·B·4 combines the effects of rising and falling outdoor temperatures. Without any other internal sources of heat, such as furnaces and stoves, the inside air temperature of the lightweight building fluctuates widely, and the air temperature of the earth-embedded building remains almost constant.

The natural conditions in combination with the mass of the building reduce the need for energy; the building's mass also aids in leveling off the demands on heating and cooling equipment. If the building does not respond quickly when the outside temperature fluctuates, the mechanical equipment will not have to be as large to satisfy the needs and will be able to operate at more constant conditions. At one extreme is a lightweight, uninsulated wood-framed building. On a cold, sunny day the furnace might not be used at all. However, at night its full capacity might be needed to maintain comfortable temperatures. The massive, earth-embedded building, on the other hand, would "see" an outdoor temperature which would be averaged over several days, and possibly as much as ten days or two weeks. A relatively small heating system would operate fairly constantly, and the comfort within the building would be fairly even throughout.

Although total overall energy consumption is reduced for some heavy buildings, a higher initial load may be placed on the heating and cooling equipment in the morning if the temperatures are allowed to drop during the night. This is primarily because the mass must be heated to room temperature before the room will be comfortable. Setting the thermostat lower at night, however, saves considerable amounts of energy.

All materials vary in their heat storage capacity. The ability of a material to store heat is measured by specific heat, the number of Btu which are required to raise one pound of the material one degree (Fahrenheit) in temperature. Water, for example, has a specific heat of 1.0, which means that one Btu is required to raise one pound of water one degree in temperature.

Material	Specific Heat Btu/lb/°F	Density lb/ft³	Heat Capacity of One Cu Ft Btu/ft³/°F
Water	1.00	62.5	62.5
Iron, Scrap	0.112	489	55
Concrete	0.27	140	38
Brick	0.20	140	28
Magnetite, Iron ore	0.165	320	53
Basalt, Rocks	0.20	180	36
Marble	0.21	180	38

Figure II·B·5. Comparison of the specific heat and density of various materials on an equal volume basis.

The specific heats of various materials which might be considered in the construction of a building are listed in Figure II·B·5. Unfortunately, the best material of those listed, concrete (specific heat of 0.27), stores only about one-fourth of that stored by the same weight of water. Fortunately, however, the density of concrete (the number of pounds per cubic foot) is considerably greater than that of water. The second column of the table shows the relative densities. By multiplying the specific heat by the density of the material, the heat capacity per cubic foot of material is obtained. These values are listed in the third column. Note that the density of water is least among the materials listed, but that the heat capacity on a per cubic foot basis is still highest because of its relatively large specific heat. In spite of the low specific heat of concrete, its heavy weight compensates somewhat, and it stores considerable amounts of heat (38 Btu per cubic foot for concrete versus 62.5 for water).

Unfortunately, designing heavy buildings goes contrary to much current thinking and design practice. Technology and design focus on trying to "do more with less," and the structural genius is the person who can use the least amount of material in the process of enclosing a space. Such thinking is usually limited to evaluating only the materials used without including the energy consumed or the longevity of the product. The visual weight of buildings is an important esthetic consideration for some people, and the trend now is to design and build a structure so that it appears to be light in weight.

Adding mass can also add to the cost of a building. Poured concrete rises and falls in favor among construction professionals. Cost, availability, ease of handling, and weight are among the factors which influence these attitudes.

More materials are required for heavy buildings than for light ones. The extra energy required for the manufacture of the added materials is difficult to compare with the heating and cooling energy which could be saved, but heavy buildings are likely to have long lives and in most cases are more durable than light ones. (Noteworthy exceptions are wood homes lasting several hundred years. Examples of massive buildings with long lives are Greek and Roman temples and Western European cathedrals.)

The task of adding thermal mass need not be a difficult one. Placing containers of water within the confines of a building (or better yet, in front of a sunlit window) is one solution, but unlikely to be accepted by very many people. Sand, gravel, concrete or water (in plastic vinyl containers) can be used to fill the voids in concrete masonry block (see Figure II·B·6). Massive fireplaces, interior partitions of concrete or brick, and even two or more inches of concrete or brick on the floor can greatly add to the thermal inertia of buildings.

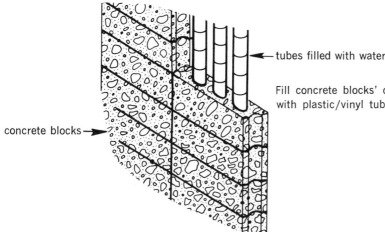

tubes filled with water

Fill concrete blocks' cavities with sand or with plastic/vinyl tubes filled with water.

concrete blocks

Figure II·B·6.
Wall construction for increased thermal mass. (An idea by Harold Hay of Skytherm Processes and Engineering.)

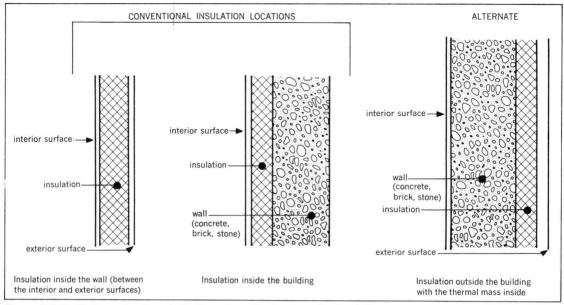

Figure II·B·7. Wall constructions indicating alternative locations for insulation.

At the University of Delaware, Dr. Maria Telkes is testing the use of phase-changing salts to increase thermal inertia. As they are warmed, the salts melt at 75°, storing large quantities of heat. They release the heat as they cool at night. Although still in the testing stage, these salts may soon be incorporated into the skin of buildings or into interior partitions, floors, and ceilings. The best location for them is in the south portions of the building which become the warmest when the sun shines.

Placing insulation on the outside of a building requires creative solutions to construction problems not frequently solved before. Insulation has customarily been placed inside of the wall (between the exterior and interior surfaces) or inside the building (see Figure II·B·7). The most common design problem in placing insulation on the outside of a concrete or masonry building is the protection of the insulation from moisture, rain, sun, and contact with people and animals. Figure II·B·8 shows 3 in. of polystyrene rigid board insulation covering the outside of a poured concrete wall. The earth covers most of the insulation, but the insulation which encloses the concrete above the surface of the ground must be protected from physical abuse and from solar radiation (especially the ultra-violet rays). Below grade, it must be protected from moisture and vermin. The styrene can be placed inside of the formwork before the concrete is poured, and the bond is extremely strong between the two materials, but it is difficult to cover and protect the insulation above ground level without going to considerable extra expense. One alternative is to plaster onto the insulation a "cementitious" material such as fiberglass-reinforced mortar. Another alternative is to devise a method of fastening rigid sheet material, such as moisture-treated plywood or cement-asbestos board.

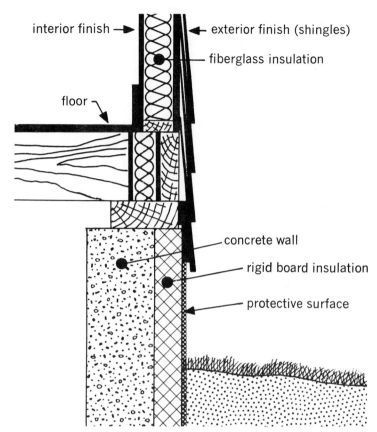

interior finish → ← exterior finish (shingles)

— fiberglass insulation

floor

— concrete wall

— rigid board insulation

— protective surface

Figure II·B·8.
Construction detail to obtain
nearly continuous exterior
envelope of insulation.

Unfortunately, the effect of building mass on the overall energy consumption of buildings is neither understood nor used as a design tool to nearly the extent that are other options such as insulation and storm windows. Victor Olgyay, in *Design with Climate,* and Baruch Givoni, in *Man, Climate and Architecture,* discuss this topic in depth. Other options for reducing energy consumption are dealt with in the next chapter, "The Building as a Heat Trap."

CHAPTER II·C

The Building As A Heat Trap

An earlier chapter, "The Building as a Solar Collector," discussed methods of bringing the sun into the building directly, without the use of a manufactured solar collector which would be placed onto the building as an "appliance." The next chapter, "The Building as a Solar Storehouse," assumed that the sun had managed to penetrate the building and that it then had to be stored as heat for later use. If the building is designed to collect and store heat for later use, it must also be designed to prevent that heat from escaping, or, at least, to slow down its escape.

The escape of heat from a building is usually called its heat loss. Conversely, during the summer, when it is hotter outside than inside, buildings absorb heat, and this is described as heat gain. (This is in addition to the solar heat gain discussed earlier in this chapter.) In effect, trying to slow down (or at least control) the movement of heat into or out of a building is the essence of energy conservation in building design. Although the reduction of heat loss during the winter is stressed here, the reduction of summer heat gain is also mentioned. Most efforts to reduce heat loss also reduce heat gain.

Apart from the use of solar energy for conserving other forms of energy, there are literally thousands of design choices to make which can increase energy savings in buildings. The exploration of these choices is in many ways more meaningful than looking to solar energy for a solution to our energy shortage. We are using far more energy than is necessary; before we look for other sources, we should first decrease the amount of energy we are using. There are almost no buildings in this country which are thermally designed for energy costs ten, or even five, years from now. For heating residential and commercial structures in the United States, over 13×10^{15} Btu (or approximately 2 billion barrels of oil) are used annually, 18% of our national total.

Fortunately, the options for saving energy in buildings are so numerous that simply listing them would require an entire volume. The tradeoffs for different types of buildings, for different types of climates, for different human needs, and for different priorities, however, make for a complicated decision-making process. Many of the energy-saving options do not relate directly to trapping heat inside of buildings, although they may have roundabout, indirect effects. Examples of such options include issues of the types and sizing of heating, ventilating, and air conditioning equipment (HVAC equipment), including methods of operation to reduce energy consumption and the controls necessary to obtain such operation; issues of using natural lighting as a substitute for artificial (electric) lighting by weighing the tradeoffs between using openings in the buildings (such as windows) for lighting, resulting in potentially greater heat loss in the winter and heat gain in the summer, against using energy in the form of electricity for lighting; issues of humidification and dehumidification; issues of the types of appliances and their operation; issues of interior planning and floor layout; and issues of moisture movement through walls, roofs, floors, and windows.

What will be discussed, although briefly, are the fundamentals of keeping heat inside a building. These include issues of insulation; window and entrance types; the shape of buildings; shutters on windows; reduction of air infiltration; wind protection; and heat recovery. There is an excellent fit between energy conservation in building design and the use of solar energy for heating and cooling. By decreasing the demand for heat, the effective size of the building is reduced as is the required size of a solar heating installation, whether of the type which uses windows and the thermal mass of the building or the type which attaches to the building and uses pumps, controls, fans, heat exchangers, and heat storage. Not only is the initial cost of the building reduced because of a smaller solar energy installation, but the long-term energy costs are also lower because the heating and cooling demands are smaller.

Apart from the solar system, other components of buildings can be reduced in size because of energy conservation. When the heating and cooling loads are sufficiently reduced, the mechanical equipment which backs up the solar system may also be reduced in size. Reduction in equipment size may mean a corresponding reduction in the size (and cost) of furnace flues, electrical wiring for mechanical equipment, and ducts. The length of ducts and pipes may also be shorter since they may not need to extend to the perimeter of the building in order to maintain comfort during the heating season.

A tighter building which uses less energy for heating also provides a more comfortable environment. Insulation reduces the cold wall effect. Interior surfaces of uninsulated walls can be as much as 8° to 14° colder than insulated walls in winter. Insulation increases the temperature of interior surfaces, and people feel more comfortable. If heat is lost too rapidly by body radiation to a cold wall, floor, or ceiling, a person will feel uncomfortable and cold. In summer, reverse conditions are in effect when warm surfaces make it more difficult for the body to remain cool.

When the interior surface temperatures of a building are uncomfortable because of the cold wall effect, people will set the thermostat higher in the winter to stay warm and lower in the summer to stay cool. This requires more energy. The effect of raising the temperature setting of a thermostat in the winter has the reverse effect of lowering it: it results in the use of more energy.

Tighter buildings also help to create more uniform room-to-room and floor-to-ceiling air temperatures. With cold walls, the air adjacent to them becomes chilled, increases in density, and falls to the floor. This displaces warm air, which rises. These continuous air movements (drafts) cause discomfort. The infiltration of outdoor air through cracks in the building skin also cause drafts, and the reduction of infiltration also improves comfort.

Extra insulation which aids in reducing energy consumption can aid in acoustical privacy between the building and the outside. Even better than insulation in this regard are storm windows and doors or double glazing.

There are many important alternatives for reducing energy consumption which do not necessarily affect the design of the building. One noted mechanical engineer who promotes energy conservation in building design cites the construction of two identical schools with identical mechanical (HVAC) designs. The schools were built in similar climates, actually within several miles of each other. One school, however, reported energy consumption for lights, heating, cooling, and ventilation to be nearly twice that of the other school. The primary reason for this great difference in performance of the two buildings was attributed to the operation of the buildings and the equipment.

Some of the basic ways in which we can reduce energy consumption involve alterations in the way we live (and operate buildings), including living at lower temperatures, and, if necessary, wearing more clothes to maintain comfort. Contrary to the belief of many people, lowering the thermostat at night does save energy. Figure II·C·1 shows the percent fuel savings with night thermostat

SETBACK 8 HRS.; 10 P.M. TO 6 A.M.

CITY	5° SETBACK	7½° SETBACK	10° SETBACK
ATLANTA	11	13	15
BOSTON	7	9	11
BUFFALO	6	8	10
CHICAGO	7	9	11
CINCINNATI	8	10	12
CLEVELAND	8	10	12
DALLAS	11	13	15
DENVER	7	9	11
DES MOINES	7	9	11
DETROIT	7	9	11
KANSAS CITY	8	10	12
LOS ANGELES	12	14	16
LOUISVILLE	9	11	13
MILWAUKEE	6	8	10
MINNEAPOLIS	8	10	12
NEW YORK CITY	8	10	12
OMAHA	7	9	11
PHILADELPHIA	8	10	12
PITTSBURGH	7	9	11
PORTLAND	9	11	13
SALT LAKE CITY	7	9	11
SAN FRANCISCO	10	12	14
ST. LOUIS	8	10	12
SEATTLE	8	10	12
WASHINGTON, D.C.	9	11	13

Figure II·C·1.
Percent fuel savings* with night thermostat setback from 75°F. (*Minneapolis-Honeywell data, 1973.) Source: (NBS.)

setback from 75° for various cities in the country. Note that a setback of 10° can reduce energy consumption more than 10% in every city shown! The thermostat should also be lowered when the building space is not being used, during the day in homes when no one is home, for example, or on weekends and holidays when commercial space is not being used.

Older homes were designed so that they could, in a sense, be reduced in size during cold periods. Sections could be closed off and only small portions heated.

Draperies should be opened during the day to allow sunlight to penetrate the building (this is not true of north-facing windows, of course); they should be closed at night. Much better than shades are interior insulating shutters which can be closed tightly over windows, in effect transforming a window into a wall. This important concept is discussed further in this chapter.

Care should be taken in the opening of windows and doors. Every effort should be made to reduce the amount of cold outside air which comes through openings in buildings. Weatherstripping around doors and windows and other openings can be the best method of energy conservation. Likewise, in most parts of the country, storm windows pay for themselves in fuel savings within a few years.

In order to understand the methods of trapping heat within buildings, the concepts of Btu and *Degree Days* must be added to our vocabularies and used as fluently as we now use the terms "horsepower" and "miles per gallon." The definition of Btu is repeated often in this book: a Btu is the amount of heat needed to raise the temperature of one pound of water one degree Fahrenheit. Thus, 100 Btu are required to raise 100 pounds of water one degree and 100 Btu are required to raise one pound of water 100 degrees.

The concept of Degree Days is somewhat more complicated. In some sense, it is analagous to a man-day of work. The amount of work a man does in one day can be termed one man-day. Similarly, if the outdoor temperature is one degree below the temperature of the building for one day, the heating load placed on the building is measured in terms of one Degree Day. Standard practice uses an indoor temperature of 65° as the base from which to measure Degree Days since most buildings do not require heat until the outdoor temperature is between 60° and 65°. If the outdoor temperature is 40° for one day, then 25 (65° minus 40°) Degree Days result. If the outdoor temperature is 60° for five days, then 25 (five days times 65° minus 60°) Degree Days also result. Likewise, if the outdoor temperature is 64 degrees for 25 days, then 25 Degree Days result. The map of the United States (see Figure II·C·2) shows total heating Degree Days. The last column of Figure II·C·5 also lists Degree Day values for various cities throughout the country; a much more extensive list is in Resource Section 2·B. Degree Days can also be obtained from local oil dealers.

Perhaps almost as critical in understanding heat loss are the concepts of R-value and U-value. R stands for *resistance,* resistance to the conduction of heat. An R-value is a measurement that indicates the difficulty (resistance) that a unit of heat has in flowing through a particular material. The higher the R-value, the better the insulating value of the material.

The term U-value also indicates heat transfer. Unlike R-values, which are applied to single materials, U-values are applied to the composite of all materials involved, such as the combination of materials comprising a wall. For example, the U-value of a typical wall would be the total insulating properties of the exterior siding, sheathing, insulation, interior finish, air spaces, and air films.

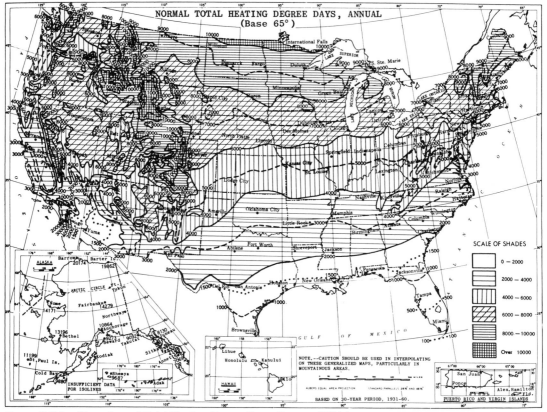

Figure II·C·2.   Mean Annual Total Heating Degree Days. Source: (NOA.)

The lower the U-value, the better the thermal resistance of a wall.

Mathematically, a U-value is the reciprocal of the sum of the various R-values of a combination of materials. If a wall has materials with R-values of 0.67, 1, 15, 1.12, and 0.18, the sum of the R-values is 17.97 (0.67 plus 1 plus 15 plus 1.12 plus 0.18). The U-value is the reciprocal of this or 1/17.97. This is equal to about 0.055.

U-values are measures of the flow of heat, in Btu, per square foot of surface during a one-hour period of time. The temperature differences between two sides of the wall (or other composite of materials) is, in a sense, the pressure which forces the heat through. If the outdoor temperature is 25°, there is more pressure on the inside of a building to push heat out than if the outdoor temperature were 60°. U-values, then, are measured in terms of Btu per hour per square foot per degree Fahrenheit. The notation used in this book is Btu/hr/ft²/°F; our sample wall above has a U-value of 0.055 Btu/hr/ft²/°F.

The vertical line charts show some typical R-values and some typical U-values (see Figures II·C·3 and II·C·4). More detailed summaries are listed in the Resource Section "Insulating Values of Materials Used in Building."

Typically, the concept of design temperatures is used by engineers to size

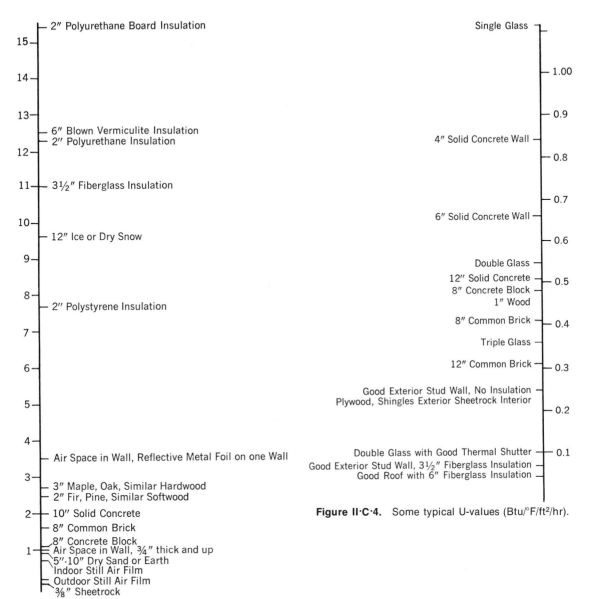

Figure II·C·4. Some typical U-values (Btu/°F/ft²/hr).

Figure II·C·3. Some typical R-values (°F/ft²/hr/Btu).

heating and cooling equipment. A design temperature is an extreme outdoor temperature which puts the greatest strain on the mechanical equipment. Since it is used so readily, buildings are often rated in their thermal performance by the number of Btu per hour which the heating (or cooling) equipment must produce in (or take out of) the building in order to keep it comfortably warm (or cool) when the outside air temperatures are at their extremes. Typical ratings for homes are 60,000 to 150,000 Btu/hr for heating, although with energy conservation measures those can be reduced by at least half. Since the design temperature

COOLING SEASON CLIMATIC DATA
Design Dry and Wet-Bulb Temperatures and Wind
Velocities in Local Use at Principal Cities

State	City	Design Dry-Bulb, deg F	Design Wet-Bulb, deg F	Summer Wind Velocity, mph	Highest Temperature Ever Recorded*
Ala.	Birmingham	95	78	5.4	107
Ariz.	Phoenix	105	76	6.0	118
Ark.	Little Rock	95	78	6.2	110
Calif.	Los Angeles	90	70	5.8	100
	San Francisco	85	65	10.7	101
Colo.	Denver	95	64	6.9	105
D.C.	Washington	95	78	5.9	106
Fla.	Tampa	95	78	7.4	98
Ga.	Atlanta	95	76	7.9	102
Idaho	Boise	95	65	5.8	112
Ill.	Chicago	95	75	9.5	105
Ind.	Indianapolis	95	76	8.9	106
Iowa	Des Moines	95	78	8.6	111
Kan.	Wichita	100	75	11.8	114
Ky.	Louisville	95	78	7.2	107
La.	New Orleans	95	80	6.9	102
Mass.	Boston	92	75	12.5	104
Mich.	Detroit	95	75	9.5	104
Minn.	Minneapolis	95	75	10.2	108
Mo.	St. Louis	95	78	9.5	110
Mont.	Helena	95	67	8.1	103
Neb.	Lincoln	95	78	9.7	115
N.M.	Albuquerque	95	70	7.8	99
N.Y.	Buffalo	93	73	12.1	95
	New York	95	75	12.5	102
N.C.	Asheville	93	75	5.6	99
N.D.	Bismark	95	73	9.5	114
Ohio	Cleveland	95	75	11.1	100
Okla.	Oklahoma City	101	77	9.8	113
Oreg.	Portland	90	68	6.5	107
Pa.	Philadelphia	95	78	9.7	106
S.C.	Charleston	95	78	9.8	104
Tenn.	Chattanooga	95	76	5.6	103
Tex.	Galveston	95	80	9.7	101
	San Antonio	100	78	7.8	107
Utah	Salt Lake City	95	65	9.8	105
Va.	Richmond	95	78	6.4	107
Wash.	Seattle	85	65	7.7	100
Wis.	Milwaukee	95	75	9.8	105
Wyo.	Cheyenne	95	65	9.2	100

*Temperatures abstracted from "Heating Ventilating Air-Conditioning Guide 1957".
Used by permission from ASHRAE.

varies with location, from 40° below zero in some cold areas of the country (lower in Alaska) to more than 30 above in some warm parts, such a rating does not give the absolute thermal performance of the building. Instead, the performance is

HEATING SEASON CLIMATIC DATA
Compiled from records of the U.S. Weather
Bureau and other sources

Average Temperature, Oct. 1–May 1	Lowest Temperature	Design Temperature Suggested by TAC	Average Wind Velocity Dec., Jan., Feb., mph	Direction of Prevailing Wind, Dec., Jan., Feb.	Normal Degree Days, Total for Year
53.9	−10	21	8.6	N	2618
59.5	16	31	3.9	E	1446
51.6	−12	21	9.9	NW	3005
58.6	28	32	6.1	NE	1390
54.3	27	—	7.5	N	3143
39.3	−29	0	7.4	S	5863
43.2	−15	14	7.3	NW	4598
61.9	10	31	8.2	NE	1161
51.4	−8	22	11.8	NW	3002
42.5	−13	—	4.7	E	4924
36.4	−23	−3	17	SW	6287
40.2	−25	2	11.8	S	5487
32.1	−35	—	12.2	NW	6909
40.2	−26	—	10.4	NW	5077
45.2	−20	9	9.3	SW	4428
61.2	7	36	9.6	N	1208
37.6	−18	8	11.7	W	5943
35.4	−24	4	13.1	SW	6580
29.6	−33	−15	11.5	NW	7989
43.3	−22	3	11.8	NW	4610
34.7	−49	−17	12.4	W	7119
37.0	−29	−2	10.9	N	6010
38.0	−13	—	7.3	NE	6124
34.7	−21	3	17.7	W	6935
40.3	−14	—	13.3	NW	5306
49.7	−2	20	7.3	SW	3281
24.5	−45	−21	9.1	NW	8969
36.9	−18	6	14.5	SW	6171
48.0	−17	14	12.0	N	3698
45.9	−2	22	6.5	S	4379
41.9	−11	—	11.0	NW	4749
56.9	7	26	11.0	N	1870
47.0	−16	—	6.5	SW	3665
53.0	−2	—	10.5	NW	2538
60.7	4	32	8.2	N	1424
40.0	−20	7	4.9	SE	5637
45.2	−7	—	5.2	NW	4082
45.3	3	24	9.1	SE	4864
33.0	−25	−6	11.7	W	7086
33.9	−38	−3	13.3	NW	7549

Figure II·C·5. Summer and winter climatic information for various U.S. cities.

based on both the quality of the building and its location. (Design temperatures are listed with Degree Days in Figure II·C·5.)

The Degree Day concept, on the other hand, evaluates the thermal perfor-

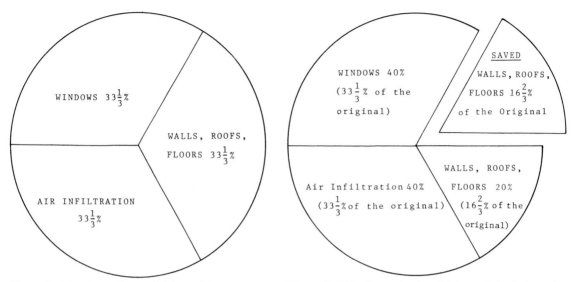

Figure II·C·6. Approximate division of heat loss for conventional housing.

Figure II·C·7. Approximate division of heat loss for conventional housing with greatly increased insulation.

mance in an absolute way by comparing buildings only with each other, independent of location. Buildings lose heat approximately in proportion to the difference between the inside and outside temperatures. Thus, the greater the number of Degree Days, the greater the heat loss. Buildings, then, can be rated according to the number of Btu which they lose per Degree Day. Typical ratings for homes range from 15,000 to 40,000 Btu/DD. But again, with energy conservation measures, these numbers can be cut at least in half.

An added advantage of the use of Degree Days is that the amount of energy which a building uses over the course of a year can be easily determined. For example, for a 20,000 Btu/DD house in a 5,000 DD climate such as New York City, the total amount of energy which the building loses through its exterior skin is about (20,000 Btu/DD) times (5,000 DD/yr). This equals 100 million Btu per year, the equivalent of about 1,000 gallons of oil per year at 70 percent furnace efficiency, or about 30,000 kwh of electric resistance heating.

> **NOTE:** Not all buildings demand heat energy based on the outdoor temperature. For example, a large office building may not need heat the entire winter because of the heat which is produced by people, lights, and machines. If these elements are eliminated from the calculations, however, large buildings can be evaluated by using a Degree Day method similar to that for small buildings.

Typical homes and most buildings lose heat in three primary ways. (A more detailed discussion of heat loss is included in the Resource Section under "Heat Theory.")

1. By conduction through walls, roofs, and floors (and to a much smaller extent by radiation and convection);

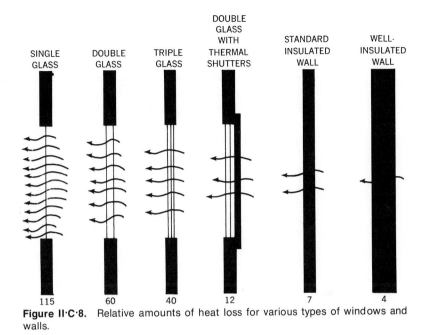

Figure II·C·8. Relative amounts of heat loss for various types of windows and walls.

2. By conduction (and by radiation and convection to a much smaller extent) through windows and other glass; and

3. By convection, the movement of air through the exterior skin of the building. This movement usually occurs through open windows, doors, and vents (intentionally or unintentionally) or by infiltration, the passage of air through cracks in the skin of the building, e.g., around the frames of doors and windows.

Depending on whether the building is well-insulated or not, whether it has many or few windows, or whether there is much air movement through it or not, each of these three components accounts for 20 to 50% of the total heat loss of the building.

Suppose that the heat loss of a building is evenly divided among the three components mentioned above. The pie diagram (Figure II·C·6) divided into thirds graphically displays this. If any one of these components is reduced by half, the total heat loss is reduced by only one-sixth (see Figure II·C·7). This fact should force us to look at all three issues at once instead of emphasizing one or the other.

In analyzing the basic sources of building heat loss, it is almost impossible to treat them as separate issues. For example, by adding storm windows to existing windows, conduction loss is reduced by half, but air infiltration is also greatly reduced (and cooling loads are lowered, too).

Note in Figure II·C·8 that different types of glass and wall construction vary greatly in the amount of heat which is lost through them. Since the values in the diagram are only conduction losses, these relative values are not absolute in their accuracy. For example, the construction details which must be used with glass result in greater losses through air infiltration than do those which use opaque materials to form walls, roofs, and floors.

Under the same indoor and outdoor conditions, a single pane of glass will conduct more than 115 Btu, while two layers of glass will conduct 60; three layers of glass, 40; two layers of glass plus an interior insulating shutter, 12; and a standard wall, 7. Very well-insulated walls and roofs will conduct 4. In other words, two layers of glass will conduct about half what a single layer will; an excellent wall will conduct about one-thirtieth (about 4%) of what a single layer of glass will. The same amount of heat will be lost through a well-insulated wall 30 ft long and 8 ft high as through a single glass window 2 ft wide and 4 ft high.

The use of insulating shutters for closing off windows at night or during periods when windows are not required greatly reduces heat loss. They can effectively reduce radiational heat loss and, depending on the construction details, almost totally eliminate air leakage. Depending on the thermal resistance of the insulating shutter, conduction heat loss through a window with a shutter can be reduced to one-tenth that of a window without. Figure II·C·8 shows loss reduced by a factor of five, from 60 down to 12.

The initial cost of shutters is high; the hardware is especially expensive. It is difficult to locate the shutters and to place them somewhere out of the way when they are not being used. Another difficulty is making sure that they are closed at night. One possibility for operation on a large scale is to use photosensitive cells which would cause the shutters to close when the sun sets and open again when the sun rises. Skylid®, discussed as a shading device in Chapter II·A, is also an effective insulating shutter, automatically opening to admit sunlight and closing to trap the heat inside the building when the sun is not shining.

As a simple approximation of the savings that can result from using shutters, compute the window heat loss by conduction and compare this to the reduced losses with shutters. When shutters are opened only during daylight hours, between 40 and 65% of the hours during the heating season, heat loss will be greatly reduced during the remaining 60 to 75% of the time by the shutters. A Degree Day load of 5,000 and single glass with a U-value of 1.15 Btu/hr/ft²/°F give a heat loss of 138,000 Btu per square foot of glass per season. If a shutter is applied which gives an overall U-value for the combination of window and shutter of 0.12 Btu/hr/ft²/°F, and if the shutter is in place over the window one-third of the time, approximately 30% of the energy will be saved. If the shutter is in place half of the time, approximately 60% of the energy will be saved. In this latter case, about 80,000 Btu, or the energy equivalent of about one gallon of heating oil at 60% efficiency, is saved each heating season per square foot of window.

This computation ignores several factors which would serve to increase savings. For instance, more Degree Days occur during the night, when the shutters would be closed, than during the day. Radiational heat loss is also most significant at night. During daylight hours, when shutters are open, net heat loss is greatly reduced, and often eliminated, by solar heat coming through the windows. The energy saving value of shutters, therefore, is significant, and their use should be strongly considered.

Zomeworks Corporation has developed a partial solution to the need for

Figure II·C·9. Manually operated shutter by Zomeworks Corporation.

shutters or for variable insulation in a building's exterior skin. It is called Beadwall®. Fundamentally, Beadwall® consists of two parallel panes of glass which are spaced approximately 3 in. apart. When the sun is shining, it penetrates the building. When the sun is not shining and when heat loss reduction is desirable, tiny, white polystyrene pellets (beads) are blown (by a vacuum cleaner type of pump) from their place of storage to the space between the layers of glass, turning the glass wall into a well-insulated translucent wall. This system is discussed further in the next chapter.

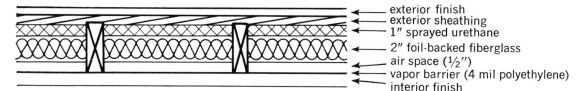

← exterior finish
← exterior sheathing
← 1″ sprayed urethane
← 2″ foil-backed fiberglass
← air space (½″)
← vapor barrier (4 mil polyethylene)
← interior finish

Figure II·C·10. A construction detail to obtain extra insulating values within a conventional wood-framed wall.

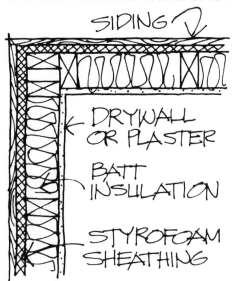

SIDING

DRYWALL OR PLASTER

BATT INSULATION

STYROFOAM SHEATHING

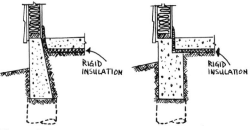

RIGID INSULATION

RIGID INSULATION

Figure II·C·12. If perimeter insulation is needed for slab-on-ground construction, install it as shown or on the exterior surface of the foundation walls. Source: (NAHB.)

Figure II·C·11.
One-inch thick styrofoam sheathing added to the exterior wall of conventional wood-framed construction.

Standard insulation techniques are also being elaborated upon as a means of tightening up buildings. One of the best references for dwelling insulation is *Insulation Manual: Homes, Apartments* by the National Association of Home Builders Research Foundation, Inc., in Rockville, Maryland (NAHB). Figures II·C·10 to II·C·16 give examples of some interesting details.

Heat loss by convection, the movement of air through the exterior skin of the building, can be a significant portion of the overall heat loss of a building. The portion is particularly large for buildings such as schools, hospitals, and auditoriums which require excessively large ventilation rates. Health codes require certain amounts of fresh, but treated, outdoor air; these amounts vary with the type and size of structure and with the number of people using the space. The rates have archaic origins in most cases, and their re-examination can result in significant savings when they are found to be excessive. A typical requirement is that a classroom, for example, have five "air changes" per hour. That is, the total volume of air within the room must be simultaneously exhausted and replaced with clean, healthy, outdoor air five times within one hour. In the winter, it may need to be heated; in the summer, it may need to be cooled. New OSHA health and safety codes require fans at many indoor chemical operations which exhaust air in such large quantities that the heat which is required to replace that air is far more significant than the amount of heat required to replace the heat which is

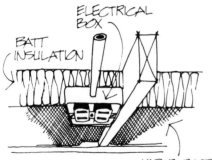

Figure II·C·13. Pack insulation around electrical boxes, switches and pipes.

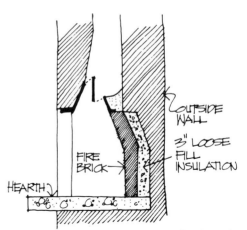

Figure II·C·14. Energy loss from fireplaces is cut by insulating behind the fireplace when it is against an outside wall.

Figure II·C·15. Provide the advantages of a sliding glass door without much of the energy waste by using a fixed pane of insulating glass and an insulated door.

Figure II·C·16. Adding a storm window to a window already having insulated glass cuts window heat loss by 30 to 50 percent.

lost by conduction through the walls of the building. Ventilation and exhaust rates for every operation should be re-examined in light of energy shortages.

For situations in which extremely large volumes of air are being exchanged between indoors and outdoors, heat recovery devices are receiving increased attention. In their simplest form, these devices transfer the heat from the exhaust air to the incoming air. During the summer, the incoming air is cooled by the exhaust air.

Smaller fans, such as those used in bathrooms and kitchens, are responsible for smaller but still significant amounts of wasted heat. Fan systems which filter and recycle rather than exhaust should be considered.

At other scales of air movement, inside and outside air are mixed when windows and doors are opened and closed. To reduce the energy needed for heating and cooling, there should be two doors at each opening. If necessary, these two doors can be next to each other, as with a storm door attached to a standard door. Better than this, however, is separating the doors from each other by a porch or vestibule so that when the exterior door is open, the interior door cannot be. This creates, in effect, a decompression chamber. Revolving doors are better than single doors for locations with intense use, and, in combination with vestibules, they are significant energy savers.

Wind is the most important factor in the moment-to-moment variation in the amount of air which penetrates a building. In *Design With Climate,* Olgyay reports that a 20 mph wind doubles the heat load of a house that is normally exposed to 5 mph winds and that a belt of sheltering trees increases in effectiveness at higher wind velocities. Fuel savings can be as great as 30% with good protection on three sides of the building. In the northern hemisphere, the north and west sides of a building are generally most exposed to wind. Therefore, buildings should be oriented away from the prevailing winds or have screens (natural vegetation or man-made) to avoid air leakage around doors, windows, and any other openings. Entrances should not be located on the north and west, but if they are, wind protection is particularly important. Cool summer winds often prevail from the west, and winter storms often blow from the northeast (see Figure II·C·17).

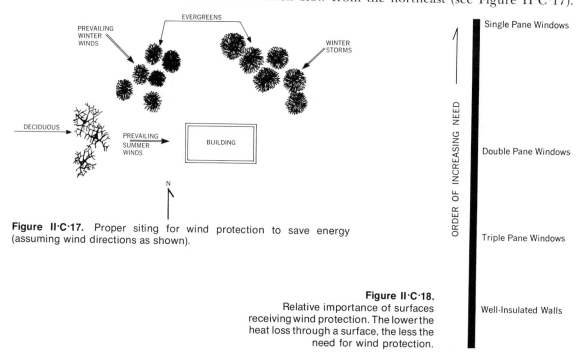

Figure II·C·17. Proper siting for wind protection to save energy (assuming wind directions as shown).

Figure II·C·18. Relative importance of surfaces receiving wind protection. The lower the heat loss through a surface, the less the need for wind protection.

The importance of wind protection on the conduction loss through surfaces of a building varies with the U-value of the surfaces. The higher the U-values, the less the insulative value and the more necessary it is to protect the surface from wind. A single-pane window, therefore, needs wind protection much more than a well-insulated wall does (see Figure II·C·18). This is due primarily to the variation in thickness of the exterior air film which clings to the outside surface. This air film and a similar interior air film are the primary contributors to the insulative value of a single layer of glass. As the thickness of this air film varies according to the variation in the velocity of the air which strikes the surface, the effective insulative value of the surface also varies. For windows, then, this variation is significant. For well-insulated walls, however, the exterior air film represents only a small fraction of its total insulative value and in fact can have a less than one percent effect on heat loss (see Figure II·C·19).

SECTION

exterior wall

outside air film with wind

outside air film without wind

inside air film

Figure II·C·19.
Effect of wind on the thermal resistance
of the outside air film of a wall.

Perhaps the most important aspect of air movement through the building as it relates to energy consumption is the penetration of air through the cracks and air spaces in the walls, roofs, and windows. The creation of dead air spaces in the walls of buildings and the tight fit of windows and doors in their frames can greatly reduce the effect of air infiltration. Air infiltration through cracks in the surface of a building is the most important consideration in wind control. This phenomenon is discussed at length in the Resource Section "Heat Theory." Some outdoor air is needed by people for ventilation and a feeling of freshness, and the natural penetration of air through the cracks is often accounted for in determining the necessary artificially-induced ventilation for complying with health codes. Every effort, however, should be made to reduce such uncontrolled air infiltration. As other factors in heat loss are reduced, the penetration of outdoor air becomes a larger portion of the remaining total. By reducing air infiltration to a minimum and, if necessary, opening a window or operating a fan to bring in fresh air, significant energy can be saved.

One of the main purposes of using building paper between the plywood sheathing and the exterior siding of houses is to reduce air infiltration. Trim details on a building's exterior are also important for reducing air penetration. Mortar joints in brick and concrete block facades should be tight and complete.

More important than cracks in the wall surfaces, however, are cracks or air spaces around the windows and doors, e.g., between the window sash and the jamb. For new construction, entrances should be designed to control air movement; weatherstripping is particularly important. As can be seen in Figure II·C·20, windows vary in their air infiltration losses, but fixed, inoperable, windows are the tightest. Not every window in a building need be operable, although there should be a sufficient number to provide natural ventilation and the necessary feeling of openness. Operable windows should be chosen for their tight fit when closed, not only when they are first installed but after they have been used hundreds of times. Pivoted and awning windows tend to be least tight and casements among the closest fitting.

ORDER OF INCREASING LOSS

— Rolled-Section Steel
Sash Industrial Pivoted

— Double-Hung, Wood Sash, Poorly Fitted
Non-Weather Stripped

— Double-Hung, Non-Weather Stripped

Double-Hung Wood Sash, Average Window,
Non-Weather Stripped
Double-Hung Wood Sash, Poorly Fitted,
Weather Stripped
Double-Hung Metal, Weather Stripped

Rolled Section Steel Sash Residential Casement

— Casement, Wood Sash

Figure II·C·20.
Relative air infiltration losses per
season of various windows.

— Fixed Windows

For existing windows which cannot be replaced, weatherstripping with durable, preferably metal, weatherstripping provides substantial reduction in air leakage. A storm window added to existing windows not only halves conduction heat loss for windows with single panes of glass but also reduces air infiltration to a similar extent. Double window sash—the standard operable window with single glass in combination with storm windows—is better than a single sash with insulating glass (see Figure II·C·21). A standard operable window with insulating glass in conjunction with storm windows (triple glazing) is, of course, an even better energy saver.

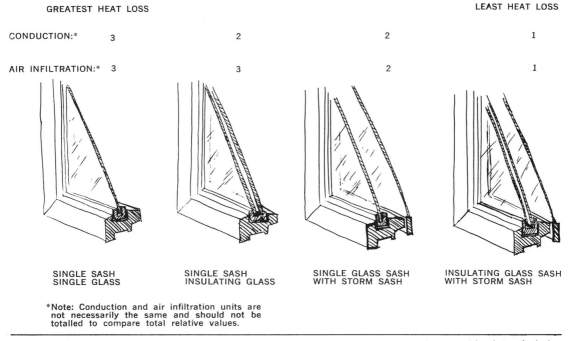

GREATEST HEAT LOSS   LEAST HEAT LOSS

CONDUCTION:*   3   2   2   1

AIR INFILTRATION:* 3   3   2   1

SINGLE SASH   SINGLE SASH   SINGLE GLASS SASH   INSULATING GLASS SASH
SINGLE GLASS   INSULATING GLASS   WITH STORM SASH   WITH STORM SASH

*Note: Conduction and air infiltration units are not necessarily the same and should not be totalled to compare total relative values.

Figure II·C·21.   Relative amounts of heat loss by conduction and air infiltration for various combinations of window sash and glass.

CHAPTER II·D

Putting It All Together: Examples

The first four sections of this chapter introduce basic concepts of designing with the sun, of constructing buildings which recognize the influence of the natural environment. They discuss how to allow the sun to penetrate at the proper time, how to use the building to store heat for later use, and how to slow down the escape of that heat from the building. Designing with these, as well as with the many other criteria in mind, is a difficult task. Indigenous architecture throughout the world has developed solutions according to local conditions. Bernard Rudolfsky's book, *Architecture without Architects,* is the best reference for examples of such design.

There are probably thousands of buildings which have placed a major emphasis on the integration of natural constraints with the many other design constraints. It is difficult to extract the best examples from this bountiful supply, and I would be pleased to receive information about other projects and buildings which have graciously and gracefully achieved this integration.

In 1935 Frank Lloyd Wright designed an entire community of what were called low-cost bermed houses. The earth was brought up to the level of the window sills. The designs were unorthodox, were not accepted by the general public, and were not executed. In the next part on low impact technology solutions involving solar energy, two houses in particular are featured. One, by Zomeworks Corporation, allows the sun to penetrate south walls of glass on sunny winter days and to warm large metal containers of water. The water stores the heat. The glass is covered with thermal shutters at night to prevent heat loss. During the summer the shutters are removed from the glass at night to cool the containers of water; they are closed during the day.

A house by Harold Hay is also featured in the next part. Instead of the sun penetrating through the walls of this house in southern California, it strikes the roof and is stored in large bag-like containers of water. Retractable shutters leave the bags uncovered and exposed to the sun during sunny winter days but close at night to prevent the escape of heat. During the summer, the shutters cover the water bags to keep the sun's heat out, retracting at night to allow the bags to cool.

These homes by Zomeworks and Hay stretch the basic concepts of the thermal performance of buildings and are reserved for discussion in the next part. Rex Roberts, however, was among the vanguard in the recent upsurge of interest in the design of basic and naturally sensitive homes. His book, *Your Engineered House,* is a primer in this respect. Another leader in the movement, Ken Kern, wrote *The Owner Built Home* and *The Owner Built Homestead.* Both publications are indispensable for laymen and professionals who are attempting to consider all areas of life-support systems in the design of homes.

In 1948 Wendell Thomas and his wife joined the Celo Community, Inc., in the mountains of western North Carolina. The first house they built there was a 32 by 24 ft shed roof dwelling called Sunnycrest. Its main feature was a 2 in. wide slot between the floor and the wall on all four sides of the house. This drained cold air from the walls down to the deep, completely dry, sealed cellar. The cool air was heated by the earth and tended to rise. A central heater on the main level induced air to rise around it from the cellar through holes in the floor. There were few windows on the east, west, and north, but much glass on the south. The earth was brought up part way around the east, north and west walls.

Thomas built a slightly smaller house in 1957, approximately 32 by 16 ft. The earth is bermed almost to the roof on the north and west and to the window sills on the south and east. The west wall of Sunnycave extends 4 ft beyond the south wall to keep winter's cold west wind off the south wall and windows. About half of the south wall above ground level consists of windows. There is little glass on the east, west, and north. Parallel to the south wall, wires are stretched on a 3 ft wide frame above the windows. Grapevines and woodbine cover this trellis, their leaves sheltering the glass from the sun's rays, and the respiration of the leaves cooling the air. In the fall, when sunlight is welcome, the leaves drop off.

The main windows are fixed, inoperable. Thomas sealed the inside pane, but left the outer one unsealed, preventing the glass from fogging. At night and on cold dark days, he covered the inside surface of glass with aluminum-painted insulation boards and then pulled heavy drapes across the front of them. Two of the large windows consisted of three layers of glass and were left uncovered for light.

In Sunnycrest, on the coldest winter morning, the temperature was 50°; on the hottest summer afternoon it was 85°. However, in Sunnycave the temperature never falls below 60° and never rises above 75° (see Figure II·D·1).

Malcolm B. Wells, the architect-conservationist, drew the sketch in Figure II·D·2. It is an elaboration of Wendell Thomas' Sunnycave, including earth on the top of the roof and natural, earth-cooled ventilation.

Mr. Wells has designed many buildings to reflect principles of ecology and energy conservation. His own office in Cherry Hill, New Jersey, is in a sense an underground office (see Figures II·D·3 and II·D·4). The office has two parts: one is bermed on three sides and covered by earth, with two doors and one skylight for openings to the outside; the other, although below grade, is bermed

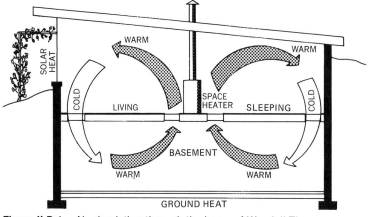

Figure II·D·1. Air circulation through the home of Wendell Thomas.

A FIREPROOF, EARTH-AND-TREE-COVERED HOUSE BUILT OF STANDARD PRECAST CONCRETE STRUCTURAL MEMBERS, THIS TYPE OF BUILDING IS SAID TO REQUIRE <u>NO</u> MECHANICAL OR ELECTRICAL HEATING OR AIR CONDITIONING! (A SIMPLER, WOOD-FRAME MODEL, BUILT IN THE CAROLINAS, HAD AN INDOOR TEMPERATURE RANGE OF ONLY 15° DURING AN ENTIRE YEAR!)

(WINTER) SOLAR HEAT THROUGH LARGE WINDOWS (H) IS SUPPLEMENTED BY EARTH-WARMING OF COLD-WALL AIR WHICH FALLS THROUGH PERIMETER SLOTS (F), IS WARMED, AND RISES THROUGH OPENING (G.). MASSIVE INSULATION (A) PREVENTS HEAT LOSS. ((B) SIMPLY INDICATES EXCAVATION.)

(SUMMER) BIG SOUTH-SIDE TREES AND ROOF OVERHANG SHADE GLASS. WARM AIR UNDER ROOF ESCAPES BY GRAVITY AT VENT WINDOWS (I.), PULLING IN EARTH-COOLED AIR THROUGH BURIED CONCRETE PIPES (D-E), WHICH CIRCULATES THROUGH HOUSE BEFORE BEING RE-LEASED AT (I) AS HEATED AIR.

Figure II·D·2. An earth-cooled, solar-heated house. (Drawn by Malcolm B. Wells, Architect.)

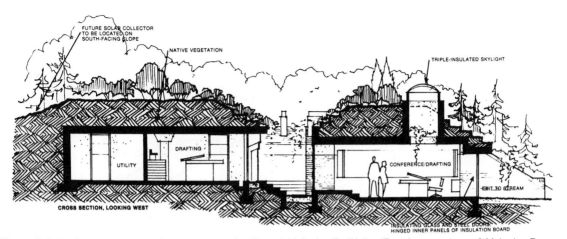

Figure II·D·3. Section through the underground office of Malcolm B. Wells. (Drawing courtesy of Malcolm B. Wells, Architect.)

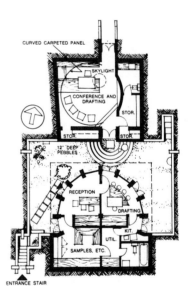

Figure II·D·4.
Plan of Malcolm B. Wells'
underground office. (Drawing courtesy
of Malcolm B. Wells, Architect.)

on only one side and covered with earth, but is exposed on three sides to the courtyard. Both units are concrete, and the earth cover is 3 ft deep. Wells uses a continuous sheet of butyl rubber between the concrete roof and the earth as a means of shedding water. The three exposed glass facades collect sunlight and natural daylight. Insulating shutters are applied at night.

A house which Mr. Wells designed at Raven Rocks in southeastern Ohio, takes his ideas even further (see Figure II·D·5). The north wall and roof is buried into the side of a hill. Its long axis, oriented east-west, provides for extensive southern exposure. The vertical glass is easily shaded in the summer by an exterior trellis. The sloping glass running the length of the top most "via" space is the primary "solar collector". The massive concrete floors and walls absorb

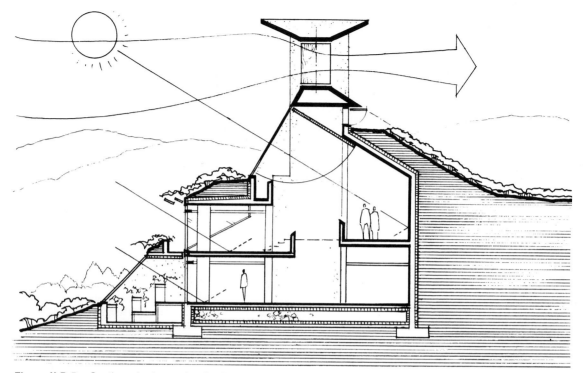

Figure II·D·5. Section of a house for the Raven Rocks Corporation showing solar heat gain and the channelling of wind through wind generators. House design by Malcolm B. Wells Architect.

a large percentage of the heat. However, solar heated room air can also be blown by a fan through a two-foot deep gravel bed under the floor for use later. Insulating panels cover the glass at night. At no point, including corners and construction joints, is concrete exposed either to air or to the ground. Instead, heat loss is reduced by the continuous application of thick rigid board insulation.

Because of the tremendous mass of the house and the large quantities of insulation, the house temperature will drop only 3% per day during average January weather with no auxiliary source of heat. For example, if the house is warmed to 75° initially, the temperature will be no lower than 61° after five days. Wind energy is converted into electricity by being funneled into wind collectors to increase its speed.

In Denver, architect Richard Crowther designed his own home for energy conservation. He combined energy conservation features with passive solar designs. Entryways are buffered from winds and storms. The east-west axis is the most favorable orientation, and the small amount of surface area of the house minimizes heat loss. Crowther attempted to minimize the need for artificial lighting by using natural daylight to light interior spaces; special skyshafts have negligible thermal heat loss. There are no west windows on the house and only a small number of north windows. All of the windows are of two layers of glass and

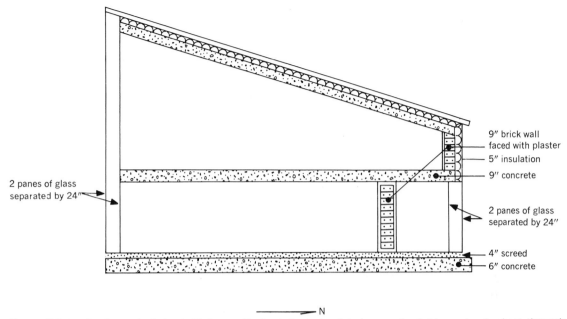

Figure II·D·6. St. George's School, Wallasey, England, heated solely by people, lights, and solar heat through windows.

are fixed in place to reduce air infiltration.

The roof and walls have a U-value of about 0.05 as contrasted to 0.075 for conventional walls in residences with 3½ in. of fiberglass insulation. Roof plenums and vent stacks are heated by solar radiation, inducing internal air convection for natural ventilation and passive temperature modulation. During the winter, solar heat penetrates through the south glass and heats up the floor slab. Large plants in the house provide additional humidity during the winter.

The annex to St. George's School in Wallasey, Cheshire, England, was built in 1961 to serve 300 students. It was designed by the late A. E. Morgan to gain directly as much as possible from solar radiation and to be of sufficient thermal mass (concrete and brick) to store the heat for long periods of time and prevent wide fluctuations in indoor temperatures. Additionally, the building insulation traps the solar heat so well that little other heat is needed. In fact, the only other sources of heat are the occupants and the electric lighting. Winters in Wallasey, located at 53°N, feature high winds and numerous cloudy days.

The building is a long, thin, two-story structure running east-west (see Figure II·D·6). The ground floor consists of 4 in. of screed over 6 in. of concrete The intermediate floor is of 9 in. concrete and the roof of 7 in., covered with 5 in. of expanded polystyrene insulation. Partition walls are of 9 in. brick.

The south, or solar, wall is 230 ft long and 27 ft high. For the most part it consists of fixed, double glazing with a 24 in. space between the layers of glass (probably for cleaning). Each classroom, however, is provided with several operable sash.

The Thermal Research Group of the Department of Building Science at Liverpool University, headed by Dr. M. G. Davies, has made extensive studies of the thermal performance of the building. For nearly all weather conditions, solar energy is the principal source of heat. On a sunny day, for example, solar energy accounted for six times the heat gain due to the occupants and three times the heat gain due to artificial lighting. The indoor temperature variations lag behind the outdoor temperatures by nearly a week. Temperature fluctuations within the building are less than 3° even on sunny days, accounted for in large part by the ability of the large thermal mass of the building to absorb the solar heat quickly.

On a much larger scale are two demonstration projects sponsored by the General Services Administration. Both projects are federal office buildings. The one in Saginaw, Michigan, designed by Smith, Hinchman and Grylls, is an environmental demonstration project. The original design includes a park for residents of the area with planting around the building and on approximately one-third of the roof. The wall surfaces near the entrances both inside and outside the building are panels made of rubble from demolished buildings. The paved surface of the parking area is also recycled rubble. Walkways are porous panels of recycled brick; the porosity eliminates the necessity for special drainage systems to accommodate water runoff. There are numerous other environmental features, including special flushing systems and water-saving requirements. The design includes some earth berming of the walls and earth covering of a portion of the roof. The double-glazed windows are protected in the summer by large overhangs. A large flat plate solar collector is built on a separate structure above the building and is expected to heat all of the domestic hot water and approximately 70% of the building's winter heating needs.

The other GSA project is an energy conservation demonstration. It was designed by Isaac and Isaac, architects in Manchester, New Hampshire, where the building is located. Dubin-Mindell-Bloome Associates was the energy consultant. The construction of the building itself is unique in that the U-values for the walls, roof, and floor are 0.06, considerably below the average (0.15 to 0.25) for buildings of this type. The mass of the wall for storing heat and cool and modulating the effects of outdoor temperatures is 100 lb/ft² compared with the normal 20 to 30 lb/ft² for similar types of buildings. The insulation is located on the exterior of this mass. There are no windows on the north wall, and windows on the east, west and south walls comprise only 10 to 15% of the wall area. All windows are double-glazed and are designed to minimize air infiltration.

In addition to achieving a high level of thermal performance by the building itself, there are other energy conservation features. Stairs, elevators, toilets, and mechanical rooms are located adjacent to the north exterior wall; since they do not require close environmental control, they act as a buffer zone. Different mechanical, electrical, and lighting systems on various floors permit direct comparison of performance and efficiency. For example, one floor has uniform overhead lighting, another floor has lights incorporated into the furniture, and other floors have non-uniform, relocatable, task-oriented lighting of various types. One floor has more windows that the other in order to determine the benefits of natural lighting for the perimeter of buildings.

The operation of the building is also designed to save energy. For example, custodial work will be done only during normal working hours so that mechanical equipment can be shut down early. The temperature will be set back one hour before closing time in the winter and will be turned down further at night.

Although these are only a few examples of buildings which are sensitive to the influences of the natural environment, they help to support some points made so far. Tremendous gains in energy savings can be accomplished through proper building design. To do so, buildings should be designed as solar collectors, as solar heat storage devices, and as heat traps. The next part discusses intermediate means for using solar energy, the direct design of a building and the actual use of an appliance-type solar energy system.

REFERENCES FOR PART II

(HUT.1) Hutchinson, F. W. "The Solar House." *Heating and Ventilation,* March 1947.

(HUT.2) Hutchinson, F. W. "Solar House: Analysis and Research." *Progressive Architecture* 28(May 1947):90-94.

(KER.1) Kern, Ken. *The Owner-Built Home.* Oakhurst, Calif.: Owner-Builder Publications, 1961.

(KER.2) Kern, Ken. *The Owner-Built Homestead.* Oakhurst, Calif.: Owner-Builder Publications, 1974.

(NAHB.) NAHB Research Foundation, Inc. *Insulation Manual.* Rockville, Maryland, September 1971.

(NBS.) National Bureau of Standards. "Technical Options for Energy Conservation in Buildings." NBS Technical Note 789. Institute for Applied Technology, Washington, D. C., July 1973.

(NOA.) U. S. Department of Commerce, Environmental Science Services Administration. *Climatic Atlas of the United States.* Washington, D. C., June 1968.

(OLG.) Olgyay, Victor. *Design with Climate: Bioclimatic Approach to Architectural Regionalism.* Princeton, N. J.: Princeton University Press, 1963.

(ROB.) Roberts, Rex. *Your Engineered House.* New York: M. Evans and Co., 1964. Available from J. P. Lippincott, Philadelphia.

(RUD.) Rudolfsky, Bernard. *Architecture without Architects.* Garden City: Doubleday and Co., 1964.

(TEL.) Telkes, Maria. "Solar House Heating - A Problem of Heat Storage." *Heating and Ventilating,* May 1974. Also Publication #19, MIT Solar Energy Conversion Project.

(WRI.) Wright, Henry N. "Solar Radiation as Related to Summer Cooling and Winter Radiation in Residences." A preliminary study for the John B. Pierce Foundation, New York, N. Y., January 20, 1936.

PART III

Low Impact Technology Solutions

As must be continually emphasized, probably the more sensible ways of using solar energy today are through processes that do not require high levels of sophisticated materials, complicated integration of components, systems, and subsystems, or many moving parts. Examples of "low impact" or "soft" technology are windows, insulating shutters for windows, shading devices, thermal inertia in buildings, and thermosiphoning solar collectors. Most of these were discussed in the previous chapter. An elaboration on the theme of the building as the solar collector is the synthesis of these low impact solutions with the other components of the building. Such solutions can be considered extensions of the building rather than appliances. A distinction between "extension" and "appliance" and, therefore, the division between these "soft" solar designs and some of those discussed in Part V are often hazy. This is necessarily so, but an effort has been made here to recognize with one, perhaps quick, hand-waving the multitudinous implications of various designs on overall energy and resource savings, pollution abatement, and reduction of general environmental disorder (entropy).

In a personal communication with this author, Dr. Hoyt Hottel, who helped to build the first MIT House back in 1939, warned that it is easy to underestimate the difficulties (and costs) of constructing a solar energy house. The reaction to these difficulties is usually to develop technologically complex solutions to the problems which plague the use of solar energy. Steve Baer, Harold Hay, Harry Thomason, and others, however, have tried to simplify their designs in low impact but sophisticated ways.

An economic comparison between low impact and high impact technologies will frequently favor the low impact route. And frequently, such systems are not only cheaper but more efficient!

For example, assume that 1000 Btu are striking each square foot of a window each day. If the window has two layers of glass, about 750 Btu will actually penetrate the space. If the average outdoor temperature is 35°, the heat loss in 24 hours is about 400 Btu. The net gain is roughly 350 Btu. The window as a solar collector has a 35% efficiency (350 divided by 1000). This is comparable to the seasonal efficiency of an appliance, the on-the-roof-type collector, but what a difference in price and overall simplicity.

Now, on the second day there may be no sun, but the heat loss through the window continues. Its net gain may be zero, so the overall efficiency is also zero. The solar collector (appliance type) has no efficiency since it does not operate. Suppose insulating shutters were placed over the windows at night, or approximately two-thirds of the time. In a two-day period, the window collects heat, 750 Btu, for eight hours and loses heat for 48 hours. During sixteen of those 48 hours the shutter is removed and the window loses about 250 Btu. During the remaining 32 hours, the shutter is in place, the per hour heat loss is reduced by a factor of five, and only 100 Btu is lost. The total heat loss for the two days is 350 Btu; the total solar heat gain is 750 Btu; and the net heat gain by the window "solar collector" is 400 Btu (750 minus 350) or 40% (400 divided by 1000) overall efficiency. This is the same or better than most fancy, well-engineered systems, but considerably cheaper and simpler. This chapter discusses solutions intermediate in cost and complexity between windows and solar energy "systems" which use an abundance of controls, valves, piping, ducting, pumps, fans, and heat exchangers.

A critical issue relating to low impact solutions is the degree of effort *(user participation)* necessary for comfort. For people accustomed only to operating a thermostat, even the simple opening and closing of shutters would be an imposition. The first chapter made it clear that we must change the way we live if we are to adapt to the new environmental imperatives. Low impact solutions usually make most sense in the context of this adaptation, but it is obvious that many of the designs can be incorporated directly into our present style of life, albeit with somewhat more difficulty.

CHAPTER III·A

Heat Collection

Thermosiphoning solar panels circulate air or water naturally without an auxiliary source of power, such as a fan or pump. This principle is paramount to an understanding of simple but sophisticated uses of solar energy. As the air or water is heated by the sun, it expands and rises through the collector. This movement draws cooler and denser air or water from the solar heat storage or from the building.

The simplest form of thermosiphoning solar collectors is shown in Figure III·A·1. With most such collectors, cool room air is drawn into the base of the wall, heated, and then expelled at the top.

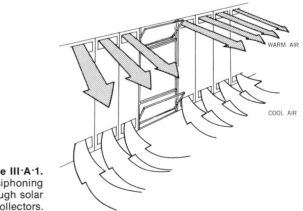

WARM AIR

COOL AIR

Figure III·A·1.
Natural thermosiphoning
of air through solar
collectors.

Some of the most significant work in the area of passive uses of solar energy and, in particular, thermosiphoning collectors, is being done at the Centre National de la Recherche Scientifique (C.N.R.S.) by its director, Professor Felix Trombe. The much-pubicized solar furnace houses the laboratories in Odeillo, France. This building itself is an excellent example of the use of solar energy for space heating done on a large scale but with no moving parts and with sophisticated simplicity. Figure III·A·2 shows the building, seen from the east. The north side of the building is concave so as to focus the sun, which strikes 63 mirrors on the hillside to the north of the building. The mirrors track the sun and reflect the light onto the parabolic reflector. The other walls, however, are compositions of windows and passive thermosiphoning solar collectors, from which the nine story building receives approximately half of its heating requirements. The solar collectors are designed so that black corrugated metal panels are located behind the glass panels which cover the east, south, and west walls of the building Figure III·A·3 is a schematic of the collector. Solar radiation penetrates through the glass and strikes the corrugated metal panel which is contained within the area defined by the glass and duct. As the metal heats, so does the air between the absorber panel and the glass. As the air heats, it rises upward and out through the vent and into the room. Cooler room air simultaneously is drawn through the same vent and down between the back of the absorber and the duct. Figure III·A·4 is an exterior view of the windows and collector.

Figure III·A·2. View of east-facing wall of C.N.R.S. office-laboratory building at Odeillo. (Photo courtesy of J. D. Walton.)

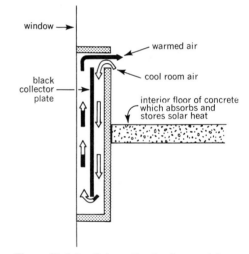

window →

warmed air

black collector plate →

cool room air

interior floor of concrete which absorbs and stores solar heat

Figure III·A·3. Schematic of a thermosiphoning solar collector at the C.N.R.S. office-laboratory building, Odeillo, France.

No provision except the mass of the building (the concrete floor in particular) is made for storing heat in this system, and it is most effective when the sun is shining. Reports indicate that the temperature in the offices and laboratories is maintained at a relatively constant level. Even during February, auxiliary heat is required only at night and on overcast days.

Odeillo has weather conditions which are particularly suitable to this design. As much as ninety percent of the daylight hours are sunny over the course of the year. Temperatures are relatively cool during the summer months, making possible the use of east- or west-facing collectors which in hot climates would greatly overheat most buildings. Also, in contrast to many large buildings, this one does not have the year round cooling load which might require solar collectors capable of obtaining higher temperatures to drive absorption-type cooling. Nevertheless, the system is a significant prototype for large-scale passive systems.

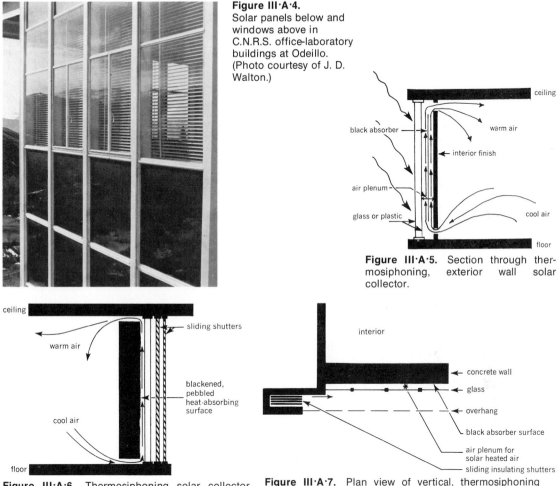

Figure III·A·4.
Solar panels below and windows above in C.N.R.S. office-laboratory buildings at Odeillo. (Photo courtesy of J. D. Walton.)

Figure III·A·5. Section through thermosiphoning, exterior wall solar collector.

Figure III·A·6. Thermosiphoning solar collector with insulating shutters.

Figure III·A·7. Plan view of vertical, thermosiphoning solar collector with sliding insulating shutters and concrete wall heat storage.

In theory, no insulation is needed in the wall between the absorber and the interior of the room (see Figure III·A·5). However, to help reduce heat loss at night, the wall should be adequately insulated; insulating shutters can be used on the outside of the glass when the sun is not shining. Figures III·A·6 and III·A·7 show sliding shutters located outside the building. The space between two transparent covers (glass or plastic) can be filled with polystyrene beads; in this Beadwall® by Zomeworks Corporation, the beads are evacuated in the morning to allow sun to penetrate and are used to fill the space at night to insulate the building.

A conventional absorber "plate" is not absolutely necessary in themosiphoning collectors. In the Chapter on air-type collectors in the last Part of the book, many alternatives to the standard flat metal sheet are discussed. Not mentioned there is the option shown in Figure III·A·6 in which the absorber surface is the actual wall surface itself, in this case, blackened pebbles cast in concrete.

An absorber surface developed by Jim Peterson and Marc Thomsen of Boulder and Jerry Plunkett of Denver is constructed of aluminum soda and beer cans cut to two inches high and attached to a sheet of plywood. The whole assembly is painted black and covered with plastic or other transparent material. Approximately 10 cans are required for each square foot of collector. Figure III·A·8 is a variation on this theme with the cans cut into halves and mounted onto the standard plywood sheathing of conventionally-framed houses. Important design criteria for air collectors are discussed at length in the last Chapter.

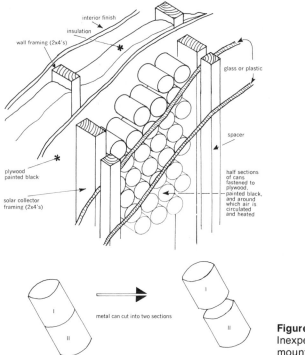

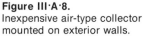

Figure III·A·8.
Inexpensive air-type collector
mounted on exterior walls.

As a means of having a view and of obtaining summer shading, insulation at night, and thermosiphoning (or fanforced) solar heated air, venetian blind-like louvers can be located between two layers of glass spaced several inches apart (see Figure III·A·9). One side of the louver is black and the other is reflective silver. There are many possible positions and functions of the louvers (see Figure III·A·10):

1. In order to obtain direct solar heat gain into the space, the louvers can be pulled up to the top of the space for maximum glass exposure, or
2. They can be left down but turned so as to parallel the light rays.
3. For controlled solar heat gain, the louvers are in the down position and

slightly tilted with the black toward the sun: this lets some light and heat in.

4. While in position 3, air can circulate between the panes of glass and come in contact with the hot black surface. This hot air can be taken to a solar heat storage area or be delivered directly to the space, e.g., to the northern side of the building which does not receive direct sun.

5. To prevent completely the sun from penetrating through the glass, the louvers can be turned to form a vertical continuous surface. During the winter, the black sides can be turned out and heat can be collected as described in 4. They would remain in this position throughout the night, reflecting the heat back into the space. If the louvers were thick enough, they could be filled with insulation, reducing heat gain during the summer and heat loss during the winter.

6. During the summer or other hot weather, the silver side can be turned to the outside, reflecting the sun's heat and keeping the space cool. They would stay in this position throughout the night, reflecting heat out and keeping cool in.

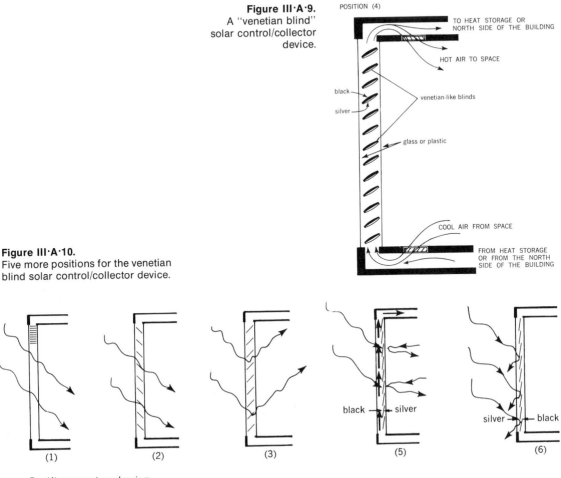

Figure III·A·9.
A "venetian blind" solar control/collector device.

Figure III·A·10.
Five more positions for the venetian blind solar control/collector device.

For (4) see previous drawing.

In order to increase the efficiency, to provide greater control of warm air movement, and to allow a smaller air plenum in the absorber, a fan can be added to the exhaust (or supply) end of the collector (see Figure III·A·11). A fan can also be used to transport the air to other parts of the building, such as to north rooms. By this means, a proper combination of windows and solar collectors can simultaneously heat both the rooms exposed to the sun (using windows) and those in shade (using the fans to transport the solar heated air).

Dampers to control air flow are sometimes needed, especially to control the cooling effect of reverse thermosiphoning. This occurs when the sun is not shining, and the air in the collector is being cooled because of conduction and radiation heat losses to the outside. As the air cools, it travels down the face of the absorber and into the room, drawing warm room air in behind it at the top of the collector (see Figure III·A·12).

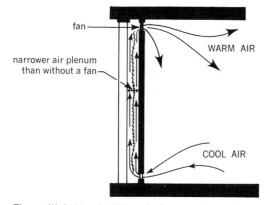

Figure III·A·11. Addition of a fan to a thermosiphoning solar collector.

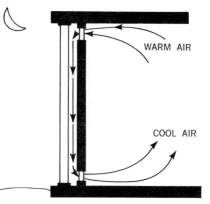

Figure III·A·12. Reverse thermosiphoning having a cooling effect.

The cooling mode can be of benefit, of course, on hot summer nights, but is a disadvantage in the winter. Dampers also control overheating of space during mild and hot weather. If properly designed, they can be used to induce natural ventilation through buildings, as shown in Figure III·A·13. Cool air can be

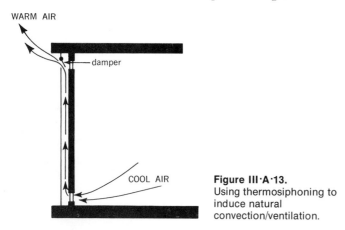

Figure III·A·13.
Using thermosiphoning to induce natural convection/ventilation.

automatically drawn into the building by the "chimney" exhaust system caused by the solar heated air.

In all cases, the dampers can be designed to operate manually or to operate automatically according to indoor and outdoor conditions. Fan pressure can also open and close them. Figure III·A·14 shows natural air pressure operating a damper. As with all air-type solar heating, the dampers should be simple in design and operation. They should close tightly, and there should be as few of them as possible.

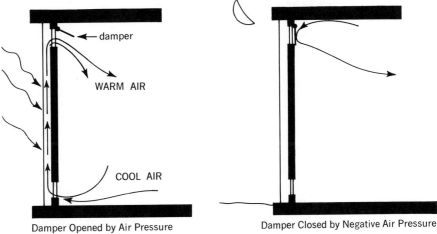

Damper Opened by Air Pressure   Damper Closed by Negative Air Pressure

Figure III·A·14. The use of dampers to prevent reverse (cooling) thermosiphoning.

CHAPTER III·B

Heat Storage

Another important and exciting element can be added to these simple heating systems: heat storage. As early as the 1940's, researchers at MIT were investigating the synthesis of solar heat storage with the solar collector itself. The resulting overall simplification of solar heating and cooling is compelling. By avoiding transport systems of ducts, pipes, fans, and pumps, as well as heat exchangers and complicated controls, significant amounts of money are saved, the operation and maintenance are simplified and reduced in cost, and comfort and efficiency can actually be increased. Performance of present-day systems may be better than when MIT did its work because of improved insulation techniques.

Figure III·B·1 shows the addition of the heat storage wall to the thermosiphoning, vertical solar collector. As the sun hits the blackened surface of the wall, the concrete in this case absorbs some of the heat while some of it simultaneously heats the air which rises and enters the room. The heat in the concrete migrates

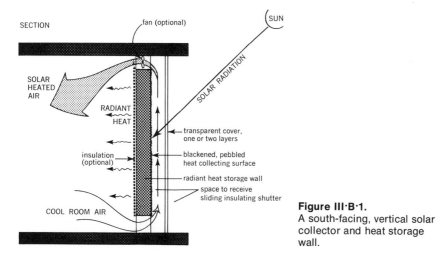

Figure III·B·1.
A south-facing, vertical solar collector and heat storage wall.

slowly inward, and, when the sun has set, radiates into the building while warm convection currents continue inside between the black concrete and the transparent cover. Such systems can be sized to maintain comfort for several days of sunless weather.

As with the cooling process due to reverse thermosiphoning, the heat storage wall can also absorb coolness at night during hot weather and store it for use during the day. Insulation can be located on the interior surface to reduce the amount of direct radiant heat from the wall and to prevent overheating. The heat stored in the wall is then retrieved primarily through the thermosiphoning air circulation. The south face of the house in Figure III·B·2 consists completely of glass, behind which is a concrete wall approximately 16 inches thick. The outside face of the concrete is rough in texture and painted black to absorb the solar radiation which penetrates through the glass and strikes the concrete. Figure II·B·3 is a schematic of the house. Not only is the concrete the solar collector, but it is also the heat storage. The concrete warms from the action of the solar energy which hits it. As the solar radiation passes through the glass (1) it is absorbed by the black coating (2) which heats the concrete wall (3). Since the long wave heat radiation cannot penetrate back through the glass as easily, the air between the concrete and the glass becomes heated and flows upward. As the hot air rises, it passes through vents in the top of the wall and across the room while simultaneously drawing cool room air in through ducts at the bottom of the wall. Excess heat is stored in the concrete for use when the sun is not shining. No means other than the natural flow of air is employed to circulate warm air to the room. To prevent cool night air from reversing this cycle, the lower inlet duct is situated slightly above the bottom of the collector so that the cool air settles there and is trapped.

Figure III·B·2.   Solar heated house built in 1967 at Odeillo, France. (Photo courtesy of J. D. Walton.)

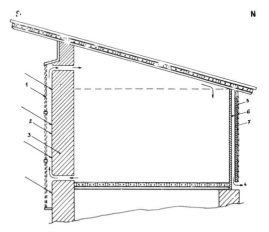

Figure III·B·3.   Schematic of solar heated house built in 1967 at Odeillo, France.

Figure III·B·4 is a closer view of the south face of the house, showing the circulation ducts at the bottom and top of the wall. Two of these houses have been occupied by engineers of the Odeillo Solar Laboratory since they were built in 1967. Approximately two-thirds of the heat used in these houses is provided by solar energy. Each of them has a floor area of approximately 1000 sq ft: the collector area is about 480 sq ft (28 sq meters). It is reported that the seasonal thermal energy provided by this system is approximately 200,000 Btu/ft² (600 kwh/m²) of collector surface (NSF.).

Figure III·B·4.
Solar heated house showing ducts for circulation of heated air. (Photo courtesy of J. D. Walton.)

Slots in the west wall of a foundation, similar to those for the first two houses, are shown in Figure III·B·5 for a three-unit dwelling showing the status of construction in February 1973. (The houses are now completed.) Figure III·B·6 is a model by C.N.R.S. of the dwellings designed by Architect Jacques Michel. Patentable aspects of the solar heating system are protected through l'Agence Nationale de Valorisation de la Recherche (ANVAR).

Figure III·B·5. West wall of foundation of three-unit solar heated dwelling showing slot in concrete for air movement. (Photo courtesy of J. D. Walton.)

Figure III·B·6. View of south and east walls of C.N.R.S. model of three-unit dwelling. (Photo courtesy of J. D. Walton.)

The schematic design of the solar heating system for this three-unit dwelling is shown in Figures III·B·7 and III·B·8. The main differences between this system and that of the first houses (see Figure III·B·3) are that windows are placed in the south wall in combination with the concrete/glass solar wall, the top air duct from the collector into the building is divided into two smaller ducts instead of one large one, and a valve or damper is provided at the top duct which can direct the flow of heated air either into or out of the dwelling. The winter heating mode is shown in Figure III·B·7. The dampers are positioned so that the heated air from the concrete/glass air space is ducted into the house. In the summer, the dampers are positioned in such a way that cool air is drawn in through an opening in the north side of the building and into the air space betweeen the concrete and glass. The air is heated (which is the drawing power for the cool north air) and is expelled outside through the damper at the top of the collector.

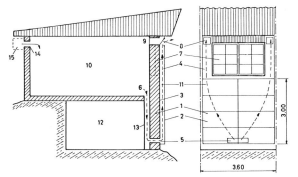

Figure III·B·7. Schematic of solar heating system of three-unit dwelling—summer operation. Source: (NSF.)

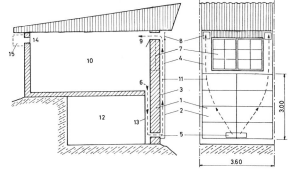

Figure III·B·8. Schematic of solar heating system of three-unit dwelling—winter operation. Source: (NSF.)

NOTES FOR FIGURES III·B·7 and III·B·8

Measurements shown are in meters

1. concrete heat storage mass
2. glass
3. blackened, rough-textured heat absorbing surface
4. air space for the movement of heated air
5. cool air inlet
6. cool room air
7. window
8. outlet and damper for the control of air flow into or out of the building

9. warm air room inlet
10. room/building
11. air flow within the collector/glass air space
12. garage or other space
13. partition wall
14. cool north air inlet and damper for summer ventilation
15. air purifier
16. insulation

The extension of the south collector wall several feet below the ground level not only increases the overall collector area but provides a trap for cool air during the winter, preventing a reverse cooling cycle.

The first solar heated house in Odeillo was built in 1962. The south facing wall, a combination of windows and solar collectors, is shown in Figure III·B·9. The primary difference between this system and that of the others at Odeillo is that water, instead of concrete, is the thermal heat storage. The solar collectors are essentially water radiators painted black and located between the south glass and the interior space. Water circulates through the thermosiphoning process (no pumps) and is stored in tanks in the ceiling area above the collectors. The tops of the radiators are just below the windows, and the storage tanks are above the wall between the ceiling joists. Figure III·B·10 is a schematic drawing of the heat collection system.

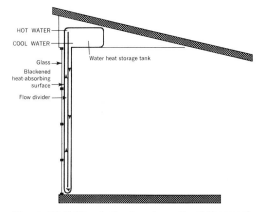

Figure III·B·9. South facing wall of solar heated house built about 1962 in Odeillo, France. (Photo courtesy of J. D. Walton.)

Figure III·B·10. Author's schematic of thermosiphoning hot water radiant solar heating system built at Odeillo in 1962.

The heat absorbing and storing wall can also be made of masonry blocks. They can be stacked and the voids filled with sand, earth, or plastic vinyl bags containing water. Unfilled voids can be used as air ducts in the thermosiphoning process. Stacks of water containers or small tubes of phase-changing eutectic salts (discussed in Chapter V·D) can also be used (see Figure III·B·11).

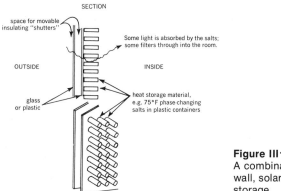

Figure III·B·11.
A combination window, wall, solar collector, heat storage.

One of the best promoters of low impact technology solar collection is Steve Baer of Zomeworks Corporation, Albuquerque, New Mexico. Baer developed a modified dome which he calls a "zome" and built a 2,000 sq ft, 11-zome cluster as a home. It has a concrete floor with adobe partitions (see Figure III·B·12). Its primary significance as a solar energy project is that its south-facing walls are made up of 55-gallon drums which are filled with water and stacked. Approximately 20 barrels comprise each wall, and they are separated from the outside by a single layer of glass.

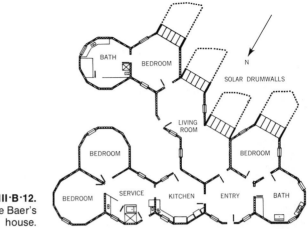

Figure III·B·12. Floor plan of Steve Baer's house.

An outdoor, reflective, insulating shutter is hinged at the base of each wall. It lies flat during the day with its reflective surface at ground level, reflecting additional sunshine onto the drums (see Figure III·B·13). At night the shutter is manually hoisted into a vertical position to reduce heat loss from the barrels to the outside (see Figure III·B·14). The shutter is aluminum-skinned and insulated with a cardboard honeycomb partially filled with urethane foam. Its estimated cost is $2/ft², and it weighs about 1½ lbs/ft².

Figure III·B·13. Exterior of Steve Baer's house showing the south-facing Drumwall solar collectors with the movable shutters. (Photo courtesy of Zomeworks Corporation.)

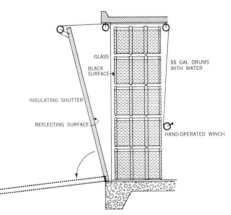

Figure III·B·14. Schematic of movable insulating shutter and Drumwall by Zomeworks Corporation.

The barrels are painted black and absorb heat, up to 1200 Btu/ft² on a sunny day. They release their heat into the living space through radiation, convection, and conduction. The rate of heat flow is controlled somewhat by a movable curtain between the barrels and the living space. One of the most striking elements of this design is the soft lighting which penetrates from between the barrels (see Figure III·B·15). Although his barrels lie in a horizontal position, Baer has also stacked them vertically in order to have fewer pounds of water per square foot of collector surface. His system can provide 75% of the heating with indoor temperatures fluctuating 10–15 degrees.

Figure III·B·15.
Interior view of Drumwall in Steve Baer's house. (Photo courtesy of Zomeworks Corporation.)

During the summer, the insulating shutters are lowered at night, and the barrels cool. During the day, the shutters are raised vertically, preventing the penetration of heat and trapping the coolness. Measures must be taken to prevent leakage of water from the barrels. Plans of the system are available from Zomeworks.

Other features of the house include honey cans filled with water which occupy space near the ceiling and are heated by skylights with shutters. A windmill pumps well water into a 5,000 gallon tank which gravity-feeds the residence.

Some of the many ways of incorporating drumwall into buildings are shown in Figures III·B·16 to III·B·19.

Total Environmental Action, Inc., a Harrisville, New Hampshire based research, education, and design firm, has followed the lead of Trombe and Baer with a house near Manchester, New Hampshire (see Figure III·B·20). The concrete walls are bermed with earth on the east, west, and north sides, and insulation is located between the concrete and the earth. The floor is also of concrete, but without insulation underneath it. The entire south side is a mix of windows and solar collectors. The only source of heat for the house in addition to solar energy is wood burned in wood stoves. The solar collectors are 12-inch-thick concrete walls exposed to the sun during the day and protected from heat loss to the outside at night by Beadwall®. Air circulates between the concrete and the Beadwall®. Solar heat is also admitted through the windows and stored in the concrete walls and floor. Insulating shutters are placed across the windows at night to reduce heat loss. The domestic hot water is preheated before reaching the hot water heater by circulating through pipes embedded in the concrete collector wall.

Of course, solar heat storage need not be built directly into the solar collector, nor need it provide all of the storage even if it is. The heat storage capacity of the building itself can be used to store excess solar heat. However, if that capacity is

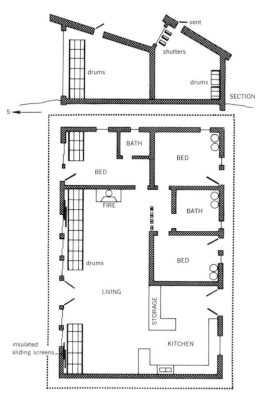

Figure III·B·16. A Drumwall house design by Jonathan Hammond. Source: (HAM.)

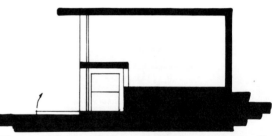

Figure III·B·17. Window seat hidden Drumwall with exterior insulating door. (Drawing courtesy of Zomeworks Corporation.)

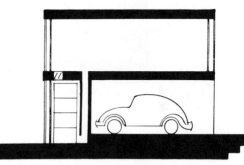

Figure III·B·18. Hidden drums with Beadwall insulation and return vent. (Drawing courtesy of Zomeworks Corporation.)

Figure III·B·19. Hidden drums with Skylid insulation. Air goes in and out through the same vent. (Drawing courtesy of Zomeworks Corporation.)

too small and if the heat storage wall cannot be designed to have enough heat storage capacity, more heat storage can be added elsewhere. Figure III·B·21 shows the auxiliary solar storage located under the floor. The figure combines several of the features already mentioned, including a fan which moves the air some distance from the collector wall.

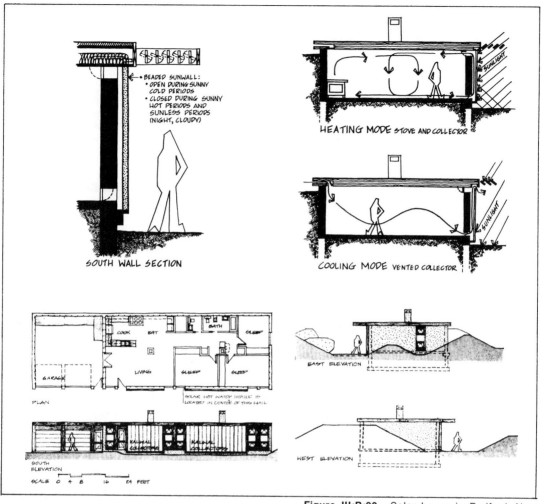

SOUTH WALL SECTION

- BEADED SUNWALL:
  - OPEN DURING SUNNY COLD PERIODS
  - CLOSED DURING SUNNY HOT PERIODS AND SUNLESS PERIODS (NIGHT, CLOUDY)

HEATING MODE STOVE AND COLLECTOR

COOLING MODE VENTED COLLECTOR

PLAN

SOUTH ELEVATION

SCALE 0 4 8 16 24 FEET

EAST ELEVATION

WEST ELEVATION

Figure III·B·20. Solar house in Bedford, New Hampshire, using concrete walls with Beadwall. Designed by Total Environmental Action, Inc. (Completed March 1975.)

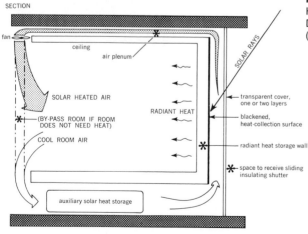

SECTION

fan

ceiling

air plenum

SOLAR HEATED AIR

RADIANT HEAT

(BY-PASS ROOM IF ROOM DOES NOT NEED HEAT)

COOL ROOM AIR

auxiliary solar heat storage

SOLAR RAYS

transparent cover, one or two layers

blackened, heat-collection surface

radiant heat storage wall

space to receive sliding insulating shutter

Figure III·B·21. A south-facing, vertical, solar collector and heat-storage wall in combination with remote heat storage.

An example of how the heat might be stored in the floor is shown in Figure III·B·22. The warm air is blown through the spaces between the floor joists. The spaces are lined with reflective foil to reduce heat loss and smooth the air flow. The heat is then transferred to containers of water in gallon size or less. The containers are plastic, glass, or metal and are arranged in irregular patterns. The concrete slab, though unnecessary, is applied as a means of storing heat gain through windows. Insulation is located between the concrete and the solar heat storage to reduce uncontrolled heat loss to the space.

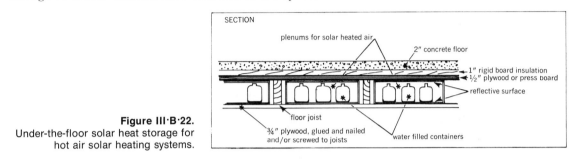

Figure III·B·22.
Under-the-floor solar heat storage for hot air solar heating systems.

Another sub-floor heat storage system has been proposed by Day Chahroudi, (see Figure III·B·23). Warm air from the collector is blown through the voids in masonry blocks lying on their sides under the floor. The warmed blocks in turn warm water contained in large waterbed-like plastic bags lying on the blocks. Warm air is drawn off the top of the bags and delivered to the building. Water can also be stored in large oil drums sitting in a crawl space. The warm air from the collector can be circulated around and through them, and the cool house air is circulated around the barrels and heated.

Of course, heat can be stored in a similar fashion in other parts of buildings such as ceilings, closets, and interior partitions. Just as the water containers were arranged between the floor joists in Figure III·B·22, so too they can be stacked on shelves in vertical walls. Similar methods have utilized tall containers of rocks (e.g., George Löf's own house in Denver).

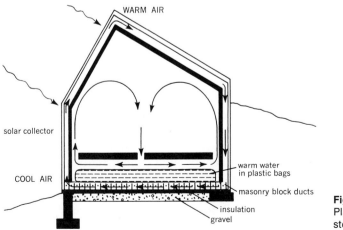

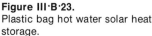

Figure III·B·23.
Plastic bag hot water solar heat storage.

If possible, the heat storage should be located within the confines of the space to be heated and cooled. The water in the storage tank in Figure III·B·24 is heated by a water-type collector. The rocks, in turn, are heated by the warm tank. Cool air from the room is drawn into the heat storage, rises, and re-enters the room as it is warmed by the rocks. When heat is not needed, the louvers on top of the heat storage are closed. Movable insulating panels can be used to separate the heat storage from the space. When heat is not needed, it is trapped by the insulation inside the heat storage. When the space requires heat, the insulating panels are removed and the heat radiates from the storage to the space.

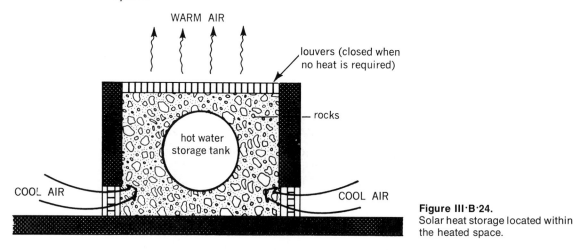

Figure III·B·24.
Solar heat storage located within the heated space.

The emphasis so far has been on south-facing vertical collectors, but with good reason, the most important of which is their relative ease of construction as compared with sloping roof-top collectors. This is particularly true of the glazing details of the transparent covers. The difference in cost between skylights and windows is testimony to this premise—due in part to the fact that it is more difficult to keep weather out of tilted and horizontal roofs and glass than out of vertical walls and glass. Another important consideration is the architectural constraint of large, steeply pitched roofs. Interior spaces under such roofs are difficult to use.

Large buildings have a great amount of wall area and less roof area. These walls can contribute significantly to a building's energy needs. Vertical walls facing south can be easily shaded in the summer. This prevents the collector from attaining high, potentially damaging temperatures. What little summer sun actually strikes a south wall can be used to help induce natural ventilation through the building.

The total amount of clear day solar heat gain on south-facing surfaces follows closely the seasonal need. In most parts of continental United States, the greatest heat gain on vertical south surfaces is in January and February, the coldest months, and the least is in July and August, the warmest months. As can be seen in the clear day insolation factors in the Resource Section, "Insolation," the mid-winter sunshine totals on vertical surfaces are less than 10% lower than on tilted surfaces, 1750 Btu/ft² per day versus 1900. When potential snow reflection of 10 to 30% is added to the 1750, vertical surfaces receive more solar energy than do tilted surfaces. Other types of reflectors, such as swimming pools, lakes, white gravel and concrete walks, work best with vertical collectors.

In some ways it seems foolish to take out a window which lets light and heat into the building directly, only to replace it with an opaque wall solar collector. By now, however, the advantages of having a mix of windows and collectors should be clear. First, usable interior wall surface can be lost if the entire south facade is of glass and windows. Second, sunlight fades and sometimes damages furniture, floors, and fabrics. Third, people are often uncomfortable when sun shines on them directly, so heat-giving direct sunlight is often cut off by draperies or shades.

Another issue of comfort is that of overheating caused by a totally glass wall, even with concrete floors and walls. With solar collectors and heat storage in the south walls, the heat can be transported to other, cool parts of the building or be trapped and stored for later use.

Heat loss to the outside is potentially reduced by having a wall solar collector rather than a window type.

CHAPTER III·C

Tilted And Horizontal Systems

Simple systems need not have vertical collectors. The building section in Figure III·C·1, for example, makes use of both windows and a tilted air-type collector. An excellent elaboration on this concept was built by Zomeworks Corporation for a neighbor, seen in Figure III·C·2 with Zomeworks in the background. This simple but sophisticated system uses no mechanical power in its operation. Air is heated in the collector and rises (see Figure III·C·3). As it rises, it comes in further contact with and actually passes through six layers of expanded metal lath painted black, and its temperature increases.

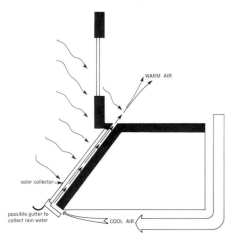

WARM AIR

solar collector

possible gutter to
collect rain water

COOL AIR

Figure III·C·1.
A tilted thermosiphoning solar collector in combination with windows.

The natural convection, thermosiphoning action pumps the warm air down through a large storage bin of fist-sized rocks. The rocks are located under the porch and house. As the air warms the rocks, it cools and is drawn through the open lath floor, into the cool air return duct, and back to the base of the collector where it is again heated.

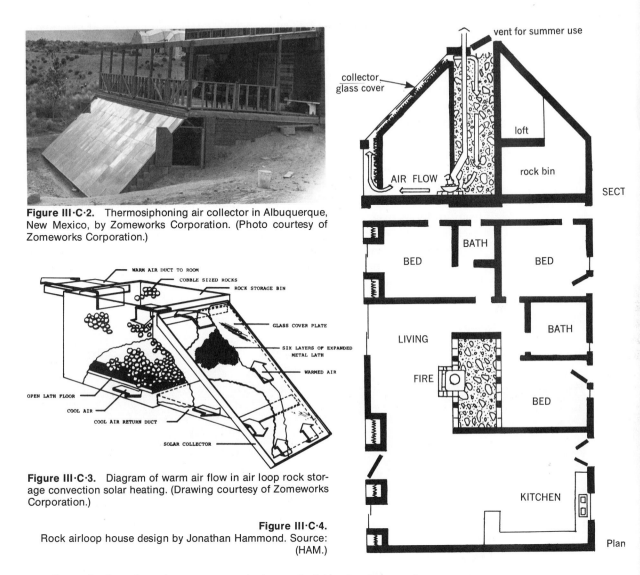

Figure III·C·2. Thermosiphoning air collector in Albuquerque, New Mexico, by Zomeworks Corporation. (Photo courtesy of Zomeworks Corporation.)

WARM AIR DUCT TO ROOM
COBBLE SIZED ROCKS
ROCK STORAGE BIN
GLASS COVER PLATE
SIX LAYERS OF EXPANDED METAL LATH
WARMED AIR
OPEN LATH FLOOR
COOL AIR
COOL AIR RETURN DUCT
SOLAR COLLECTOR

Figure III·C·3. Diagram of warm air flow in air loop rock storage convection solar heating. (Drawing courtesy of Zomeworks Corporation.)

Figure III·C·4.
Rock airloop house design by Jonathan Hammond. Source: (HAM.)

vent for summer use
collector glass cover
loft
rock bin
AIR FLOW
SECT

BATH
BED   BED
BATH
LIVING
BED
FIRE
KITCHEN
Plan

When, during the winter, warm air is needed by the house, dampers are opened which permit cool air to travel to the duct under the rock storage (not shown in diagram), to be warmed by the rocks as it rises, and to re-enter the house.

Figure III·C·4 is another natural convection loop utilizing hot air collectors and rock heat storage. The air flow from the storage to the collector passes through the house. The fireplace and flue help to heat the storage. During the summer, a vent at the top of the storage induces natural ventilation through the house at night, simultaneously cooling the rocks for daytime use.

Another low impact technology solar project, designed by Harold Hay of Sky Therm Processes and Engineering in Los Angeles, is in Atascadero, California (see Figure III·C·5). The solar collector of this one-story house is horizontal rather than vertical or tilted. The flat roof supports large, waterbed-like, black plastic bags full of water (see Figure III·C·6). The water bags, eight inches deep, contain 7000 gallons of non-circulating water. The water has a heat capacity effect of 16 inches of concrete, but the weight of only four. The earthquake-resistant steel roof deck measures 52 by 36 ft. The water bags are exposed to the sun during the day; at night insulative sliding panels are rolled on aluminum rackways into place above the bags to prevent them from losing their heat to the night air. Instead, they radiate their heat down into the space below by warming the metal ceiling of the house (see Figure III·C·7). The sliding "shutters" are insulated with two inches of polyurethane foam. A one-quarter horsepower motor responds to a thermostatic signal and, running only two minutes morning and night, slides the insulation panels on tracks. They move from positions over the roof ponds to be stored above a carport or patio area, stacking three deep (see Figure III·C·8). They can also be operated by hand. In the summer, the reverse operation occurs: during the day, the insulative shutters cover the bags to protect them from the hot sun. At night they are rolled back and the black bags radiate their heat to the cool night sky and, in turn, cool the space below. Figure III·C·9 shows the various modes of operation.

Figure III·C·5. Exterior view of solar heated and cooled house in Atascadero, California, by Harold Hay. (Photo courtesy of Harold Hay.)

Figure III·C·6. Plastic roof-top solar collector containing water; Atascadero. (Photo courtesy of Harold Hay.)

Figure III·C·7. Interior; Atascadero. (Photo courtesy of Harold Hay.)

Figure III·C·8. Sliding insulating shutters in a retracted position; Atascadero. (Photo courtesy of Harold Hay.)

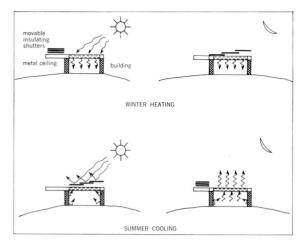

Figure III·C·9.
Various operating modes of
Harold Hay's naturally heated
and cooled house.

In Persia, Hay observed an ice-making structure known as a yakh-chal. For centuries desert people there have used the yakh-chal to produce ice through the process of heat radiation into the night sky. The ice was stored in hive-shaped structures. It was possible to manufacture ice in shaded basins when the ambient temperature of the night air was as high as 48°F. This same principle is the basis of Mr. Hay's house-cooling process. He points out that "over 60% of the residential use of energy in the Los Angeles area could be handled by solar energy or night sky radiation. Potential savings can be 70% if solar water heaters are added."

In preliminary work in Phoenix, Arizona, Mr. Hay and John I. Yellott investigated this natural air conditioning system. With ambient temperatures ranging from subfreezing to 115°F, the system maintained room temperatures between 68 and 82° throughout a normal year of Phoenix weather without supplementary heating or cooling. Figure III·C·10 shows cross-sections of Hay's first ceiling ponds. The Atascadero house differs primarily in its use of plastic bags to encase the water.

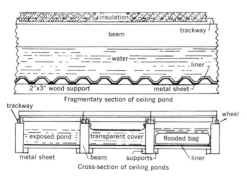

Figure III·C·10.   Flat, rooftop solar collectors on an Arizona solar house by Harold Hay and John Yellott. Source: (HAY.)

Hay's effort at Atascadero is being made in cooperation with the Solar and Night Sky Radiation Research Group at the School of Architecture at Cal-Polytechnic University in San Luis Obispo, California. The team, headed by

Professor Kenneth Haggard, evaluated the architecture, construction, maintenance, thermal and acoustical performance, economics, and occupancy reaction to the house. HUD allocated $40,000 to help with this evaluation. During its first year of operation, comfort was maintained without any source of energy other than solar and nocturnal radiation. Hay's system works best in latitudes between 35°S and 35°N (where the sun is high in the sky) and where winters are moderate and freezing weather is rare.

Walls of earth, brick or stone can support roof ponds, the cost of which becomes nominal because the usual ceiling and roof expenditures are deductible from the total cost. Savings can be realized in heating and cooling equipment and in such supporting items as large investments in generation and distribution of electricity.

Natural air conditioning can be installed, operated and maintained by relatively unskilled persons. It does not need motors, compressors, condensers, gases and ducts, nor does it require materials other than those essential to modern construction. Natural air conditioning provides gentle and uniform comfort with no noise, drafts, or disease-spreading air circulation.

"It is ironic that a highly developed country is now developing a system having greatest merit for developing countries," Hay has said. "Where a narrow comfort zone is not required, movable insulation can be used without roof ponds in much simpler and lower-cost ways. Such means could improve the health and productivity of people with low incomes. For higher priority needs, the more costly roof ponds could substitute for present imported thermal control equipment except in situations with high ventilation requirements, such as theaters. Natural air conditioning has important advantages for hospitals and clinics in developing countries."

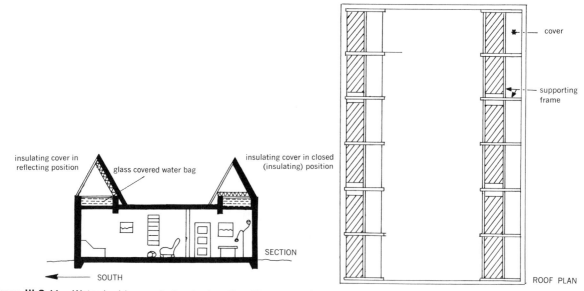

Figure III·C·11.   Water bed house design by Jonathan Hammond. Source: (HAM.)

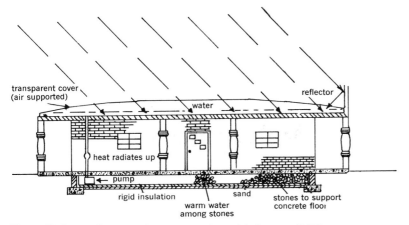

Figure III·C·12. Thomason's sunny south model. Source: (NSF.)

The modified version of the Sky Therm® concept shown in Figure III·C·11 was designed for conditions in Davis, California. The movable insulation reflectors are open during sunny winter days and closed at night and during cloudy weather. During the summer, the water bags are exposed to the cool night sky, allowing the water to soak up the heat from inside the house during the day, thus keeping it cool.

Figure III·C·12 is another variation of Harold Hay's theme. This one is by Dr. Harry Thomason in Washington, D.C. Instead of providing movable insulation to cover the roof ponds, the water is drained by gravity to an under-the-floor reservoir which is also filled with fist-sized stones. During the winter, water is pumped to the roof when the sun is shining and drained to the reservoir when it is not. The solar heated water warms the stones, and together they warm the house by gravity flow of heat through the concrete floor. When the water is on the roof, the stones continue the heating process. During the summer, the water can be pumped to the roof at night to be cooled for use by the house during the day.

CHAPTER III·D

Variations And Hybrids

The space between the transparent covers and the heat storage wall works best when it is only two to five inches thick, but there are advantages in making the space large enough for human use. For example, the space can be used as a hall, porch, or vestibule (see Figure III·D·1).

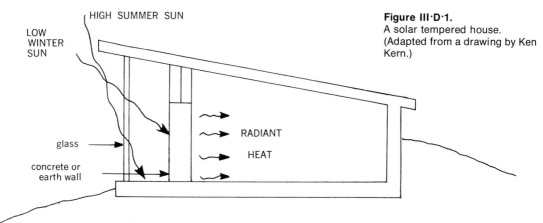

HIGH SUMMER SUN

LOW WINTER SUN

glass

concrete or earth wall

RADIANT

HEAT

Figure III·D·1.
A solar tempered house.
(Adapted from a drawing by Ken Kern.)

The *biosphere,* Figure III·D·2, is an integration of a house, a greenhouse, a solar heater, and a solar still. Conceptualized by physicist Day Chahroudi, the space between the solar collector and the heat storage wall is large enough to be used for growing food. The north wall of the greenhouse stores the heat and serves as the south wall of the house. Together with architect/engineer Sean Wellesley-Miller at the Massachusetts Institute of Technology, Chahroudi is developing a thin membrane whose solar transmittance varies with its temperature. When it is cool, it transmits about 95% of the solar radiation which strikes it at right angles. When it is warm, it is almost totally opaque. This results in high solar heat gain into the greenhouse during sunny, cold winter weather and in almost no solar heat gain during excessive summer heat (enough radiation will penetrate to nurture plants). By using many additional layers of transparent membrane, each spaced an inch or so apart, the "wall" is also a good insulator. The combination of high solar gain and good insulation has not been achieved before, nor, of course, has the combination of transparency and opacity without the use of mechanized controls or moving parts.

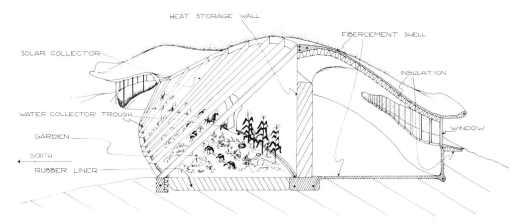

Figure III·D·2. "Biosphere" by Day Chahroudi.

Zomeworks has also enlarged the space between the transparent covers and the heat storage to accommodate a greenhouse. The prototype, shown in Figures III·D·3 through III·D·6, is located in Monta Vista, Arizona. It has a ground area of about 240 square feet and a glazed roof area of 270 square feet, facing south. The end walls are solid and insulated, as is the north wall, and 55-gallon drums of water, which are used in the drum wall of Steve Baer's home for heat storage, are stacked along the north wall. The sun's rays penetrate through the glass and heat the vertically-stacked drums. Excess heat is exhausted through hinged vents along the north side of the roof ridge. They are automatically opened and closed by freon canisters which are activated by the sun. These automatic shutters were first developed as the product called Skylid® for solar shading of windows (see Part II).

The most unusual feature of the greenhouse is its double-glazed roof which uses Beadwall®. The two layers of glass are spaced 3″ apart, and tiny polystyrene beads fill the space when the sun is not shining, greatly reducing heat loss from the greenhouse (see Figure III·D·4). The beads are stored in bins at the rear of the building during sunny hours (see Figure III·D·6) and are bown into the inter-glass space when the solar intensity is low (at night and during cold, cloudy periods, for example). The movement of the beads between the storage bins and the inter-glass space is automatically accomplished in about one minute by thermostatically-controlled vacuum cleaner-type blowers. The heat loss through this bead-filled, double glazing is approximately equal to that of a wall with the equivalent amount of fiberglass insulation. The reports are that after several cloudy days of near zero weather, the room temperature stayed above 50°.

Figure III·D·3.
Interior of greenhouse by
Zomeworks Corporation which
combines Drumwall with
Beadwall. (Photo courtesy of
Zomeworks Corporation.)

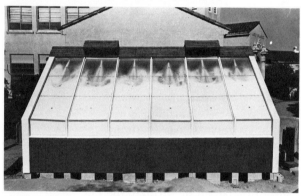

Figure III·D·4.   View of south facade. Beads filling the Beadwall
cavity. (Photo courtesy of Zomeworks Corporation.)

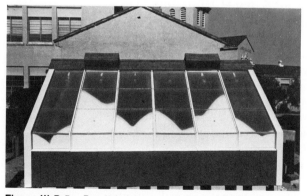

Figure III·D·5.   Beads being evacuated from cavity. (Photo courtesy of Zomeworks Corporation.)

Figure III·D·6.   North wall of greenhouse
showing containers of beads. (Photo courtesy
of Zomeworks Corporation.)

CHAPTER III·E

Retrofitting Buildings

Applying solar heating and cooling to existing buildings should be one of our nation's top priorities, providing immediate reduction in our fossil and nuclear energy demands. Unfortunately, our inventory of tens of millions of buildings and related structures is of less interest to professional designers and energy administrators than new buildings are. For building owners, and particularly the homeowner whose heating and cooling bills are doubling, and then doubling again, the retrofitting of existing buildings must have top priority.

As with new buildings, retrofitting old ones can be done on varying levels of technological complexity, monetary and energy expense, and common sense. At one end of the scale of complexity and expense are the five schools which were solar powered through the initiatitve of the National Science Foundation. In the middle range is one of the first solar heated houses, located in Boulder, Colorado, and designed by Dr. George Löf in 1950.

The end of the scale presently applicable for most home retrofitting includes the simpler and often more efficient methods. There are three basic ways of retrofitting buildings (see Figure III·E·1). One way is to apply the collectors to existing, or slightly modified, exterior surfaces of the building, that is, the walls or roofs. Another way is to attach them to an addition onto the building, such as a porch, garage, or a new wing. A third way is to build a structure separated from the building. This might be an auxiliary out-building, such as an unattached shed, garage, or barn, or a structure, built for the sole purpose of supporting the collector. The last chapter discusses this option.

Because of the unusual constraints of existing buildings, the size orientation and tilt angle (discussed in the last chapter) of the collector may be predetermined. Often the economic constraints of trying to alter existing conditions restrict the optimization of the design. For collectors which heat domestic hot water, the design is somewhat more flexible because the collectors are smaller and are used year round; the sun position varies much more during twelve months than during the shorter heating season. Collectors for cooling have difficulty attaining adequate efficiency under the best of conditions and should conform to optimum design as much as possible, making application to existing buildings difficult. For space heating, the size can be as small as 100 square feet or larger than half the floor area of the building. For domestic hot water heating, it can be as small as necessary or as large as 30 square feet per person.

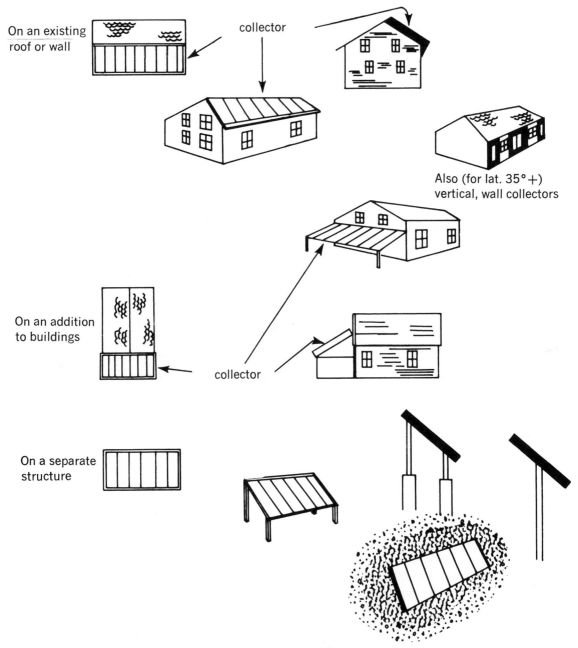

On an existing roof or wall

collector

Also (for lat. 35°+) vertical, wall collectors

On an addition to buildings

collector

On a separate structure

Figure III·E·1. The application of solar collectors to existing buildings.

The orientation of the collectors can range from south-southeast to south-southwest for space heating and from southeast to southwest for domestic hot water heating. Dr. Doug Taff at Garden Way Labs is experimenting with systems which combine east-facing with west-facing collectors.

For space heating collectors, the tilt angle (measured from the horizontal) can vary from an angle of the latitude to an angle of latitude plus 55°. For 40°N, the range then is from 40° to 90° (vertical). For domestic hot water heating collectors, the tilt can range from latitude minus 10° to latitude plus 25°. For 40°N, this allows a range of from 30° to 75°.

In all of the above ranges, the seasonal or yearly overall efficiency will not vary more than 10 or 20% from the optimum. One of the easiest ways of collecting solar heat with existing roofs is to pass water or air over the shingled surface. The surface should be as black as possible, painted if necessary, and free from debris. Frames for two layers of glass or the equivalent (such as fiberglass-reinforced polyester) are attached to the rafters, taking care to prevent leaks.

The roof could also be covered with corrugated aluminum painted black and covered by glass. Water is released through a perforated pipe along the ridge and collected in a gutter, or the equivalent. Dr. Thomason experimented with this method and found it to be relatively inefficient. However, if the roof is an existing one, the small cost involved in converting it to a solar collector might justify a low efficiency.

Figure III·E·2 shows some possible design details. Portions of south-facing walls could be converted to air-type collectors in a fashion similar to that for a roof. Water-type collectors would be less practical in this case because of the absence of a sloping surface over which the water can trickle.

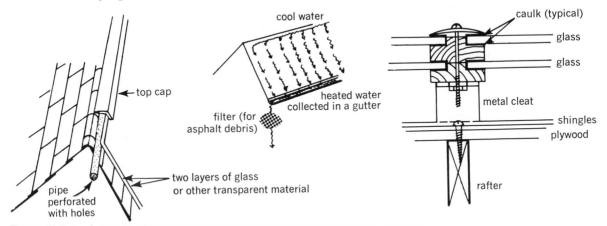

Figure III·E·2. Converting an existing roof to an open flow water-type solar collector.

Separate structures can support collectors in yards. Such collectors, their advantages and disadvantages, are discussed in the last chapter. An example of such a device which can be easily set up and dismantled is shown in Figure III·E·3. Cool air from the house is blown out through the bottom of the window to the solar collector and back through the top of the window. The installation is similar to a window air conditioner. Better control is obtained by ducting the cool air from one window and returning the warm air through another.

Figure III·E·3.
A portable, in the yard air-type solar collector. (Design and drawing by Lea Poisson.)

A quick but modest effort can be made in retrofitting existing buildings by constructing simple windowbox air-type solar collectors. Figures III·E·4, III·E·5, and III·E·6 are modifications of the vertical, thermosiphoning solar collectors on the office building at Odeillo (see Figure III·A·3). They are designed to be incorporated into the openings of existing windows. Figure III·E·4 is a design credited to Buck Rodgers of Embudo, New Mexico. The cool air from the room is drawn into the collector by the warmed air leaving it. The vertical variation on this Figure III·E·5 is particularly applicable to large buildings.

Although the windowbox collector can be of almost any size, its effectiveness, although substantial on a per square foot basis, really is almost immeasurable unless it is significantly larger than the window. If a collector size of one-quarter to one-half the building floor area is required to provide 50% of the heating, it should be clear that large collectors are required for a noticeable effect on overall energy savings. Figure III·E·6 shows one method of making the collector larger than the window.

The difficult task of adding heat storage to existing buildings has been practically solved by J. P. Gupta and R. K. Chopra of the Defense Laboratory, Jodhpur, India. They developed a simple solar room heater which requires no mechanical power and which can be incorporated into existing buildings. As can be seen in Figure III·E·7, the solar collector is resting against the building and faces south. A tall, uninsulated hot water tank stands inside the room with its back against, but well-insulated from, the outside wall. Water circulates by thermosiphoning action from the flat plate collector, up to the tank, and back down to the collector. For climates with freezing weather, anti-freeze is added to the water. Heat radiates from the front of the tank to the room.

The range of low impact alternatives has only been skimmed here. Innovations and new designs are desperately needed, as is the development of mechanisms which will encourage quick adoption by every segment of the construction industry.

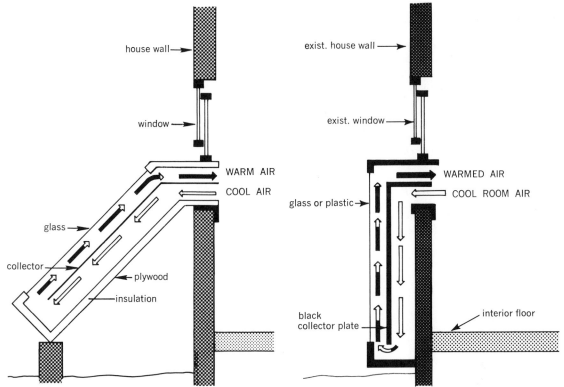

Figure III·E·4. Window box solar collector, invented by Buck Rodgers, Embudo, New Mexico. (Taken from *Alternative Sources of Energy*, No. 13.)

Figure III·E·5. A variation of the window box collector.

Figure III·E·6. The window box: a frontal view.

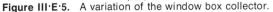
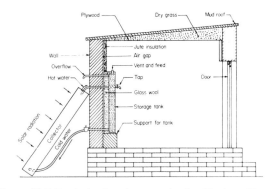

Figure III·E·7. A simple solar room-heater. (Designed by J. P. Gupta and R. K. Chopra of the Defence Laboratory, Judhpur, India.)

The following Part on solar domestic water heating is a prelude and an interim step to Part V on higher impact technology solutions for using solar energy for space heating and cooling.

REFERENCES FOR PART III

(**HAM.**) Hammond, Jonathan; Hunter, Marshall; Cramer, Richard; and Newbauer, Loren. "A Strategy for Energy Conservation: Proposed Energy Conservation and Solar Utilization Ordinance for the City of Davis, California." Prepared for the City of Davis with the support of The Case Institute, August 1974.

(**HAY.**) Hay, Harold. "New Roofs for Hot Dry Regions." *Ekistics* 31 (February 1971): 158-164.

(**KER.**) Kern, Ken. *The The Owner-Built Home.* Oakhurst, Calif.: Owner-Builder Publications, 1961.

(**NSF.**) National Science Foundation and Research Applied to National Needs. *Proceedings of the Solar Heating and Cooling for Buildings Workshop.* Washington, March 1973. Department of Mechanical Engineering, University of Maryland.

PART IV

Solar Hot Water

by Douglas Mahone,
Associate,
Total Environmental Action, Inc.

The heating of water for domestic uses (bathing, cleaning, etc.) presently accounts for about 12% of the energy consumed nationally for residential and commercial purposes. This enormous demand for energy is being met primarily by such conventional sources as gas, oil, and electricity. A considerable part of this demand could be met by solar heating, using existing and well-proven technology. Except for Florida and southern California, solar hot water heaters have not found a wide market in the United States, but they are quite common in Israel, Australia, Japan, and other parts of the world. The low cost and steady supply of conventional fuel in the United States until now has been a major factor against the acceptance of solar hot water heaters. Recent changes in fuel availability and the realization that the bottom of the barrel is in sight force us to give further consideration to the use of solar energy for heating water.

The use of solar energy for heating water is in many respects quite similar to its use for heating buildings. There are, however, several aspects of solar hot water (SHW) that make it a potentially better investment of energy, money, and effort than solar building heating.

For one thing, the demand for hot water is relatively constant throughout the year. Thus the collector and other parts of the solar water heater will be working harder and longer to produce the savings in fuel that eventually must pay for the higher initial cost of the system. The solar building heating system, on the other hand, is fully operational only during the coldest months of the heating season.

A SHW heater can also be sized to fit the demand more accurately. Although heating systems deal with extreme loads only a few days of the year, they still have to be large enough to meet those extremes. A SHW system, on the other hand, will have roughly the same load day in and day out; except in unusual applications, the design load (expected maximum load) should be close to the normal daily load. Without the problems of widely fluctuating demand, SHW heaters can be relatively cheaper and simpler than solar building heaters.

A problem common to all types of solar heating is the variable nature of sunshine. SHW heaters, however, often have an additional advantage over solar space heating systems because the requirements for hot water are less rigid than those for space heating. If there is an extended sunless period during which the supply of hot water runs out, the consequences will probably be less severe than if the building were to lose its heat. It is the difference between letting the laundry wait a bit longer and having the pipes freeze and burst; between going over to a friend's for a shower and having plants and foods freeze. Of course, many commercial and industrial installations cannot risk being without hot water, and these would need a system with reliable carry-over capacity and a substantial auxiliary heater. If a certain amount of inconvenience or variability of use is accepted, however, a SHW system can be greatly simplified in design.

For relatively constant hot water demands, the common solution is to provide an auxiliary heater. The simplest is a conventional hot water heater, which is reliable, automatic, and readily available. It has only to make up the difference between incoming water temperatures from the solar heater and the constant supply temperature to the user. If the solar system is providing full-temperature water, the auxiliary remains off. Its controls therefore are very much simpler than those required by most space heating systems. Many SHW systems require no other auxiliary controls than the thermostat that comes with the conventional hot water heater. Of course, there are SHW systems that have special requirements or use more sophisticated auxiliary heat sources, but as a general rule, there are fewer problems in providing auxiliary heat to SHW systems than to solar space heating systems.

All of these features of SHW heaters (similarity to solar space heating systems, year-round demand, constant load, less rigid demand, ease of control, and auxiliary supplement) make them a logical starting point for anyone interested in capturing some of the sun's heat for his own use. In addition, the scale of a SHW heating system is considerably smaller than that of most solar space heating systems. By starting out with a SHW heater, it is possible to learn a great deal about using the sun's energy and to gain experience that is directly applicable to the solar heating of buildings. At the same time, capital outlay is low and useability of the finished product high and immediate.

The use of solar energy for heating water involves several specific issues. One of the first concerns variations in the amount of water to be heated. It is standard practice to figure "average" daily consumption per person at 28 gallons in public housing and 41 gallons in luxury apartments. In Australia, on the other hand, 10 gallons/person/day is figured for residential purposes. In general, one-third of total daily water consumption in a home will be hot water, and the maximum probable hourly demand will usually be one-tenth of this total. These figures are based on current use patterns, which tend to be sporadic: a big demand in early morning and again in the evening, corresponding primarily to bathing and dishwashing uses. Commercial and industrial buildings usually have their own special requirements. Consumption must be evaluated according to varying conditions.

A corollary to the issue of hot water consumption is that of conservation. Even though a SHW heater is an energy conservation device in itself, resources are required for its manufacture and for auxiliary heat. Additional savings, however, can result from the prudent use of hot water, in contrast to the more common use patterns of waste and casual consumption. Simple changes in habits can produce significant reduction in demand: using a basin of hot water rather than a steady flow from the tap to rinse dishes; taking showers rather than baths; providing spray nozzles on faucets rather than the conventional steady stream nozzles; washing clothes in warm (or cold) rather than hot water; or using "suds saver" washing machines that recirculate hot wash water rather than disposing of it. These and other methods used by water- and energy-conscious individuals will reduce hot water demand, simplifying the cost and design problems of SHW systems and saving both water and energy.

In order to meet the demand, the method of heat collection has to be balanced against the pattern of use. This can be a process filled with compromise. For example, one of the simplest SHW heaters is a black rubber hose sitting out in the sun. By adjusting the rate of flow through the hose, one can get a steady stream of hot water. If this water is used directly as it comes out, the result is hot water that is virtually free. If it is to be used for showers or dishwashing in the evening, however, the water would flow through the hose into a storage tank. If the water is to be used in the morning, the storage tank must be insulated to keep the water hot overnight. If hot water is needed on days when there has been no sun, the storage tank must be larger and better insulated. An auxiliary heater is usually necessary to insure the availability of hot water.

It is readily apparent that the more stringent the demands on the SHW heater, the more complex and costly the system becomes. In addition to providing larger and better storage for the hot water when demands are less flexible, more hot water must be produced and stored on those days when the sun does shine. This could involve either more hose or a more efficient type of collector.

Another aspect of SHW heater use is that of user participation. For example, a simple and satisfactory solar hot water heater consists of a shallow trough of water with a transparent cover sitting in the sun (see Figure IV·1). But this heater has to be filled in the morning and drained in the afternoon or early evening. Someone has to fill the tank, to cover it if the sun clouds over, to decide when the water is sufficiently hot, and to drain the hot water for use. This simple heater would be unacceptable to most people as an alternative to their conventional water heaters because most people are used to getting hot water with no more thought or effort than it takes to turn on a faucet. SHW heaters that are as automatic and trouble-free as the conventional kind can be built, but these are at the other end of the scale of complexity from the simple trough. In between these extremes are many varieties of heaters, with varying degrees of user participation required, an important determinant of the final system choice.

Two factors which have prevented the wide acceptance of SHW heaters in the U.S. are the cheap availability of conventional fuels and the expectation that hot water will always be available at the turn of a faucet. Consequently, the only heaters that appear acceptable are the more complicated and self-operating kind. In Florida and southern California, the market for SHW heaters, interestingly enough, peaked ten to fifteen years ago and had been declining until recently when the "energy crisis" aroused new interest. These heaters have evidently been acceptable from the point of view of the person who pays the hot water bills. For the owner, however, who has to pay the user cost, they have required a bit more maintenance than their conventional counterparts. In addition, the initial cost of the SHW heaters has not, until recently, been significantly balanced by energy savings on a short-term basis. On a longer-term basis, however, the savings have been significant, and they will become increasingly so as energy costs continue to rise.

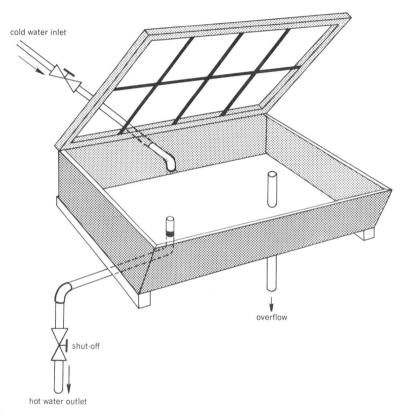

cold water inlet

overflow

shut-off

hot water outlet

Figure IV·1.
Open trough solar water heater.

Before describing SHW systems capable of meeting U.S. hot water demands and use patterns, some simpler SHW heaters should be mentioned. The characteristics of the previously cited hose and water trough—simplicity, low cost, high user participation—are common to several other kinds.

The Japanese have built water heaters that are essentially plastic water bags set on a level platform (see Figure IV·2). Some have reflectors below to reflect additional energy up to the bottom side. A variation of this kind of heater is a simple wooden box with a sheet of plastic tacked to the inside to hold the water (see Figure IV·3). The heater doesn't require a transparent cover, but it will be more efficient if it has one. In these heaters, the collector and the storage are one and the same.

A major disadvantage of flat-basin heaters is that they must be horizontal. In the tropics, where the sun is high overhead all year, this disadvantage is minor, but in higher latitudes, where the sun is lower in the sky, a horizontal collector becomes less efficient, partly because much of the energy is reflected off the collector. In the winter, when the sun is lower, efficiency decreases even more. A more efficient heater would be one tilted toward the sun.

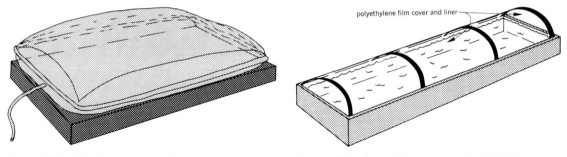

polyethylene film cover and liner

Figure IV·2. Plastic bag solar water heater.

Figure IV·3. Trough type solar water heater.

Simple flat, metal tanks tilted to receive optimum sunlight have been built and used as water heaters in Japan, Algeria, India, and elsewhere. These are basically two pieces of metal sealed at the edges to form a container from four to eight inches thick. Construction is rigid enough to resist the increased water pressure caused by tilting the tank up. The side toward the sun is painted black, a glass cover plate is added, and, as with the flat-basin heaters, the collector and storage are the same.

The depth of water in both types of integral collector/storage heaters can be determined by simple experimentation. The shallower the tank, the quicker the whole volume will heat up, but the less volume there will be for a given surface area. An advantage of the flat basin types is that the depth of the water is easily variable. On a less than ideal day, the basin might be filled only half way, heating less water to a higher temperature. For either type, the size of the container would be determined by the volume of hot water the particular application would require. The best depth would be chosen by experimentation, and the area would follow from the volume requirement.

An interesting variation on these collector/storage tanks was developed in the West Indies (see Figure IV·4). The collector surface is a flat metal container, painted black, with a glass cover plate. At one end, the container bulges up to form a large space for holding water. A baffle plate within the container enables a thermosiphoning circulation pattern to carry warmed water from the collector area up into the storage space and cool water from the bottom of the space over to the collector area. The storage space heats up during the day and is insulated to cut down losses at night. The biggest single disadvantage of integral-type collector/storage heaters is their inability to prevent loss of heat when clouds cover the sun or night falls (although the West Indies design is a big step towards solving this problem). Because of this, virtually all commercially successful SHW heaters have a separate collector and storage tank.

One particular arrangement used in Israel, Australia, Japan, and the United States (see Figure IV·5) consists of a tilted collector with transparent cover plates, a separate, highly insulated water storage tank, and well-insulated pipes connecting the two. The bottom of the storage tank is at least a foot higher than

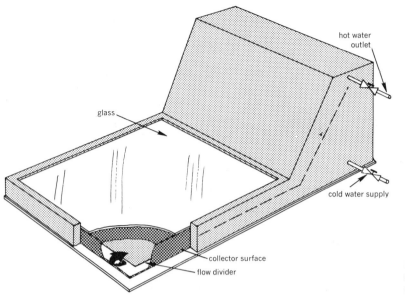

Figure IV·4.
Solar water heater with accompanying storage.

the top of the collector, and no auxiliary energy is required to circulate water through it. Circulation occurs through natural convection, or themosiphoning. When water in the collector is heated by the sun, it expands (becomes less dense) and rises up the collector, through a pipe, and into the top of the storage tank. This forces cooler water at the bottom of the tank out another pipe leading to the bottom of the collector. This water, in turn, is heated and rises up into the tank. As long as the sun shines, the water will quietly circulate, getting warmer.

Reverse thermosiphoning is an effect that can cause potential heat loss. Just as thermosiphoning is driven by warmed water rising through the collector, cooling, or reverse thermosiphoning, can be driven by nighttime cooling of the collector fluid. If the tank is elevated above the collector, the effect is negated because the cool water simply settles to the lowest point of the system (the bottom of the collector) while the warm water stays up in the tank. This is not the case when the tank and collector are on the same level. One commercial SHW made in France uses a carefully designed valve that allows water flow in only one direction. The valve closes when reverse thermosiphoning begins. Most systems, however, simply rely on a height differential to avoid the problem.

Several design variations of collectors have been used (see also Chapter V·B). All are basically an insulated box with one or more transparent covers, containing a blackened metal absorber plate and some arrangement for circulating water past the plate. One type has a long, continuous tube bent in a sinusoidal shape so that water flows from bottom to top (see Figure IV·5). The pipe is soldered to the surface of a flat metal plate which provides a greater absorption area than the tubing alone, and which readily conducts heat to the tubing. Another type has a series of parallel pipes running vertically between a supply header at the bottom and a return header at the top (see Figure IV·6). This is likewise soldered to a flat plate.

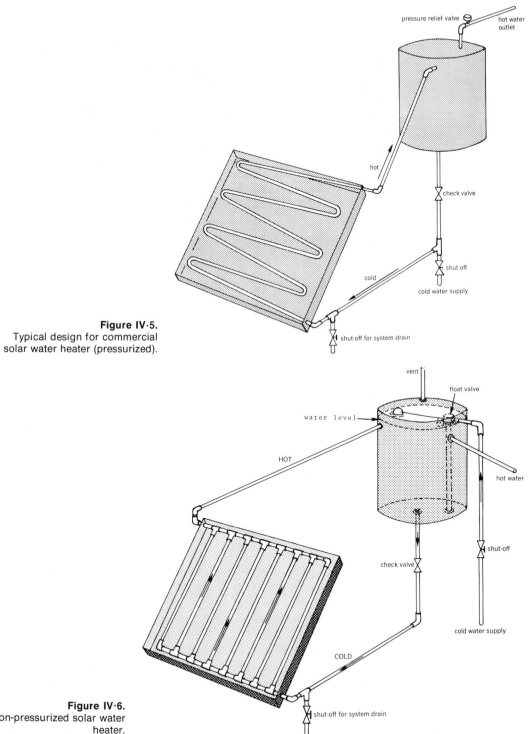

Figure IV·5.
Typical design for commercial
solar water heater (pressurized).

Figure IV·6.
Non-pressurized solar water
heater.

There are several construction variations of each of these two basic circulation patterns. By attaching baffles between two metal plates, one can create either a sinusoidal or parallel path for the water to follow. There is a technique for bonding two metal plates together with a circulation pattern embossed between them called roll-bonding, which can also be used to provide either type of circulation pattern. A variation of this technique is the so-called tube-in-plate absorber plate which produces parallel tubes integral with the metal plate and connected by headers top and bottom. Both techniques produce highly efficient heat transfer plates. Another common technique involves welding or riveting a corrugated metal plate to a flat metal plate, creating channels for water flow. The difficult task here is making a header to interconnect the channels.

No matter what the construction details, the basic goals of any absorber plate are designed to provide: a maximum energy absorbing surface; an efficient means of transferring heat quickly and evenly to the water flow; a steady, uniform distribution of water; and an inexpensive unit that is leakproof and reliable. This last requirement is the most challenging, because the unit will be subject to extremes of temperature, intense thermal expansion and contraction, strong sunlight, corrosion, pressure, and possibly freezing.

The most common materials used for absorber plates are copper and aluminum, with galvanized iron running a poor third. The costs of all these materials, especially copper, are rising rapidly. Although cost is a major factor in material selection, even more important is ease of fabrication, which depends largely on the technology available. Soldering copper is relatively easy, but expensive. Roll-bonding requires elaborate and sophisticated machinery, but can be done quite cheaply on a large scale. If the unit is homemade, the economics change somewhat (no labor cost), and fabrication is determined by available materials and skills. It is still quite expensive to buy a ready-made unit, but manufacturing efficiencies are expected to bring down costs dramatically for the more sophisticated techniques, unless energy and material shortages become more severe.

The absorber plate is coated with a substance that enhances its absorbing ability. The simplest and cheapest (and in many ways the best) coating is flat, black paint. The ideal coating, however, is any one of a number of substances called selective blacks, which are discussed in more detail in Part V. These blacks have high absorptance and low emissivity (regular black paint has about equal absorptance and emissivity) to minimize reradiation of energy. Several of the commercial units available around the world have good selective blacks. Unfortunately, the techniques for applying the coatings are not simple. They involve acid baths, electrolysis, sometimes even vacuum deposition. There are also additional complications because most of the commercial coatings are easily damaged by water. This means the insulated box holding the absorber plate must be moisture-sealed and airtight, and the attachment of the transparent cover becomes much more difficult. Selective blacks will become better and more available as current research and improved manufacturing techniques simplify the technology. Until then, simple black paint has the advantages of easy application, durability, and cheapness, factors which outweigh its non-selectivity for most applications.

The box that holds the absorber plate prevents the loss of most of the energy that hits the unit. These losses are by reflection and absorption at the cover plate; reflection, convection, and reradiation from the absorber plate; and conduction out the back of the absorber. The first loss is difficult to prevent, although work is being done to produce high transmittance and low reflectance glasses and plastics.

The second loss is handled by the cover plate(s) and the absorber plate coating. The cover plate is more effective at preventing losses. Most covers are glass or plastic; all provide the greenhouse effect of letting in the short-wave energy and sharply reducing the out-going long-wave reradiation of heat. Convective losses are sharply reduced by the creation of a small air space and by eliminating outside air movement across the absorber plate. In cold climates, where convective and conductive losses through the cover plate are increased by greater temperature differences between the absorber and the outside air, two and sometimes three layers of glass or plastic are often used. This reduces energy transmission to the absorber, but the loss is offset by the insulating value of the additional air space.

There is no "best" material for cover plates. The most commonly used material is glass, which is readily available, durable, and has good transmittance; its disadvantages include cost, weight, brittleness, and difficult glazing details. Several plastics are used, and more are being developed in an effort to offset some of their disadvantages. Two disadvantages are ultra-violet degradation of the plastic (often within a few years) and its inability to withstand high temperatures. Even if the plastic is cheap, the labor and inconvenience of replacing a quickly deteriorated cover are considerable. More durable plastics, such as plexiglass and Lexan®, tend to be quite expensive. Fiberglass-reinforced polyester sheets are less expensive, but their long-term durability may be limited by their lower resistance to the effects of heat. The principal advantages of plastics are light weight, good workability, relatively easy glazing details, toughness, and, potentially, low cost.

The third area of heat loss, conduction out the back, is reduced through the use of conventional insulation, but direct contact between the absorber plate and the insulation must be prevented. When the sun is shining and the collector is not operating, temperatures of the plate can get high enough (250°—400°F) to damage the insulation. By placing the collector flat against a sloping roof, the insulation serves the double function of insulating both the absorber and the building. In addition, conduction heat lost from the absorber will enter the living space, where it is still useful.

Piping from the collector to the storage tank has to be well-insulated and should present minimal interference with circulation. With most of the commercially available units, the tank is close to the collector so as to minimize the length of piping, and joints and bends are kept smooth and to a minimum. Insulation around the pipes helps to conserve the heat collected and to maintain the temperature difference between incoming and outgoing water so that the thermosiphoning action will not be impaired. The actual engineering design and sizing of the piping to prevent this impairment is subtle and goes beyond this discussion. In general, the goal is to have large diameter pipes, gradual bends, and short runs. An example (MOR. 1) is a 45 square foot collector, with a height difference of 9 feet from the bottom of the collector to the hot water inlet at the top of the tank, and with 16 feet of piping. With a one inch tube (outside diameter), the collector will have to produce a 11° temperature rise (from 125° to 136°) in order for thermosiphoning to occur. The required temperature rise is greater at lower temperatures: A rise of 13° (from 10° to 113°), or of 35° (from 40° to 75°) is required to produce the same circulation. This means start-up is a little slower in the morning than one might hope, but the problem is not a serious one because most of the useful heat collection occurs in the middle part of the day. If the pipe is too small, thermosiphoning will still occur, but it will require greater temperature differences; this means lower efficiency and should be avoided.

The storage tank is generally large enough to hold about two days' supply of hot water. The advantage of greater storage capacity (a larger tank) can be offset by the increased losses due to more surface area and longer storage time. The judgment is basically one of economics: a larger tank will carry over longer sunless periods but will cost more; a smaller tank will be cheaper, but auxiliary heat will have to make up for inadequate sun and will probably involve consumption of a non-renewable fuel. An auxiliary heat source of some kind will almost certainly be needed (a system that could carry over the longest sunless period would be so over-designed for average conditions as to be economically ridiculous), so the tradeoff is between more frequent use of the auxiliary or a larger, more expensive system.

Since thermal gradients between different height levels of a thermosiphoning system are essential to circulation, it follows that the locations of pipe connections to the tank are important. The supply pipe to the bottom of the collector should feed from the coolest part (bottom) of the tank. Heated water from the collector should feed to the warm part (top) of the tank. About a half day's worth of storage should be left above the hot water inlet for the hottest water to accumulate, because water coming from the collector will not always be fully hot (cloudy day, early morning) and would otherwise cool down the hottest water. If the auxiliary heat is an electric immersion heater (as is common practice with the commercial units), it should be placed below one full day of storage capacity. This allows it to heat up a useable amount of water without doing more than is necessary. (Fuel-fired auxiliaries, a different matter, are discussed below.) The hot water outlet to the user should be from the top of the tank, and the cold water refill should be at the bottom. Insulation should be as thick as is economically reasonable: the equivalent of four inches of fiberglass insulation is not excessive, and more should be used if the tank is outside in cold weather. One distinct advantage of placing it inside, besides lower insulation requirements, is that whatever heat is lost will go to a living space rather than to the outdoors.

Determining the size of the collector is, like so many other decisions in solar heater design, as much an exercise in economics as in engineering. As with the storage tank size, it is a tradeoff between higher initial capital cost or higher auxiliary costs. Beyond the point where the unit is large enough to provide the minimum daily consumption of hot water, a larger collector provides increasingly poorer payback in terms of conventional heating dollar saved per dollar of capital outlay for solar equipment. A larger system is capable of carrying over longer sunless periods, but it does not work to full capacity during normal sunny periods.

Calculations to size the collector can involve some sophisticated engineering, but generally, under favorable conditions, a good collector will heat between one and two gallons of water per square foot per day. Based on this range, a rough sizing for preliminary design can be readily determined from daily demand figures. For example, assuming an efficient collector and a moderately good location, what size collector would be needed for a family of four using 80 gallons of hot water per day? At 1½ gallons per square foot per day, they would need 80 × 1.5 = 54 square feet, or a collector roughly 7 feet × 8 feet. When solid performance predictions are called for, more elaborate engineering calculations would be made as the design process progressed.

The location and tilt of the collector and its relationship to the tank are important considerations which are strongly influenced by the architectural design and layout of the building to which they are attached. The collector should be directed towards true south, and should be tilted up from horizontal at an angle equal to the latitude of the location. These orientations are theoretically optimum for year-round collection of heat. The actual ideal for a particular location depends on a good many other factors. For instance, if there is a roof on the building that could take the collector, it would probably be far cheaper to put it there than to build a separate structure. If the azimuth of the roof is within 15° of true south, and if the tilt is within 10° of the latitude angle, the loss in efficiency from the ideal would be no more than 10%-15%, and it would be better simply to add on a little more collector. Tilt and orientation are less critical at latitudes closer to the equator.

If the collector is on the roof, the fact that the tank in a thermosiphoning system has to be higher than the collector has often meant mounting the storage tank high in the ridge of the roof or sometimes even in a false chimney constructed on top of the roof. The roof structure, of course, must be strong enough to hold it. An alternative is to build a support for the collector on the ground, feeding an elevated tank next to it or inside the building. The design constraints of thermosiphoning collectors are demanding, so flexibility of installation is limited. Because of this, other methods of circulating water are used in applications where the arrangement demands cannot be readily satisfied.

The plumbing involved in any water heater presents some problems, most of them common to all plumbing, to the designer. The primary one is the choice between pressurized and non-pressurized systems. The pressurized variety is able to supply hot water at locations above the storage tank because system pressure from the water main or well pump is maintained (see Figure IV·5). This imposes considerable stress on the water channels in the collector, which must be designed accordingly. The non-pressurized systems supply hot water by gravity flow only to users lower than the tank (see Figure IV·6). If pressurized hot water is required (for showers or appliances), the difference in height will have to be large enough to meet the requirement (about 0.43 psi results from each foot of elevation between tank and user). If the height difference cannot be accommodated, the only solution is to install a separate pump and pressure tank. The stresses within a non-pressurized system are lower, which allows cheaper and easier construction.

It should be noted that virtually all of the commercially successful thermosiphoning water heaters are produced in warm climates where freezing does not occur. The danger of pipes or channels in the collector freezing and bursting is obvious, but less obvious is the problem caused by radiational cooling. Copper pipes in collectors have frozen and burst on clear, windless nights when air temperatures never dropped below 40°F; radiational heat loss was simply greater than heat gain from the surrounding air, so the water temperature dropped below freezing.

If bursting should occur, repairing an absorber plate with soldered-on pipes would be difficult; it would be almost impossible on a Roll-Bond® plate. Plastic piping would solve the freezing problem because of its elasticity, but it is a poor material in other respects. Most plastic has poor heat transfer qualities, loses strength when hot, and is subject to ultraviolet degradation. Plastics are being developed, however, to have long term durability in collectors while undergoing frequent freeze/thaw cycles.

A SHW heater in a cold climate has to have positive safeguards against freezing, and this makes it a more complicated design than the typical, warm weather thermosiphoning system. There are three basic ways to protect a collector: (1) provide movable insulation to cover it (see Figure IV·8); (2) provide means for draining it (see Figures IV·5, 6); or (3) circulate an anti-freeze solution rather than water (see Figure IV·7). The use of movable insulation is much more feasible with SHW heaters than with building heaters because they are relatively small. But problems still exist. The movement would have to be automatic in some sense (even if only a simple sensor that sounded an alarm bell to notify the operator of freeze danger), because humans are fallible and the consequences of freeze-up are so severe. The moving mechanism would have to be well designed and reliable because the whole thing would be out in the weather constantly. All of this can be costly. More is said about movable insulation in Part III, "Low Impact Technology."

Provision for drainage is difficult in thermosiphoning systems because their operation depends on a closed, fluid-filled circuit. Most of the simpler systems sidestep the freezing problem entirely because they are filled only when heating (and they usually do not have small water channels that would freeze and burst). The drainage problem is basically the same in pressurized and unpressurized thermosiphoning systems. The tank must be shut off from the collector, a vent at the top and a drain at the bottom of the collector opened, and the water allowed to flow out. The water channels have to be designed to drain completely so that no water pockets remain to cause problems. To refill the collector, the drain must be closed, water allowed in, the vent closed after all air had escaped, and finally the shut-off to the tank reopened. This process, while not difficult to operate, is difficult to automate. Manual operation requires considerable user time (especially on days of intermittent sun), and automation may require a large increase in capital expenditure. Previous designers have also encountered vapor lock problems caused by remaining air pockets in the refilling process. Occasionally steam results from the filling of an extremely hot collector and adds to the air lock problem.

The most widely used protection against freezing is the use of an anti-freeze/water solution instead of water in the collector. Since the anti-freeze solution should not mix with the domestic hot water supply, provision for heat exchange between the solar heated solution and the water in the storage tank is usually accomplished by running the solution through a coil of metal tubing immersed in the storage tank (see Figure IV·7). This coil connects at either end to the supply and return pipes from the collector, and the fluid thermosiphons through in the regular manner. By requiring a larger system and increasing capital costs, this heat exchanger introduces a small inefficiency, but it is relatively trouble-free in operation. An alternative to the coil of tubing in the anti-freeze loop is the use of a small tank within the large tank to provide the necessary heat transfer surface.

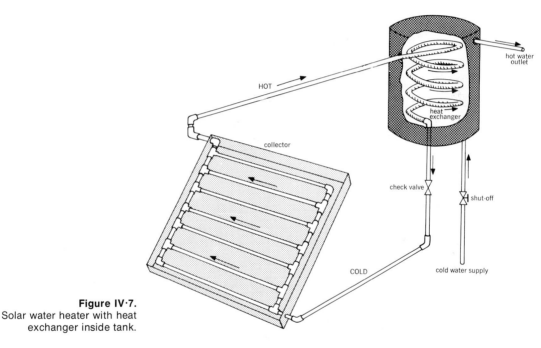

Figure IV·7.
Solar water heater with heat
exchanger inside tank.

A fourth and less widely used method of preventing freezing is the introduction of heat to the absorber when it drops to temperatures close to freezing. The two primary ways in which this is done is through the circulation of warm liquid from the storage tank or the operation of electric heat strips attached to the absorber.

There is a wide variety of other systems for SHW heating that were designed in response to particular situations or problems. An electric immersion heater is an auxiliary heat source that can be more readily integrated with a standard thermosiphoning system than a fuel-fired auxiliary can be. One way around this problem is to use the SHW storage tank as a preheater for water flowing into a conventional water heater (see Figure IV·9). In this system, the cold water supply flows through the SHW storage tank on its way to the conventional heater. The heat exchanger (immersed coil or small tank) for the anti-freeze solution preheats the water as it moves through the tank. When the SHW system is fully heated, there will be no need for temperature boosting by the conventional heater. If boosting is required, the thermostat included in the conventional heater actuates it, reducing the complication of automatic controls.

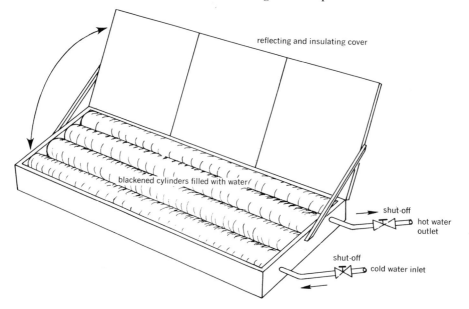

Figure IV·8. Solar water heater with movable insulation.

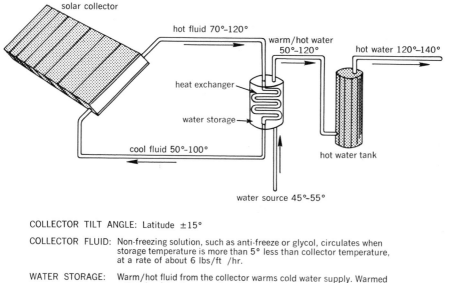

COLLECTOR TILT ANGLE: Latitude ±15°

COLLECTOR FLUID: Non-freezing solution, such as anti-freeze or glycol, circulates when storage temperature is more than 5° less than collector temperature, at a rate of about 6 lbs/ft /hr.

WATER STORAGE: Warm/hot fluid from the collector warms cold water supply. Warmed water circulates to conventional hot water heater and is heated additionally, if necessary.

Figure IV·9. Solar domestic water preheat system.

The constraint placed upon architectural design by the thermosiphoning arrangement can be alleviated through the use of pumps. Where the required arrangement simply cannot be accommodated because piping runs would be too long or because there is no room for an elevated tank, a pump arrangement is commonly used. The cost of the pump and controls and the auxiliary power necessary to run the pump present disadvantages, but the freedom of system layout is considerable. A pump, for instance, allows the collector to be on the roof, the storage tank in the cellar, and the faucet on the first floor. The collector and its piping form a circulation loop, with its own pump, tied into the hot water heater. If it is a closed loop (running through a heat exchanger in the tank), it could be filled with an anti-freeze solution to prevent freezing.

Another variation made possible with a small pump is a Thomason-type collector. Instead of flowing in closed channels, water in the collector is allowed to trickle freely down over the surface of a corrugated metal sheet. It is collected in a gutter at the bottom and flows by gravity into a storage tank. Since the system is not pressurized, flow to the user would also have to be by gravity, unless it were mechanically pumped into a pressure tank. The advantage of an open trickle collector is that the water drains into the storage tank simply by shutting off the pump. Thus, freezing problems are eliminated. A variation of this principle is the Kawai collector, developed in Japan, which consists of a layer of fabric or fibrous material sandwiched between two metal plates. Water is allowed to trickle slowly through and drain out the bottom.

Pumps have also proved useful with once-through heaters. The collector, a long network of blackened iron pipe lying flat on a black roof, is really an elaborate variation of the garden hose sitting in the sun. When the sun is shining brightly, a continuous flow of hot water comes out of the pipe and flows into a storage tank. If the water supply is a city main, the pump is eliminated. A variation of this system has been successfully used to heat swimming pools. Water being pumped through the pool filter is diverted through a pipe network on a roof and warmed a bit before it re-enters the pool. (Because of the low operating temperatures of collectors for swimming pool heating, transparent cover plates are often unnecessary.)

There is one more class of heaters that lies between SHW systems and solar building heating. These heaters are all basically preheaters for conventional water heaters that use the building's heat storage (heat from the large building collector) as the heat source. For most air or water systems, a simple heat exchanger transfers heat from the storage to water on its way to a conventional hot water heater. There the temperature is boosted, if necessary, to useable levels. In more sophisticated systems, the heat exchanger is replaced by a small heat pump that concentrates heat from the storage area into a smaller hot water tank, boosting the temperature high enough for domestic use. All of these indirect systems are subject to greater inefficiencies due to increased mechanical demands and heat exchanger inefficiencies, but in the context of an entire building energy system, they may make sense. Most of the collector and storage problems mentioned earlier are handled by the heat systems of larger buildings, where they have to be solved anyway. All of the cost and efficiency considerations are then applied to the system as a whole, and the incremental cost of the SHW subsystem is more easily justified in terms of savings, despite a possibly lower efficiency.

The Bibliography has a separate section called "Solar Domestic Hot Water Heating and Swimming Pool Heating." For the information on technical aspects related to solar water heating refer to Part V as well as to *How to Design and Build a Solar Swimming Pool Heater* by Francis deWinter.

REFERENCES FOR PART IV

(**HAM.**) Hammond, Jonathan; Hunter, Marshall; Cramer, Richard; and Neubauer, Loren. "A Strategy for Energy Conservation: Proposed Energy Conservation and Solar Utilization Ordinance for the City of Davis, California." Prepared for the City of Davis with the support of The Case Institute, August 1974.

(**MOR.**) Morse, R. N. "Solar Water Heaters." *Proceedings of the World Symposium on Applied Solar Energy*, Phoenix, Arizona, 1955, pp. 191-200.

(**STO.**) Stoner, Carol Hupping, ed. *Producing Your Own Power: How to Make Nature's Energy Sources Work for You.* Emmaus, Pennsylvania: Rodale Press, Inc., 1974.

PART V

Systems For Indirect Use Of The Sun's Energy

At this time, most attention concerning utilization of solar energy for heating is directed toward so-called solar collectors, large expanses of glass on a tilted surface. The heat is transported by a fluid (either liquid or air) to a storage area for later use or delivered immediately to the space to be heated. Direct methods of using the sun's energy require few controls and little complication. Indirect systems, however, are often encumbered by fluid transport systems, elaborate collectors, large heat exchangers, control systems, valves, pumps, expansion tanks, and heat storage containers. In order to design these systems properly, close coordination is required among the client, the building designer, the engineer, and the contractor. In this way, effectiveness can be achieved and undue complexity avoided. Figure V·1 is a summary diagram of a system which indirectly uses the sun's energy.

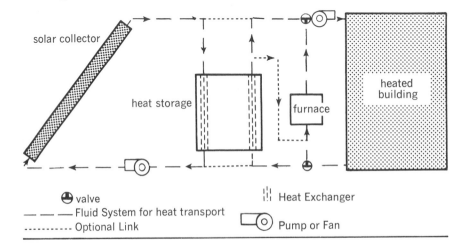

Figure V·1. Basic solar system diagram for indirect use of the sun's energy for heating buildings.

CHAPTER V·A

Fluid Systems For Heat Transport

One of the first decisions to be made in choosing a solar energy system is what type of fluid should be used to transport the heat energy. There are two basic fluid transport systems: one links the solar collector to the solar heat storage; the other delivers the heat (or cool) from the storage to the building. Secondary systems may accompany both.

Liquids and gases are included in the definition of fluid. To date, liquid systems, which include water, ethylene glycol solutions/water, propylene glycol/water solutions, and oil, have predominated. Air is the only gas which is used.

In deciding on the type of fluid, the following areas should be considered:
- human needs and comfort;
- compatibility with the building design;
- compatibility with other mechanical devices (e.g., the backup systems);
- climate;
- relative cost (initial, operating, maintenance);
- relative complexity; and
- long term reliability.

When human comfort requires only heating, air transport systems should be strongly considered because of their relative simplicity. (This is also a major consideration, of course, in the selection of the even simpler passive systems discussed earlier.) When domestic hot water must be provided in addition to heating, however, the choice between liquid and air systems becomes more difficult. Water can be preheated enroute to the water heater, where it is raised, if necessary, to its final usable temperature. When this preheating is done in combination with space conditioning, rather than alone as its own separate system, the water supply line is usually diverted through a heat exchanger within the solar heat storage. For liquid systems, the heat storage usually consists of a water tank, which is relatively compatible with conventional heat exchangers. Less conventional alternatives are used with storage systems for air-type collectors. For example, a relatively small tank, perhaps 30 to 60 gallons for a

residential installation, can be buried within a rock-type heat storage. The warmed rocks in turn warm the water inside of the tank. From there the water continues on to the hot water heater for additional heat if necessary.

> **CAUTION:** For liquid systems which utilize an anti-freeze solution, every attempt must be made to prevent poisonous contamination of the domestic hot water. This is accomplished either by total isolation of the ethylene glycol/water mixture from the domestic hot water or by the use of less toxic propylene glycol. (Many experts, though, also discourage the use of propylene glycol because of its toxicity.)

If provisions for cooling are required, an air system is a less likely choice than a liquid. Although some research has been done with air, the majority of work in solar powered cooling is being done with liquid systems. The same (or at least similar) thermo-dynamic and physical properties which in the past dictated the use of liquids for conventional cooling systems also dictate their use in combination with solar energy.

Air systems, however, can be used successfully for cooling. For example, in some parts of the country cool night air can be blown through the rock bed and stored for use the following day. If necessary, refrigeration compressors can cool the rock bed even further, using off-peak electricity at night.

The heating and cooling distribution systems for the building can also affect the choice of fluid. For example, many people are uncomfortable with forced-air heating or cooling, the system most compatible with conventional air-type solar systems.

Radiant heating systems are usually of the hot water type, although there are heating systems which use warm air circulated through radiant wall, ceiling, and floor panels. Hot water radiant systems, such as baseboard radiant convectors, are relatively compatible with liquid solar energy systems. Hot liquid from the collectors circulates through the heating system or through heat exchangers in the solar heat storage tank(s). The main disadvantage of liquid heating systems is that usually they must be sized for relatively high (140°—190°F) water temperatures. Unfortunately, the higher the required water temperature, the lower the overall efficiency of solar energy systems. Steam systems (212°F plus) therefore are generally incompatible with solar energy.

Forced air distribution systems can be compatible with liquid-type solar collectors. The warm or cool water from the storage tank is circulated through heat exchangers, or fan coil units, and the air is blown across them and heated or cooled in the process. The air is then delivered to the building. In the case of the Solaris System®, designed by Dr. Harry Thomason, heat from the water tank warms the rocks which encircle it. The rocks in turn heat the house air which circulates through them.

Certain types of buildings may limit the amount of space which can be allotted to heat storage. Until phase-changing salts are readily and commercially available, the main choices for heat storage are large tanks of water and rock (pebble) beds. Tanks consume one-half to one-third the volume of rocks, and this fact alone may dictate the choice of liquid-type solar collectors. Figure V·A·1 summarizes the options for collector fluids.

Although people may prefer various types of heating and cooling systems, the luxury of choice is usually possible in residences but not in larger buildings. Since collectors operate most efficiently at low temperatures, this in turn causes the heat distribution fluid of the building to circulate at low temperatures. The lower the fluid temperature, the larger the amount of fluid which must be circulated in order to provide enough heat or cool. The larger the building, the more likely, therefore, is the use of liquid systems since distribution piping consumes relatively small amounts of space. For air to be as efficient as liquid, large ducts and/or rapid air velocities must be used in combination with large fans. The increased duct sizes result in higher costs and in the use of greater amounts of valuable space. The large fans required to provide higher air speeds through the ducts involve higher initial and operating costs and greater energy consumption. Higher air speeds also require higher air temperatures so that people will feel warm drafts rather than cool ones. One way to use air-type collectors effectively is to feed the air directly to exhaust fans as make-up air or to ventilation equipment as preheated cold intake air.

COLLECTOR FLUID OPTIONS	HEAT STORAGE OPTIONS	SPACE CONDITIONING OPTIONS
air	rocks	forced warm or cool air
	small containers of water	air-warmed (or cooled) radiant panels
	small containers of phase-changing salts.	
water or other fluid such as oil or a water-antifreeze solution	large containers (usually tanks) of water	forced hot or cold water, for example baseboard or fan coil units
	large containers of water encircled by pebbles	hot or cold water radiant panels
		forced warm or cool air (this requires a heat exchanger to remove the heat or cool from storage and transfer it to the air)

Figure V·A·1. Summary of collector fluid options.

Building designs which limit the location of solar collectors to the roof (rather than to the south-facing walls) are more likely to require the use of liquid systems because of the extra distance involved in transporting fluid to and from heat storage, which is usually located close to the ground because of its weight. The longer distances can also mean that more building space is consumed by air ducts. The choice of liquid, which uses small pipes, thus becomes preferable. For all fluid transport systems, the network of ducts and piping should be kept simple, short, and heavily insulated.

Climate may dictate the choice of fluid transport systems. For example, in cold climates, where a building may require only heating, air collectors is often the

likely choice. When a liquid system is subject to freezing conditions, an anti-freeze/water solution may be necessary to prevent freezing in the collector when the sun is not shining. Alternatives are to completely drain the water from the collector when the sun stops shining or to use an absorber which will not burst under repeated freezing and thawing. Freezing due to temperatures a few degrees below 32°F can be circumvented through the use of an additional cover plate.

Water and air are much less expensive than oil and anti-freeze to use in heat transport systems. In cases where water is at a premium, air is least expensive. In cities or dusty regions, however, great expense may be required to filter, clean, and purify the air.

An air-type collector can usually be made more cheaply than a liquid type. Other components of the system, including storage and heat exchangers (or lack of them), also cost less, and local labor economics often favor the installation of air ducts over water pipes. In addition, maintenance costs can be lower for air systems since air leaks are not nearly as destructive as water leaks. In liquid systems, anti-freeze solutions deteriorate and must be changed every two to four years, an expense not often considered. It is true that millions of gallons are changed each year in cars and trucks, but a residential solar energy system requires 10 to 50 times more anti-freeze than a car does. If used on a large scale, both availability and disposal are problems. Anti-freeze solutions also have lower heat transport capabilities than water. Air systems, on the other hand, can have higher operating costs because a greater amount of electricity is required to move heat with air than with water.

The potentials and design tradeoffs of various fluid options are also discussed later in relation to the design, construction, and performance of specific systems.

CHAPTER V·B

Flat Plate Solar Collectors

The major component of a solar energy system is the solar collector which converts the sun's light energy into useful heat energy and then transfers that heat to a fluid. The fluid transports the heat to the building or to storage for later use. It can also be used to actuate a refrigeration (air conditioning) cycle or to heat domestic water.

Numerous examples of wonderfully simple, yet sophisticated, collectors have already been discussed. These have included such devices as windows and vertical collector/storage wall combinations. Indirect solar energy systems are distinguished by the separation of the various functions of solar heating and cooling into distinct components. Harold Hay's rooftop "waterbeds" combine the functions of absorbing the heat, storing it, and transporting it to the space to be heated. In traditional terms, however, a solar collector denotes a very particular piece of equipment, not unlike an appliance, which is attached to a structure. Most collectors used for the heating and cooling of buildings are described as "flat plate" collectors. These collectors can be categorized as either liquid-types or air-types, according to the fluid which circulates through them.

LIQUID-HEATING FLAT PLATE COLLECTORS

The absorber plate, the main component of a collector, stops the sunlight, converts it into heat, and transfers it to the liquid. Its surface is usually painted black. To reduce heat loss, cover plates, such as glass which transmits sunlight, are placed over the absorber. Heat loss out the back side of the absorber is reduced through the use of insulation. All of these components are often containerized (wooden or metal box) for shipping, ease of installation, or protection from moisture. Figure V·B·1 shows a typical collector with these various components.

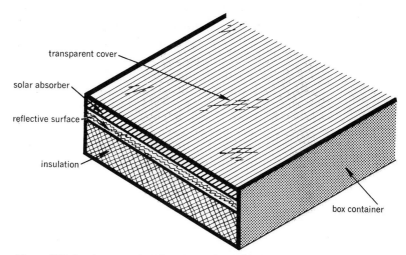

Figure V·B·1. A prototypical flat-plate solar collector.

Absorber Plates

Traditionally, absorber plates for liquid systems have been made of copper, aluminum, or steel. The characteristics to consider in choosing an absorber are:

- availability
- energy and resources required in manufacture
- conductivity (thermal performance)
- durability
- ease of handling
- cost

Availability. Copper is available, but it is in short supply on Earth and its price continues upward. Aluminum and steel are relatively easy to obtain, but even aluminum resources are being depleted. Steel absorber plates are less available commercially than aluminum, particularly those which are treated for corrosion.

Energy and Resources Required in Manufacture. Since our energy supplies are dwindling and energy conservation is more and more vital to the survival of Earth, the choice of metal must also be based on the amounts of energy required for its manufacture. The manufacture of aluminum, for example, requires considerably more energy per pound than does steel.

There is simply not enough copper on Earth to provide every building with an all-copper solar collector, even if the absorber lasts 50 years and the copper is recycled. The same limitation may hold true for aluminum. All metals, in fact, should be used only after alternatives have been considered. Systems which do not require metal absorber plates should be used whenever possible.

Conductivity (thermal performance). Metal need not be used as the absorber plate if the fluid comes in direct contact with every surface which is struck by sunlight. With almost all liquid systems now in use, however, the liquid is channelled, usually through tubes attached to the plate. Heat must be conducted to the tubes from those portions of the plate which are not in contact with the fluid. If the conductivity is not great enough, the heat will escape from the collector before it is conducted to the tubes. With metals of high conductivity, such as copper, the sheet can be thinner and the tubes can be spaced further apart. The most conductive metal is copper, followed by aluminum. To obtain similar results with all other conditions being the same, an aluminum sheet would have to be twice as thick and a steel sheet nine times as thick as a copper sheet.

Durability. The durability of metals is limited by the threat of corrosion. Copper resists corrosion better than aluminum and steel, but under certain conditions, water and anti-freeze can corrode all three. The high operating temperatures of solar collectors also accelerate corrosion. There are, however, methods of reducing the effects of corrosion. If the absorber is drained to prevent freezing, for instance, the admission of air into the system raises the potential for corrosion. Thus, corrosive oxygen from the air should be prevented from entering. (gases such as nitrogen are being investigated as alternatives to air in self-draining systems.)

Particular care must be taken to prevent corrosion of aluminum; swimming pool water can cause aluminum absorber plates to leak within months of installation. Steel must be either galvanized or stainless. Corrosion can also be reduced by adding so-called *inhibitors*, some of which are chromate-based, to the water or to the anti-freeze. The use of soft water with a low mineral or metal content is recommended. The inside surfaces of aluminum tubes or other liquid passages can be treated with a zinc lining process which is expensive and not generally available at this time.

The pH of the system is the most critical aspect of corrosion. For aluminum, the solution must be approximately neutral, with a pH around 6 or 7. Any deviation, either lower (acidic) or higher (basic), greatly increases corrosion. The pH must be monitored carefully to prevent deviation.

Other treatments, such as the circulation of corrosion inhibitors through the system for several days prior to the introduction of the inhibited liquid, are under investigation, but their estimated cost—several hundred dollars for a 4,000 gallon system—is high.

All of the treatments mentioned only reduce the corrosion problem; they provide no guarantee of success. Until metal corrosion is guaranteed and proven, there are very few alternatives to copper for liquid absorber plates. Notable exceptions are glass and the open-faced, corrugated aluminum or steel of Dr. Harry Thomason. Non-metal absorber plates, e.g., glass and plastic, are free of corrosion but have shorter lives.

The following advice on corrosion protection was issued by Olin Brass Company, and is reprinted with their permission.

CORROSION PROTECTION FOR ALUMINUM ROLL-BOND®
SOLAR ENERGY COLLECTOR PLATES

We believe that it is desirable to recognize that precautions should be taken to maximize the corrosion resistance of solar energy collector plates which will typically operate with a water-based heat transfer fluid circulating within the ROLL-BOND® passages. The following suggestions and discussion are offered for consideration by the user as he develops his own specific designs and establishes his own specific operating characteristics for the solar energy collection system.

We anticipate that "typical" solar energy collector systems will be "multi-metal" systems in which the water-based heat transfer fluid circulates not only through aluminum but also through other non-aluminum base metals.

We do not have specific operating experience with solar energy collectors in multi-metal systems. They may, however, be compared with automotive cooling systems, particularly those in which an aluminum radiator is used, as in the Corvette automobile, and those in which an aluminum block or aluminum head is used, as in the Vega automobile. These systems do operate successfully and may be presumed to offer satisfactory corrosion performance. They are indeed multi-metal systems in that the heat transfer fluid (water-anti-freeze mixture) circulates not only through aluminum (the aluminum radiator or the aluminum block) but generally also circulates through steel components, copper and brass components, and comes in contact with soft solder used in the joining of copper and brass components. This type of multi-metal system appears to be comparable to that which we would regard as "typical" in a solar energy collector circuit. The coolant used in this type of automobile system is a water-ethylene glycol mixture and the ethylene glycol contains various inhibitors and buffers which are intended to make it compatible with the multi-metal cooling system. In the case of General Motors' vehicles, it is our understanding that the coolant is made to their Specification GM-1899-M, "Antifreeze Concentrate—Ethylene Glycol Type."

We understand that ethylene glycol degrades in such service and that organic acids are among the degradation products. Since a significant reduction in pH, which could accompany the formation of the organic acids, would serve to make the mixture more corrosive, we understand that one function of the buffers used in typical automotive anti-freeze formulations is to control pH to a desirably high value. When and if the buffers are exhausted, pH may drop. The answer to the potential problem represented by this in the case of automotive cooling systems lies in the usual recommendation of the automobile manufacturer for periodic replacement of the coolant mixture. In most cases, this is recommended at twelve month intervals.

Without lengthy experience with solar energy collector systems, we cannot say whether or not replacement of the water-ethylene glycol mixture will be required at longer or shorter intervals than is required in the automobile cooling system. We do note that the operating temperatures which seem to be typical of most solar collectors tend to be somewhat lower than automobile cooling system operating temperatures. On the other hand, we also note that the total operating time to be expected of a solar collector during a calendar year is undoubtedly greater than is normal for an automobile cooling system. This suggests that, until such time as the system manufacturer and his customers have gained sufficient experience with a specific system, in order to predict heat transfer fluid life, provisions be made for monitoring pH of the solution so that it may be changed when necessary. It is suggested that the recommendations of the anti-freeze manufacturer be sought as to the pH range to maintain.

The usual automotive coolant mixture uses ordinary "tap water." Of course, the composition and corrosivity of tap water varies quite widely, and there are geographical areas in which normal tap water is known to be quite corrosive. There is some indication that, under some circumstances, the corrosivity of the tap water used is a factor in the eventual corrosion performance of the automobile multi-metal system. Since a solar collector system is a stationary system, it seems reasonable to suggest that the variable of possibly unsuitable "tap water" be avoided through the use of distilled and/or de-ionized water in the mixture with the commercial ethylene glycol anti-freeze. This appears to be a worthwhile and not too costly precaution.

In some "multi-metal" systems, it may be desirable and/or necessary to use water rather than a water-ethylene glycol mixture as the heat transfer medium. In these applications, it will be necessary to treat the water in order to insure the integrity of the protective oxide films. Various types of inhibitors are available, and instructions for their use can be obtained from the manufacturer. Chromate inhibitors are popular and perform admirably in service.

A typical solution would be 600 ppm chromate made up as follows:

$$0.9 \text{ gpl } Na_2Cr_0O_4 \cdot 10H_2O \text{ (.120 oz. per gal. } Na_2Cr_0O_4 \cdot 10H_2O)$$
$$0.082 \text{ gpl } Na_2Cr_2O_7 \cdot 2H_2O \text{ (.011 oz. per gal. } Na_2Cr_2O_7 \cdot 2H_2O)$$

The powders are dry mixed and dissolved in approximately twice the weight of water. This solution can then be added to the storage tank. The pH should be between 6 and 7. The inhibitor level should be maintained at 500 ppm chromate. Glass phials of chromate solutions containing known amounts of chromate could be used to check the level in the system on a color comparison basis. The pH level is important in color determination and should be kept within a range of 6 to 7.

A word of caution: Chromate inhibitors should not be considered for use with water/anti-freeze mixtures.

Aluminum components in the system should be galvanically isolated from components made from other metals.

The use of a "getter" column, through which the heat exchange fluid circulates before entry into the aluminum portion of the system, has been suggested. Such a "getter" column would consist of a plastic cylinder containing a series of aluminum sheets. The water would circulate over this surface and the function of the surface would be to pick up "heavy metal" ions which may have gotten into the system as the result of corrosion of other metal components or as the result of "tramp" material within the system.

It is desirable that heat transfer fluid design velocity through the system be limited to two or three feet per second.

Although we understand that normal operating temperatures for most solar energy collectors will be below 212°F, we must also consider the possibly higher temperatures which will be encountered during shutdown of the collector system. Of course, collector systems intended to provide space heating only will be shut down in the summer months. Further, systems intended for use in providing both space heating and domestic hot water heating will operate only intermittently during the summer months because of reduced heat demand. A survey of current literature indicates that, under shutdown conditions (no flow of heat transfer fluid) and at times of high ambient temperature, collector temperatures may reach 300°F or more. We understand that, at temperatures over 250°F, most organic foam insulations will degrade. Further, degradation of the chlorinated and/or fluorinated hydrocarbons, which appear to be widely used in "blowing" organic foams, may also occur at these temperatures. The potential degradation products include hydrochloric acid (HCl) and hydrofluoric acid (HF), both of which are highly corrosive to all common metals, including aluminum. Therefore, we suggest that the use of foamed plastic insulation on solar energy collector plates be avoided.

As indicated above, all of these suggestions which are offered for the consideration of the system manufacturer or system user are intended to help to optimize the corrosion performance of the aluminum ROLL-BOND® components in the system. We recognize that there is substantial diversity of system design and this may introduce other factors which ultimately affect corrosion performance. We recognize that the design of collector systems is not our function and is beyond our control. Certainly, we will be glad to discuss specific design features with the designer and offer for his consideration additional suggestions as may be appropriate. We also recognize that the subsequent operation of the system is altogether out of our control.

In view of all the circumstances outlined above, we do want to point out that what has been offered in the foregoing discussion are suggestions intended to assist the designer and user in optimizing the performance of the final product. They should not be considered as constituting an implied warranty and are advisory in nature. The only warranty which we can reasonably provide is that the ROLL-BOND® panels which we produce will conform to final specifications and/or drawings, and will be free from defects in material and workmanship, provided, however, that we will not be responsible for any corrosion failure, regardless of the cause.

Although the preceding discussion has related to the use of aluminum ROLL-BOND® in solar energy collectors, we do want to point out that it is probable that much the same precautions would be relevant to the use of other metals, such as steel. Appropriate suggestions should be sought from material manufacturers if other metals are used in the collector.

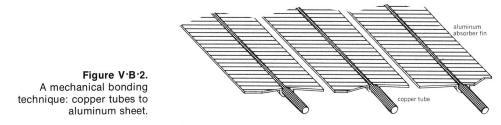

Figure V·B·2.
A mechanical bonding
technique: copper tubes to
aluminum sheet.

Ease of Handling. The weight of the absorber is not crucial in the design of a solar collector, but it does play a role in ease of construction and handling. The total weight of collectors is usually less than five pounds per sq ft, the absorber usually accounting for about one pound of this.

Copper can be difficult to work with because it hardens as it is formed and bent; steel, of course, requires special tools. All metals require careful cleaning, perhaps with a cleaning acid, before the application of the black surface. Copper is especially difficult to paint.

If tubes are to be attached to the plate, soldering or welding difficulties should be considered. Soldering of copper tubes to copper sheets is relatively easy, but expensive, for skilled labor. Aluminum, except with difficulty in some cases, cannot be soldered or welded to any metal. Special mechanical bonds, such as in Figure V·B·2, are proving successful.

Cost. Cost must be weighed against relative thermal performance. The performance of the various metals, discussed fully later, shows that for present disparities of cost among copper, aluminum, steel, and stainless steel, aluminum is the best buy. Unfortunately, its totally unsolved corrosion problems preclude its widespread use. Since copper is a nearly depleted resource, stainless steel, at approximately the same cost, may be the most logical long term choice among metals.

Types of Absorber Plates for Liquid Systems. There are three basic designs used to bring liquids in contact with the absorber plate for the purpose of taking away heat. One is the Thomason-type, open-faced corrugated sheet over which liquid flows (see Figure V·B·3). A second is the tube-in-plate method, used by the refrigeration industry, in which passages are incorporated into the absorber sheet itself. The third and most popular method prior to the advent of the tube-in-plate is the application of the tubes onto the sheet, either on the back side protected from the sun, or on the front side, exposed to the sun.

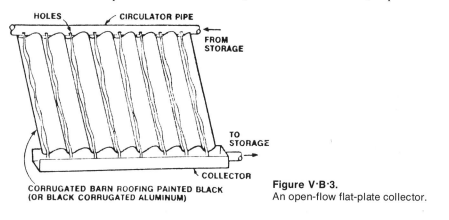

CORRUGATED BARN ROOFING PAINTED BLACK
(OR BLACK CORRUGATED ALUMINUM)

Figure V·B·3.
An open-flow flat-plate collector.

The first method has many advantages, as Dr. Thomason has demonstrated. For low-temperature applications (under 110°F), its performance appears to be competitive with tube-type absorbers, but efficiency drops off sharply at higher temperatures. Its clearest advantage is that it is self-draining and needs no protection against corrosion or freezing. Cool water from storage is pumped to a header at the top of the collector. Holes of 1/32 inch diameter are drilled into the pipe at the valley locations in the corrugations. A gutter collects the warmed water at the base of the collector, and it drains back to the storage tank. Although the destructive power of the sun, in combination with the water trickling over the surface, erodes the black paint, Thomason reports relatively little decrease in overall efficiency. He also suggests that condensation of water on the cover plate does not decrease efficiency.

Efforts have been made to obtain a continuous and even laminar flow of water over a flat surface so as to obtain maximum contact with the absorber surface. Such an even flow is almost impossible to obtain, but its accomplishment could eliminate the need for an absorber plate of high conductivity; the use of non-metals could then be possible.

Tube-in-plate absorbers are best represented by the aluminum product, Roll-Bond®, by Olin Brass. Tubes are incorporated into the plate when two sheets are fused together. The tube pattern is embossed on the sheets prior to the fusing process; when the pattern is expanded under pressure, the tubes are formed. With steel, two sheets are spot-welded, and the water flows through the space between the two sheets. Two variations are shown in Figures V·B·4 and V·B·5.

Most of the original experimental work with flat plate collectors was done with absorber plates consisting of tubes attached to a metal sheet. The performance of tube-in-plate absorbers is based on this work. While working on the first experimental solar house at MIT in 1940, Hottel and Woertz wrote "The Performance of Flat Plate Collectors" (HOT.1), the classic work on collector performance.

Later, Hottel said: "The use of tubes spaced six inches apart and in good thermal contact with a blackened copper sheet 0.02 inches thick (or blackened aluminum sheet 0.04 inches thick) is 97 percent as good as the extraction of heat by a completely water-cooled black-sheet; consequently there is no expectation of an invention which will improve the performance of the black absorber plate as a device for the transfer of heat to a fluid. (Its improvements as an absorber and radiator by use of spectrally selected surface treatment is another matter.)" (HOT.2).

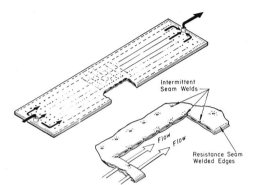

Figure V·B·4. Expanded metal solar collector plate. Source: (MOO.)

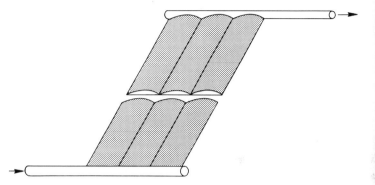

Figure V·B·5. A welded steel, closed-liquid solar collector absorber plate.

Numerous other studies have indicated the optimum tube spacing for various types and thicknesses of metal plates. Figure V·B·6 is adapted in part from one of the most significant, a study by Bliss (BLI.2). The *efficiency factor* shown on the graph is used in thermodynamic calculations of performance: the higher the number, the better the efficiency. The actual values vary with laminar or turbulent flow, the latter producing the higher values. In making comparisons between cost and overall efficiency, Bliss found a less than 10 percent variation from the optimum tube spacing in the range of three inches to seven inches. He also found that in 1959 the cost of using an increased thickness of copper (from 0.010″ to 0.030″) rose faster than the resultant increased efficiency. In the tradeoff between cost and efficiency, optimum tube spacing was four to six inches for a 0.010″ thick sheet. Four to five inch spacing on 0.040″ steel was comparable. Again, these values were computed in 1959 and are based on the method chosen by Bliss. See his paper, however, for assistance in computing values with present prices.

Tube sizing is another matter. Most of these considerations revolve around pressure drop, fluid flow rates, and cost. Because of the increased cost for increased pipe size, the tubes should be as small as possible. Typically the tubes on the absorber plates (the risers) are ⅜ to ⅝ inches in diameter; the top and bottom headers are ¾ to one inch in diameter.

The smaller the tube diameter, the lower its cost, but also the faster the same amount of fluid must travel to obtain optimum flow rates. The faster the fluid flows, the greater is the pressure drop through the absorber's "plumbing system." Velocities of less than 4 feet per second are recommended to reduce the rate of corrosion. The required pump size is proportional to the pressure drop, as is the amount of energy (and the cost of that energy) for moving the fluid and picking up the heat energy from the collector. Figure V·B·7 graphs the relationships among tube size, flow rate, and pressure loss. (More viscous fluids, such as anti-freeze solutions and oil, have higher pressure drops.) DeWinter reports (DEW.) that capillary effects in tubing smaller than ⅜ inch diameter might prevent proper emptying and thus present potential freezing problems.

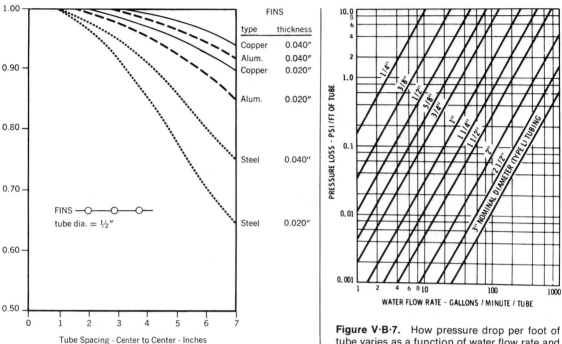

Figure V·B·6. The variation in absorber plate efficiency according to metal thickness, metal type and tube spacing. Based on data in "The Derivations of Several 'Plate-Efficiency Factors' Useful in the Design of Flat-Plate Solar Heat Collectors" by R. W. Bliss. Source: (BLI.2)

Figure V·B·7. How pressure drop per foot of tube varies as a function of water flow rate and tube diameter. Source: (WIN. 1)

The configuration of the tubes on or in the absorber plate is important to overall performance. Uniform fluid flow, low pressure drops, ease of fabrication, and low cost are all design considerations.

Uniform fluid flow is perhaps the most important consideration. Flow is considered turbulent in headers and laminar in risers for the common situation of water entering the bottom header at one side of the collector and leaving the top header at the other side. Duffie and Beckman (DUF.) summarize the studies of Dunkle and Davey, who found that the pressure drops from the bottom of the collector to the top are far greater at the ends of the absorber plate than at the center. This means that flow rates are higher in the end risers than in the middle ones. Figure V·B·8 shows measured temperatures during experimentation for a bank of twelve collectors, all connected in parallel; temperature differences of 22°C (40°F) were found across the collector surface. Figure V·B·9 shows alternative connection arrangements to obtain more uniform flow distribution and temperatures.

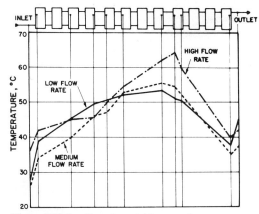

Figure V·B·8. Experimental temperature measurements on plates in a bank of parallel flow collectors. From Dunkle and Davey (1970). [Reprinted from (DUF.)]

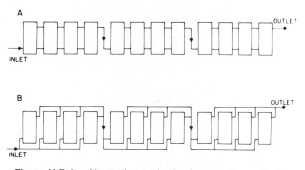

Figure V·B·9. Alternative methods of connecting collectors in banks. a) Series-parallel, b) parallel-series. From Dunkle and Davey (1970). [Reprinted from (DUF.)]

Dunkle and Davey also recommend that the headers be large enough to cause the major pressure drop in the risers, not the headers. Banks up to 24 risers can be satisfactory for either forced or natural circulation. Duffie and Beckman report further that "for forced circulation banks of over 24 risers, no more than 16 risers should be connected in parallel, and for larger banks, series-parallel or multiple-parallel connections can be used."

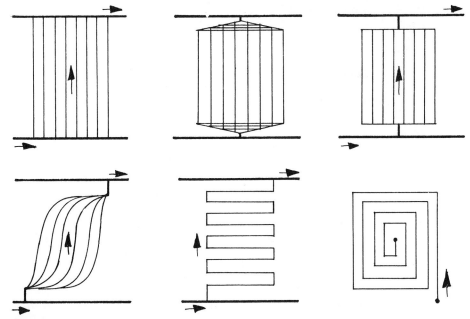

Figure V·B·10. Various tube configurations for solar absorber surfaces.

Tube configurations for individual collector units are shown in Figure V·B·10. *A* through *D* include the grid patterns and some variations. *E* and *F* are the sinuous types which eliminate fluid flow distribution problems but have larger pressure drops. They are also significantly easier to fabricate since they eliminate most of the numerous plumbing connections of the grid types.

For self-draining systems, the tubes and flow pattern must be arranged to allow for complete draining. Past systems have reported the trapping of air and, thus, incomplete refilling after draining. Sinuous patterns are best at minimizing this problem.

The tubes must be strong enough to withstand pressures due to overheating. Up to 4 psi must be provided for and pressure relief valves must be included to accommodate any steam produced when water refills a hot, empty collector.

Absorber Surfaces

Absorber surfaces (or coatings) and cover plates should be considered simultaneously. Their functions are similar, and the choice of cover plates is related to the absorber surfaces. For example, it has been shown that the use of a selective surface in combination with one cover plate is more efficient than a flat black paint in combination with two cover plates.

The primary function of the absorber surface is to increase the percentage of sunlight which is retained by the absorber plate. By definition, a black body is a perfect absorber of radiation; all wavelengths at every incident angle will be absorbed by a black body. Real substances, however, will always reflect some of the radiation which strikes them and will do so to an increasing extent with increasing angles of incidence (see Figure V·B·11).

A black body is also a perfect emitter of thermal radiation. Although there are no perfect emitters in nature, most black colors absorb as much energy as they simultaneously emit—about 90 or 95%. The ideal for an absorber surface is that it reflects no shortwave light radiation (it absorbs it all) and reflects all longwave radiation (it emits none of it). Such an ideal surface is called a *selective surface*.

Unfortunately, selective surfaces are still expensive and are not readily available. For general application, however, hardware store, flat, black paint can be used. The absorber surface must be thoroughly cleaned first, even to the extent of using an acid bath. Various black paints can be tested by measuring the temperatures of surfaces coated with different paints and exposed to sunlight. Asphaltum paints have been used with success as have high temperature black paints made for wood stoves. PPG Industries uses a black "PPG Duracron enamel." DeWinters (DEW.) recommends a coating sold by Sears Roebuck and Company, "Tar Emulsion Driveway Coating and Sealer" (not the "Coating and Filler").

The performance of a selective surface is measured by its absorptivity (α) of solar energy, by its emissivity (ϵ) of longwave thermal radiation, and by the ratio of absorptivity and emissivity (α/ϵ). These concepts are discussed in Appendix E, "Emittances and Absorptances of Materials."

Selective surfaces must be evaluated on their applicability to the particular absorber plate material, their cost, their availability, and their durability. Each selective surface is designed for application to particular materials; selective surfaces appropriate to copper will not necessarily be applicable to aluminum. Cost is a consideration since their use presumably should either decrease the cost of other collector components (for example, eliminate the need for a second cover plate), or increase the performance of the collector sufficiently that it justifies the expense, either by raising the usable temperature attainable from the collector or by increasing the total amount of energy collected. Present costs range from \$0.25 to \$2.50 per square foot of surface.

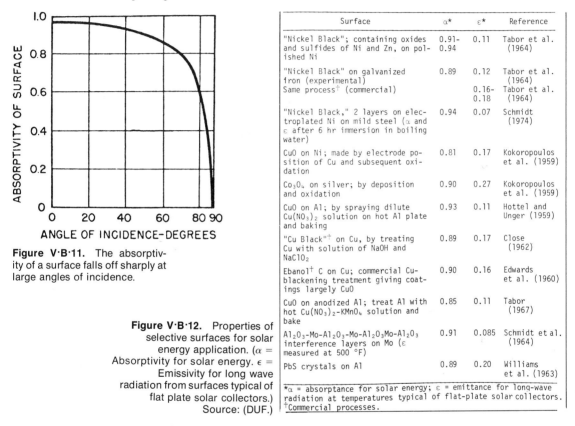

Figure V·B·11. The absorptivity of a surface falls off sharply at large angles of incidence.

Figure V·B·12. Properties of selective surfaces for solar energy application. (α = Absorptivity for solar energy. ϵ = Emissivity for long wave radiation from surfaces typical of flat plate solar collectors.) Source: (DUF.)

Surface	α^*	ϵ^*	Reference
"Nickel Black"; containing oxides and sulfides of Ni and Zn, on polished Ni	0.91-0.94	0.11	Tabor et al. (1964)
"Nickel Black" on galvanized iron (experimental)	0.89	0.12	Tabor et al. (1964)
Same process[†] (commercial)		0.16-0.18	Tabor et al. (1964)
"Nickel Black," 2 layers on electroplated Ni on mild steel (α and ϵ after 6 hr immersion in boiling water)	0.94	0.07	Schmidt (1974)
CuO on Ni; made by electrode position of Cu and subsequent oxidation	0.81	0.17	Kokoropoulos et al. (1959)
Co_3O_4 on silver; by deposition and oxidation	0.90	0.27	Kokoropoulos et al. (1959)
CuO on Al; by spraying dilute $Cu(NO_3)_2$ solution on hot Al plate and baking	0.93	0.11	Hottel and Unger (1959)
"Cu Black"[†] on Cu, by treating Cu with solution of NaOH and $NaClO_2$	0.89	0.17	Close (1962)
Ebanol[†] C on Cu; commercial Cu-blackening treatment giving coatings largely CuO	0.90	0.16	Edwards et al. (1960)
CuO on anodized Al; treat Al with hot $Cu(NO_3)_2$-$KMnO_4$ solution and bake	0.85	0.11	Tabor (1967)
Al_2O_3-Mo-Al_2O_3-Mo-Al_2O_3Mo-Al_2O_3 interference layers on Mo (ϵ measured at 500 °F)	0.91	0.085	Schmidt et al. (1964)
PbS crystals on Al	0.89	0.20	Williams et al. (1963)

*α = absorptance for solar energy; ϵ = emittance for long-wave radiation at temperatures typical of flat-plate solar collectors.
[†]Commercial processes.

Not all selective surfaces are generally available. Sometimes the limit on availability is due to high transportation and handling costs to and from the plant at which the surface is applied. Many surfaces, too, are protected from widespread, general use, by patents. The difficult application process, because of the necessity for quality control, limits their use. Electroplating, chemical baths, and vapor depositions are common methods of application. Microscopic layers of about one-half micron must be applied in uniform thicknesses. Figure V·B·12 summarizes some selective surfaces which have been or are being investigated. Figure V·B·13 gives some "recipes."

Durability is a key factor in prescribing the type of surface. Destructive forces include moisture, high temperatures, and sunlight. An approximate relationship between the performance of flat blacks and selective surfaces is shown in Figure V·B·14. A flat black absorber plate with two transparent covers has approximately the same performance as one which has a selective surface and only one cover. For collector temperatures below 150°F or so, a second cover plate over the selective surface does not significantly affect performance. However, at temperature ranges sufficiently high for actuating absorption cooling equipment (above 180°F), a second cover glass may be necessary. At low operating temperatures (below 100°F), on the other hand, a selective surface does not necessarily

---

(1)   Nickel-Black

The metal base must be perfectly clean: this is effected using standard chemical cleaning techniques as used in the electro-plating industry.

The black coating is then obtained by immersing as a cathode in an aqueous electrolytic bath containing, per litre:[7]

   75 gms nickel sulfate ($NiSO_4 \cdot 6H_2O$);
   28 gms zinc sulfate;
   24 gms ammonium sulfate;
   17 gms ammonium thiocyanate;
    2 gms citric acid.

The pH of the solution should be about 4 and a pure nickel anode is used. The bath is operated at 30°C. Electrolysis is carried on for 2-4 minutes at 2 ma per cm^2 the exact time depending upon the nature of the base metal and the temperature.

In cold weather there is danger of precipitation from the solution given above so that in recent work the same solution but half as concentrated has been employed instead - with the same results. (The citric acid may be omitted).

Better results are obtained by the two-layer technique. Thus on galvanized iron the electrolysis is carried on at 1 ma per cm^2 for 1 minute followed by 1-2 minutes at 2 ma per cm^2.

(2)   Copper Oxide on Aluminum

An aluminum base is first covered with an oxide layer by anodizing. For this purpose the aluminum body is immersed as cathode in an aqueous solution containing 3% by volume of sulfuric acid and 3% by volume of phosphoric acid, with carbon as an anode. An electric current of 6 milliamperes/cm is passed during 20 to 30 seconds through the solution, then the current is reversed for a few seconds to give partial anodizing. After rinsing, the aluminum body is immersed for 15 minutes at 85-90°C in an aqueous solution containing per litre:

   25 gms of copper nitrate $Cu(NO_3)_2 \cdot 6H_2O$;
    3 gms of concentrated nitric acid;
   15 gms of potassium permanganate.

After this treatment, the aluminum body is withdrawn, dried and heated to about 450°C for some hours, until the surface color has become almost black.

This treatment is rather sensitive to the type, composition and grain structure of the aluminum so that the results are not equally good on all grades of commercial aluminum.

(3)   Copper Oxide on Copper

The formulation is quoted from Ref. 5 and is very similar to that of Salem and Daniels.[9]

Before blackening, the copper is buffed to remove dirt and oxide layers to yield a clean bright surface. After being degreased in a boiling bath of metal cleaner it is washed in clean water and rubbed with a soft wire brush to remove gritting particles.

It is then treated, for various times - between 3 and 13 minutes* - in the blackening bath at a solution temperature of 140-145°C.

The bath comprises:

   16 oz sodium hydroxide (NaOH);
    8 oz sodium chlorite ($NaClO_2$).

per gallon (imperial) of water.

---

* At 3 minutes the measured $\alpha = 0.79$; $E = 0.05$. At 8 minutes $\alpha = 0.89$, $E = 0.17$. Longer times increase E with little increase of α.

Figure V·B·13. Several recipes for selective black surfaces. (Reprinted by permission of ASHRAE.) Source: (JOR.)

---

increase efficiency. To date, the expense of selective surfaces has only occasionally been justified by the increased cost. Miromit solar hot water heaters from Israel use a surface developed by Dr. Harry Tabor; the Beasley solar hot water heaters from Australia also use a selective surface, as do the collectors available from Sunworks, Inc. (they use one called Ebanol C).

Cover Plates

A cover plate is the generic term applied to a rather large group of transparent materials which are used to cover the absorber plate and which are usually located about one inch from it. The short-wave sunlight penetrates through the "transparent" material (transparent to sunlight), hits the absorber plate, stops, and turns into long-wave infrared. The previously transparent material is now relatively opaque to these long waves, and the heat is trapped.

The materials, from glass to plexiglass to fiberglass reinforced polyester to thin plastic films, vary in their abilities to transmit the short-wave sunlight and to trap the long-wave heat. They also vary in their other properties: cost, weight, resistance to degradation by solar radiation, break resistance, ease of handling, scratch resistance, and so forth.

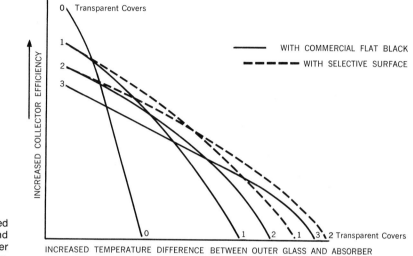

Figure V·B·14. Generalized effect of absorber coating and numbers of transparent cover plates on collector efficiency.

A basic knowledge of the solar spectrum makes an understanding of the function and choice of cover plates somewhat easier. Figure V·B·15 shows the solar radiation (insolation) which impinges on the extraterrestrial layer of the atmosphere. Beneath this top line is the relative amount of that insolation which penetrates through the atmosphere to the sea under typical clear day conditions. The bottom line represents the radiation which penetrates through (is transmitted by) one sheet of glass which is held perpendicular to the movement of the light. Ultra-violet light is less than .4 microns in length and infrared is longer than .7 microns. Visible light is the spectrum between these two. Long-wave infrared (heat) is usually considered to be 3 to 20 microns in length or longer. Glass is almost totally opaque to these long waves.

Solar radiation is either transmitted, absorbed, or reflected. The sum of the actions of these three properties accounts for 100% of the radiation. Glass, one of the most popular choices for cover plates, varies in its solar transmittance according to the angle of incidence of the solar radiation. Figure V·B·16 shows how the transmittance capability of very clear glass varies for different numbers of layers.

As a tried and tested cover plate, glass is one of the most favored choices. It is readily obtainable, it has good solar transmittance (85 to 92%, depending upon iron content—lower for glass thicker than ⅛ inch), is opaque to long-wave heat, is thermally stable at high temperatures, and is relatively scratch and weather (except hail) resistant. Most importantly, however, it is familiar to everyone.

Some of its disadvantages are that it must be handled in small pieces and with great care to prevent breakage. It is often applied to the collector on site rather than shipped long distances because of the breakage problem and its large weight. Its inability to span distances, as well as its size limitations, require that many difficult and expensive greenhouse-type glazing details must be used in fastening it to the collector. Glass is also relatively expensive compared to many of the other glazing choices.

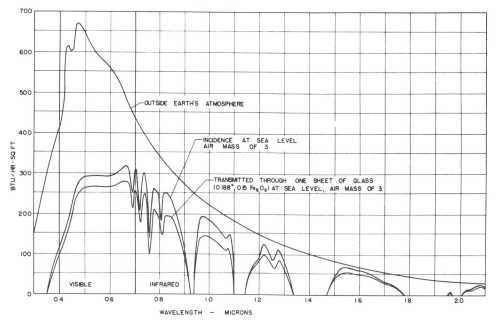

Figure V·B·15. Curves showing solar energy transmittance of one sheet of glass (0.188 in. thick, 0.15 Fe$_2$O$_3$) at sea level, for solar radiation transmitted through standard air mass of 3. Curves are compared with standard solar energy curve outside earth's atmosphere. Source: (MIT.)

In the use of glass, care should be taken to allow the glass to expand and contract within its glazing frame while simultaneously insuring a tight fit against moisture and air leakage. It should never rest on metal because of the potential excessive heat. Glass should also be chosen for low iron content. When looking on edge, the greener the color, the higher the iron content and the lower the solar transmittance. High quality glass absorbs 3–4% of the insolation rather than 7%. In areas of frequent storms of large hailstones, thick or tempered glass must be used, or it may be covered with screen (up to 4% of the solar energy will be prevented from hitting the collector).

In double glass systems, the space between the two layers should be vented to allow potential water leakage and moisture condensation to escape. Desiccant-containing chambers have been used by PPG Industries and others to absorb the moisture in their collectors.

MIT found in the late 1950's that the inner layer of insulating glass (e.g., thermopane) would break under intense thermal stresses, primarily because the inner layer became much hotter than the outer layer. The tightness of their sealed common edges caused the internal stresses. All insulating glass used in collectors should be *properly* tempered to reduce this problem. Ordinary thermo-pane can crack at temperatures as low as 300°F. PPG Industries uses their ⅛ inch "Herculite K" insulating glass, which is available without the collector, if desired.

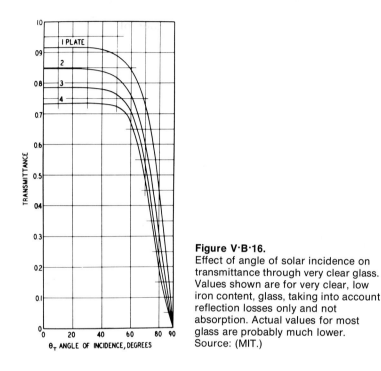

Figure V·B·16.
Effect of angle of solar incidence on transmittance through very clear glass. Values shown are for very clear, low iron content, glass, taking into account reflection losses only and not absorption. Actual values for most glass are probably much lower.
Source: (MIT.)

Anti-reflection coatings have been added to glass to reduce reflection and increase the transmittance. Up to 95% transmittance has been achieved. These coatings, which vary in effectiveness according to different wavelengths and angles of incidence, are not yet generally available and are relatively expensive for the somewhat small (2 to 4%) increased performance. The deposits, on both surfaces, are approximately one-fourth wavelength in thickness.

Takao Kobaysaki and Stephen Sargent (KOB.) include the following alternatives to glass for collector covers:

- Plastics
  polymethyl methacrylate or acrylic (Acrylite®, Lucite®, Plexiglass®)
  polycarbonate (Lexan®, Merlon®)
  polyethylene terephthalate or polyester (Mylar)
  polyvinyl fluoride (Tedlar)
  fluorinated ethylene-propylene, or fluorocarbon (Teflon, FEP)
  polymide (Kapton®)
  polyethylene
- Other Materials
  fiberglass-reinforced polyester (Sun-lite®)
  plastic-plastic laminates
  plastic-glass laminates

In general, plastics have higher solar transmittance because most of them are thin films. They are usually cheaper and come in large sheets, potentially reducing the number and expense of glazing details. Many of them are more resistant to breakage and are lighter than glass. Their maleability makes some plastics candidates for new and innovative covering techniques.

Unfortunately, most of them are partially transmittant of the long-wave infrared heat from the absorber plate and are, therefore, less efficient as thermal traps. Long-wave transmittance is as high as 86% for 2 mil polyethylene and 73% for 2 mil fluorinated ethylene-propylene (KOB.). For some plastics, however, the increased solar transmittance mitigates the increased heat transmittance, as does the use of a selective surface on the absorber plate. Efficient thermal traps become increasingly important, however, at higher collector operating temperatures, and many plastics, therefore, become less desirable.

Many plastics degrade when exposed to the ultraviolet rays of the sun. The films are particularly vulnerable because they are so thin. Some of the thicker plastics become yellowed, thereby losing some of their solar transmittance.

In addition, many plastics soften at the high temperatures which they reach on collectors: on cooling, some of them return to their original shape, but others, such as the acrylics, remain deformed. The low scratch resistance of some of the plastics is a critical issue in dirty or dusty regions. Hard, scratch-resistant coatings are available at increased cost for some plastics (e.g., acrylics).

Teflon has a very high solar transmittance (95% plus) and good weatherability (as plastics go), but it is expensive and not as strong as some of the other plastics. At elevated collector temperatures, Tedlar has a life-span as short as two years. Mylar is relatively strong to begin with but degrades fairly quickly when exposed to ultraviolet.

Sun-lite, by Kalwall Corporation, is emerging as a particularly good alternative. Although its solar transmittance is over 90%, it decreases slightly (and permanently) at high temperatures and under exposure to the sun, but levels off above 80%. Its relatively low cost, however, coupled with its ease of handling, its durability, and its light weight, make it a very attractive choice. It is available in rolls of unlimited lengths in standard widths of four and five feet, making it a good candidate for at least the cooler outer layer of a two-cover system.

Suntec is developing a membrane with a very high solar transmittance to be used in multi-layer systems (up to six). In this way, relatively high transmittance is retained and heat loss through the front is greatly reduced. Another membrane by Suntec turns opaque at high temperatures, protecting the absorber plate from overheating.

The number of covers is a critical factor in the design of the collector, the rest of the solar system, and the heating and/or cooling system. Generally, the lower the temperature required from the collector, the fewer the number of cover plates required. Solar swimming pool heaters, for example, may require no cover at all.

The larger the number of covers, the more solar energy is absorbed and reflected by them, and the smaller the percentage of sunlight which actually strikes the absorber surface. This is particularly true at sharp angles of incidence. The cooler the outside air temperature, the greater the number of covers needed to obtain the required operating temperatures of the collector without sacrificing collector efficiency. For example, assuming the same solar weather conditions, two covers may be required in New England but only one in Florida to obtain the same relative collector performance.

In order to use the collector for nocturnal radiational cooling, in which the heat storage fluid is circulated through the collector at night as a means of cooling by radiation to the night sky, no cover glass is used. Although cooling is thus obtained in the summer, winter collector efficiency is substantially reduced.

Figure V·B·14 illustrates that for different collector operating temperatures, various numbers of cover plates result in more optimal efficiencies. Although an additional cover plate can result in higher efficiencies, the extra cost of installation might not warrant its use. Besides the cover installation itself, there is, for example, the differential cost of storage. Two covers will bring the storage temperature higher, increasing the effective storage capacity without increasing installation expense. One cover glass may result in larger sizing, not only of storage but of the heating system (larger ducts and fans, or larger pipes and pumps, for example) in order to make use of lower storage temperatures. Tybout and Löf (TYB.) found that the use of two covers is most economical for heating throughout most of the U.S.

Hottel states (HOT.2): "The optimum number of glass panes is greater the higher the desired temperature of operation of the black plate. For domestic water supply one pane is generally the optimum; some Russian collectors have one pane at the water-entrance end of the absorber and two at the exit. House heating generally justifies use of two panes. Absorption refrigeration may require three panes or the use of a selective blackening, whichever is cheaper, because of the higher temperature level of heat required. Consideration of use of more than three panes, to obtain heat at a still higher level, should in any case be accompanied by consideration of surface treatment of the glass to reduce reflection losses or treatment of the absorber to make it selective or both. . . ."

Whillier (WHI.2) gives the following table as a guide to the selection of the number of cover plates.

COLLECTOR TEMPERATURE ABOVE AMBIENT TEMPERATURE	NUMBER OF COVERS
−10° to 10°F	none
10 60	1
60 100	2
100 150	3

These are only guidelines, of course, The actual decision must be based on cost and on other considerations: the configuration of the collector, materials used in the collector, and the design of the heating/cooling equipment and storage mentioned above.

An additional consideration is that the use of a selective surface in combination with a single cover may be more efficient than two covers in combination with flat black. Although multiple-layered cover plates may produce higher winter efficiencies, collectors are more easily damaged because of the higher temperatures during the summer when the collector may not be operating. Where outdoor temperatures of only a few degrees below freezing are expected, a second cover may solve the collector freezing problem.

Cover plate spacing is controlled primarily by its effect on collector efficiency and the cost of installation. Collector efficiency is affected (2 to 4%) by the shading of the glazing framework on the absorber surface and by the difference in the insulating values of various thicknesses of air space: spacings of ½ to one inch (0.04 to 0.08 ft) are the most efficient.

The net effective area of the collector is controlled in part by the size, number (frequency and length), and type of glazing details. Care should be taken to minimize their reduction of the gross, useable square footage of absorber surface. Glazing gasketing and sealants should also be considered for their resistance to ultraviolet and infrared radiation and to high temperatures. The glazing/fastening details should provide for drainage and should resist penetration by water, snow, and ice. They should also allow for movement of the cover plates as they expand and contract with temperature change.

Some Other Collector Design Factors

Insulation. In order to reduce the loss of heat from the absorber plate, insulation is added to the back side. If the collector is applied to a surface of the building, such as a wall or roof, heat is not "lost" but is transferred to the building itself. This is an advantage during the winter, but a disadvantage during the summer. Except in climates with cool summer temperatures, the collector should be insulated to minimize this heat gain and to raise its overall efficiency. Six inches of fiberglass or its equivalent is the usual choice for roof top collectors; as little as four inches can be used for vertical collectors. Where the collector is its own structure and does not cover a building surface, six to eight inches of fiberglass, or the equivalent, are generally used.

Fiberglass insulation is favored over styrofoams and urethanes because of its stability at high temperatures. The choice of a particular type of urethane, in particular, must be based on its stability at high temperatures: some urethanes will deform, expand, and give off potentially toxic gases. Flammability is also a consideration.

Whenever possible, and particularly for urethane, the insulation should be separated from the back side of the absorber plate by an air space. A reflective foil should cover the insulation, separating it from the air space. This reflects heat back to the absorber, decreasing the temperature of the insulation and increasing the efficiency of the collector.

The perimeter, or edges, of the collector surface must be insulated to reduce edge losses. The perimeters of absorber surfaces are generally at lower temperatures, with resultant lower overall efficiencies because of these edge losses. Collector test work shows that edge losses can be reduced by massively insulating the edges and by making the collector surface area large relative to the length of perimeter. Insulation should not be added unduly if it reduces the potential absorbing surface.

For collectors which are available as components from manufacturers, metal, plastic, or wood pans are sometimes added as a protective casing around the outside (back) of the insulation. Galvanized steel is common, but pans of even better weather resistance should be used in corrosive atmospheres.

Neither dead loads nor wind loads are problems for vertical wall collectors or for collectors which are integrated into the roof, since the building must withstand wind conditions anyway. For collectors which are separated from buildings or which are attached with their own structures to the tops of roofs, wind loads are the main consideration in the structural design of the support system. In general, to reduce the cost of the structural system and to reduce the sail effect of a large collector, surface areas should be kept small to maintain a low profile. This can be done by placing a series of long, low collectors one behind the other. Collectors weigh from 1 lb per sq ft for plastic collectors to 6 lbs per sq ft for glass and metal collectors. This is well below the design loads for the roofs of most buildings.

Snow loads have not been a problem in past solar projects. In most cases, collectors have been designed at angles steep enough to maintain natural snow run-off. When this method fails, warm storage heat travels through the collector to warm the collector surface and melt the snow. Of course, the snow loads on collectors would be no greater than those for ordinary roofs and, in fact, because the snow generally slides off, could be less.

Accessibility. Large collector surfaces may require more maintenance attention than ordinary walls and roofs, and they are more difficult to maneuver on because of their relatively slick and fragile cover surfaces. Long, low collectors, catwalks, and provisions for ladders can reduce the problems of access. Where local atmospheric and climatic conditions require frequent window washing, accessibility is even more important. In any case, the weight of maintenance personnel must be considered in the design of the cover plate details.

Thermal Capacity of Collectors. Klein, Duffie and Beckman (KLE.) showed that the heat capacity of collectors has almost no effect on performance. However, in an unpublished essay, "Thermal Capacity of a Flat-Plate Collector for a Solar-Heated House: Does it Make Much Difference Whether the Capacity is Large or Small?" (December 4, 1973), William Shurcliff goes to great lengths to show that, under steady-state conditions of morning start-up and under dynamic conditions of intermittent sunshine, collectors can vary by 25 % in their overall efficiency as their heat storage capacity varies. The smaller the capacity, the greater the efficiency. This will be tempered somewhat by sunshine conditions, of course, but the more intermittent the sunshine, the more important it is to have the smaller capacity with shorter heating up periods.

Intuition suggests that Shurcliff is right to some extent (perhaps not to 25%, however) since the lower heat capacity enables the collector to reach operating temperature faster. When sunshine is intermittent, the low heat capacity allows fuller utilization of short bursts of available sunlight.

Hottel and Woertz (HOT.1) showed that the heat loss caused by heat capacity is between 0.30 and 0.60 Btu/ft²/day per degree F temperature difference between the collector and the outdoor air. For their collector, which consisted of two covers of glass, a copper absorbing surface, and fiberglass insulation, the glass accounted for 52%, the copper plate for 20%, and the insulation for 28% of the total heat capacity. They did not view this heat loss as excessive, particularly when compared with collectors in which the liquid does not drain from the absorber at night. Systems in which the liquid does not drain at night as much as double their effective collector thermal capacity.

Collector Examples and Construction Details

The prototypical solar collector is shown in Figure V·B·17. It was on this basic model that Hottel and Woertz (HOT.1) performed their vanguard research on solar collector performance. Thousands of innovators around the globe have struggled to develop efficient but inexpensive methods of fastening the tubes to the plate. Figure V·B·18 shows three tube locations: below the plate, above the plate, and integral with the plate. John Yellott (ASH.) reports that Austin Whillier studied the effect of bond conductances and concluded that steel pipes are as good as copper if the bond conductance between tube and plate is good. Bond conductance can range from a high of 1000 Btu/hr/ft/°F for a securely soldered tube to a low of 3.2 for a poorly clamped or badly soldered tube. Bonded plates with integral tubes are among the best alternatives for performance, but they require mass production facilities to obtain cost benefits.

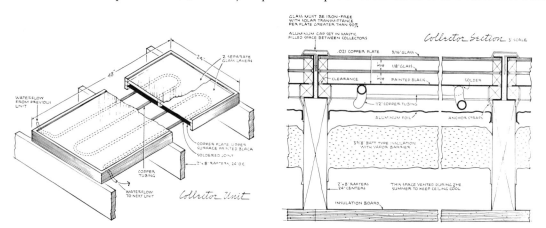

Figure V·B·17. A solar collector with two cover glasses and a copper tube-in-plate absorber. Source: (WHI.2)

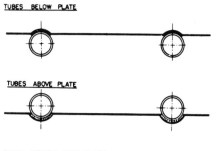

Figure V·B·18.
Tube-in-plate configurations.
(Reprinted by permission of ASHRAE.) Source: (JOR.)

Figure V·B·17 shows the basic construction details of some of the early solar collectors at MIT. Copper tubes are soldered to the back side of black copper sheets. Two clear layers of glass cover the assembly.

Thomason's Solaris System, in which the water flows down the surface of corrugated aluminum or galvanized steel, avoids many of the freezing problems associated with tube-type collectors. The ridge and gutter details of his system are shown in Figure V·B·19. Two variations of the tube-in-plate absorber surface are proposed in Figures V·B·20 and V·B·21. Both collectors use polyurethane foam insulation to back the absorber plate. Once again, extreme care must be taken in choosing the urethane which will be stable at the high temperatures which can be attained by collectors when not in use (e.g., during the summer).

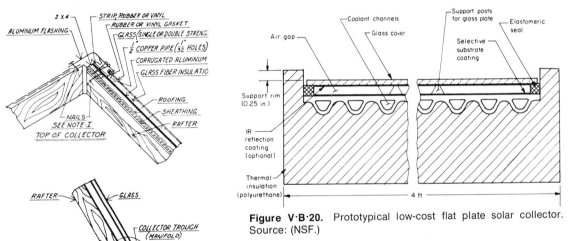

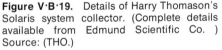

Figure V·B·19. Details of Harry Thomason's Solaris system collector. (Complete details available from Edmund Scientific Co.) Source: (THO.)

Figure V·B·20. Prototypical low-cost flat plate solar collector. Source: (NSF.)

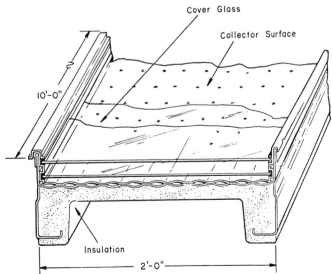

Figure V·B·21. Structurally integrated solar collector unit. Source: (MOO.)

The collector in Figure V·B·21 is itself a structural member. It could replace every component of typical roofs, including the rafters, the insulation, and the shingles. The absorber plate is composed of two sheets of metal welded to each other; the high pressures within the tubes require that this weld be of high quality.

One of the first commercial solar collectors available in this country for the heating of buildings was designed by Everett Barber in Guilford, Connecticut, in 1973. The first designs, shown in Figures V·B·22 and V·B·23, were used on

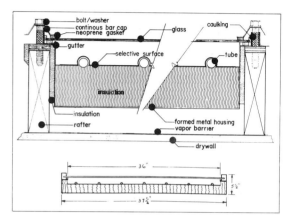

Figure V·B·22.
A first-generation flush-mounted collector design by Sunworks Corporation. (Drawing courtesy of Sunworks Corporation.)

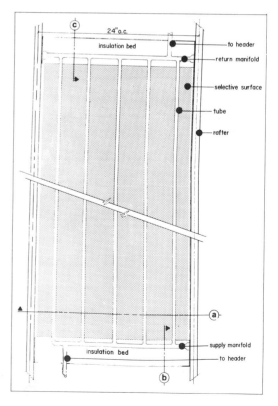

Figure V·B·23.
Details: flush-mounted module by Sunworks Corporation. (Drawing courtesy of Sunworks Corporation.)

several houses. They are available through Sunworks, Inc. The "flush-mounted module" was designed to fit between roof rafters and to replace the insulation, sheathing, and weatherproofing roof surface. This model was replaced by the "surface-mounted module," which relies on the conventional roof to give the building rigidity and the required water and weather protection.

Barber believes that tried and tested materials produce the best long-term results, and, therefore, the best long-term economic payback. Both the fluid tubes and the absorber plate to which they are soldered are totally of copper.

LIQUID HEATER
FLAT-PLATE SOLAR HEAT COLLECTOR* -- SURFACE-MOUNTED MODULE

This module is intended for installation on any structurally-sound surface or specially constructed framework. It can be mounted on a surface, such as an existing roof, and connected end-to-end or side-by-side. We recommend that no more than three collectors be connected in series.

Technical data on materials
    cover: single glass, 3/16" thick, tempered, edges swiped 92% solar trans-
        mittance
    absorber container: sides - aluminum extrusion; rear - aluminum sheet
        0.05" thickness
    air space between cover and absorber: 1" - 1 1/8"
    absorber:
        -copper sheet: 0.010" thick (7 oz.)
        -selective black: minimum absorptivity 0.90; maximum
            emissivity 0.12 manufactured by Enthone, Inc.,
            durable to 400°F.
        -copper tubes: 1/4" Ø (0.375" OD) M-type
        -tube spacing: 6" on center
        -tube pattern: grid
        -manifolds: 1" Ø (1.125" OD) M-type copper
        -tube connections to manifold: silver solder
        -bond between tube and sheet: soft solder
        -connection to external piping: 1" Ø (1.125" OD) K-type
            copper, extending 2" beyond
            collector ends; supply, top right;
            return, bottom left (when viewed
            from the top)
        -manifold/tubes pressure-tested to 15 atm.
    insulation behind absorber: 2 1/2" thick fiberglass, 1.5 lb./ft.^{3} density
        R= 10.4
    gasketing material: glass: aluminum extrusion and neoprene 'U' gasket
    weatherproofing: this module can be placed out in the weather without need
        for further weatherproofing

Dimensions of the surface-mounted module
    -outside dimensions overall: 36" wide by 84" long x 4" thick
    -effective absorber area = 18.56 ft.^{2}
    -ratio of usable absorber area to total surface covered: 0.884
    -thickness of module: 4"
    -glass area: 18.96 ft.^{2}

Method of anchoring
    Continuous angle clips are fastened to each end of the frame for anchoring;
    holes to be drilled by installer as required

Weight per module
    120 pounds, filled

Recommended flow rate through the collector:
    -1 gph/ft.^{2} of collector (F_r= 0.9)
    -flow resistance at this rate is negligible

Collector Coolant

The coolant can be inhibited alcohol-water mixtures such as standard automobile anti-freeze made by Union Carbide or duPont. In areas where regular tap water is used as a coolant, it is important that the pH be controlled between 6 and 8. These collectors can be used with other coolants but the user must contact the manufacturer for approval of specific liquids. (See guarantee statement.)

The manufacturer reserves the right to change design details without notice.

*Patents Pending.

This choice will yield long life in comparison to every other material now available for this type of absorber surface. A selective surface is applied to the copper sheet, and this is covered with a layer of glass. Galvanized sheet metal contains the entire assembly. Independent tests by NASA and others show this collector to have very good performance through a wide range of collector temperatures.

PPG Industries followed closely on the heels of Sunworks, producing a less expensive collector of lower quality (see Figure V·B·24 and V·B·25). The lower quality is primarily due to the use of aluminum, rather than copper, as the absorber surface. Many of the first panels were installed without solving the corrosion problem and subsequently leaked.

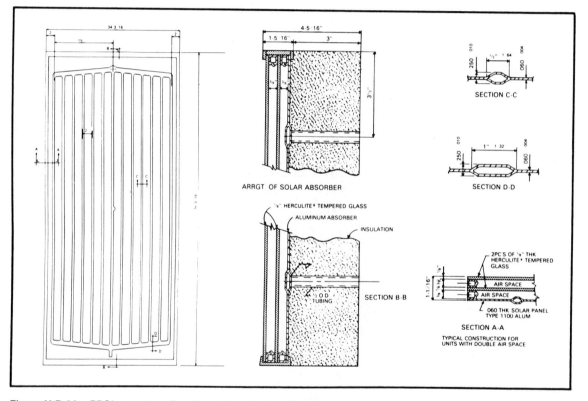

Figure V·B·24. PPG's prototype Baseline solar collector. (Drawings courtesy of PPG Industries.)

On the other hand, the PPG panel has apparent advantages. Instead of tubes fastened to a sheet, the aluminum absorber plate is of the tube-in-plate type (Roll-Bond®). Although the surface is covered with a flat black rather than with a selective surface, it is covered by two sheets of glass—PPG's "Herculite K" tempered insulating glass. The assemblies are available with or without insulation or containerization on the back of the absorber plate. PPG has since switched to copper absorbers in their basic collector line.

A first design of a collector developed by Sun-Earth, Inc. is shown in Figure V·B·26. Its most notable feature is the incorporation of the header piping into the component which is also available as the collector. For collectors of other designs, the large diameter (1 to 1½ inches), the correspondingly long header lengths, and the numerous connections which must be made to the many panels significantly increase the total effective cost of the collector area.

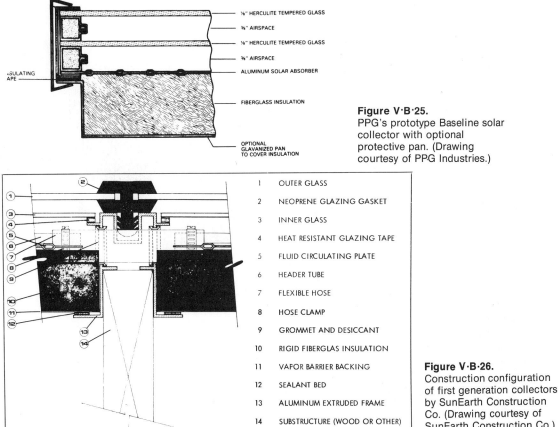

Figure V·B·25.
PPG's prototype Baseline solar collector with optional protective pan. (Drawing courtesy of PPG Industries.)

1	OUTER GLASS
2	NEOPRENE GLAZING GASKET
3	INNER GLASS
4	HEAT RESISTANT GLAZING TAPE
5	FLUID CIRCULATING PLATE
6	HEADER TUBE
7	FLEXIBLE HOSE
8	HOSE CLAMP
9	GROMMET AND DESICCANT
10	RIGID FIBERGLAS INSULATION
11	VAPOR BARRIER BACKING
12	SEALANT BED
13	ALUMINUM EXTRUDED FRAME
14	SUBSTRUCTURE (WOOD OR OTHER)

Figure V·B·26.
Construction configuration of first generation collectors by SunEarth Construction Co. (Drawing courtesy of SunEarth Construction Co.)

Sun-Earth is also attempting to eliminate the expensive mullion which is used to fasten the collectors to the roof and to each other.

Revere Copper and Brass, Inc., developed a solar collector by adding rectangular copper tubes to their laminated panel system, in which a thin copper sheet is laminated to plywood to make a copper composite building panel used primarily as a roof surface (see Figure V·B·27). Each standard 2 feet by 8 feet panel uses from two to five tubes, depending on efficiency requirements. Although special clips, battens, and couplings have been designed, the performance and durability of this collector are often questioned. The Copper Development Association built their "Decade 80" house in 1974 to demonstrate the applicability of the product and to promote the use of copper in the solar energy industry.

Some of the durability questions include possible de-lamination of the thin copper sheet from the plywood at high temperatures; the possible discontinuous and loose fit between the rectangular tubes and the copper sheet; and the "bleeding" of heat from the absorber plates to the outside through the copper "solar batten" which secures the single or double cover plate to the collector.

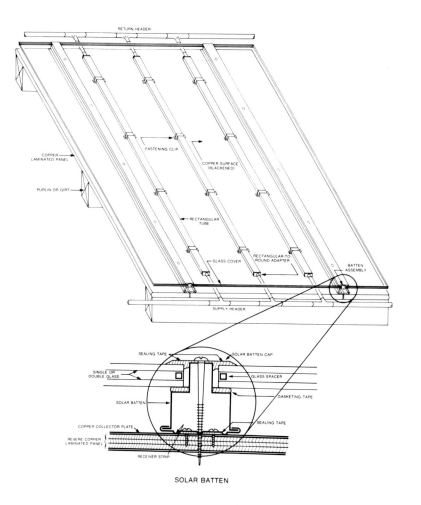

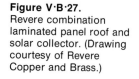

Figure V·B·27.
Revere combination laminated panel roof and solar collector. (Drawing courtesy of Revere Copper and Brass.)

The solar collector shown in Figure V·B·28 was developed by William Edmondson, editor of the *Solar Energy Digest*, a monthly newsletter on solar energy. Although Solarsan® was conceived as a solar water heater, it is readily adaptable to house heating. In fact, since it was designed to be incorporated as part of the roof, it almost makes more sense to build this collector to the full dimension of a roof, using it for heating both the building and the domestic hot water. Since the cost of several of the components can be applied to the roof, the actual cost of the collector materials can be relatively small, possibly around $1.00 per square foot. Many of the necessary details, such as the bond of the copper tubes to the aluminum, are not shown in the drawing. To provide the best performance possible, *Solar Energy Digest,* published in San Diego, California, makes plans available.

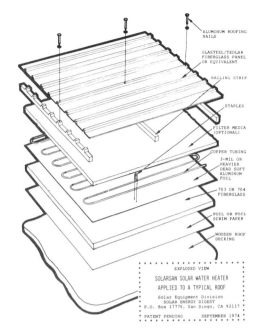

ALUMINUM ROOFING
NAILS

GLASTEEL/TEDLAR
FIBERGLASS PANEL
OR EQUIVALENT

NAILING STRIP

STAPLES

FILTER MEDIA
(OPTIONAL)

COPPER TUBING

3-MIL OR
HEAVIER
DEAD SOFT
ALUMINUM
FOIL

703 OR 704
FIBERGLASS

FOIL OR FOIL
SCRIM PAPER

WOODEN ROOF
DECKING

EXPLODED VIEW
SOLARSAN SOLAR WATER HEATER
APPLIED TO A TYPICAL ROOF
Solar Equipment Division
SOLAR ENERGY DIGEST
P.O. Box 17776, San Diego, CA 92117
PATENT PENDING   SEPTEMBER 1974

Figure V·B·28.
William Edmondson's Solarsan solar water heater. (Drawing courtesy of *Solar Energy Digest.*)

AIR-TYPE FLAT PLATE COLLECTORS

Solar collection and utilization systems which use air as the heat-collecting medium should be considered for all space heating applications. This is particularly true when cooling and/or domestic water heating are not being considered or are relatively unimportant.

Air systems are more appealing than liquid systems because they require less plumbing and are therefore less costly. The complications of liquid systems stem from potential freezing problems within the collector; the need to allow for the expansion of liquid as it heats up in the system, including the possibility for the flashing of the liquid into gas (such as water into steam); the possibility of leakage anywhere within the system; and the corrosion of the metal plumbing. The relative simplicity of air systems is attractive to people who wish to build their own system, but as with all systems of collecting, storing, and using the sun's energy, their precise design is difficult and all except the simplest of systems must be designed by someone knowledgeable in simple mechanics and heat transfer. Air collectors are, however, relatively easy to maintain and repair. Fans, damper motors, and controls may fail, but the large components, including the collector, the heat storage, and the ducting, should last indefinitely.

The construction of air collectors and related components and systems is relatively simple when compared to the tasks of plumbing and of trying to find and utilize an absorber plate compatible with liquid systems. Except for Dr. Harry Thomason's design, in which water flows over corrugated metal, most absorber plates have tubes attached to or incorporated into them and they are not easily handled even by skilled workers. More easily handled are absorber plates compatible with air systems: since they are not connected into a plumbing system that must be leak-proof and since they do not have to have great care taken in allowing for expansion and contraction, they do not have to be built with as much precision.

In fact, for air-type collectors, the absorber surface need not be metal. Since, in many collector designs, the air comes in contact with every surface heated by the sun, heat does not have to be conducted from one area of the absorber surface to another, as is the case with absorber plates for liquid collectors. Almost any blackened surface which is heated by the sun will transfer heat to air when the air is blown on it. This heat transfer mechanism opens up numerous possibilties for absorber surfaces.

Raymond Bliss and Mary Donovan used four layers of black cotton screening to construct absorber plates for the Desert Grassland Station in Arizona and Dr. George Löf used sheets of glass painted black in his houses in Colorado (see Figure V·B·29); the glass plates in Löf's design had a two-thirds overlap of 18 inches. Each of the plates consisted of two pieces, a black part and a clear part. The black was obtained by applying black frit to ordinary window glass and firing it in an annealing furnace. The assembly was covered by two layers of glass. Four 4 ft sections were arranged in series and at a tilt of 60° from the horizontal. At first, because of a faulty method of supporting its edges, the glass spalled itself as it expanded and contracted. The method of support was changed, however, and the breakage stopped. Unless the edges of the glass are protected, the plates will chip and eventually break. Under the direction of Dr. Erich Farber, at the University of Florida, the black portion of the Löf designed plates were replaced by blackened aluminum, overlapped in the same manner.

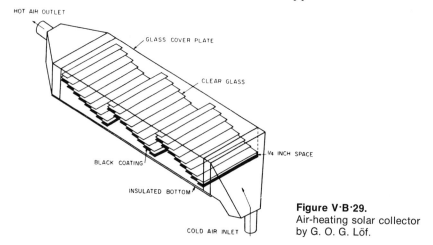

HOT AIR OUTLET

GLASS COVER PLATE

CLEAR GLASS

¼ INCH SPACE

BLACK COATING

INSULATED BOTTOM

COLD AIR INLET

Figure V·B·29.
Air-heating solar collector
by G. O. G. Löf.

Other materials for absorber surfaces include metal scraps (or pieces of metal cans) attached to a rigid board, metal lath, fiberglass meshes (e.g., air filters), crushed glass or rock, cloth, and even paper. Many of these can be obtained cheaply or as recycled or reused materials; however, the entire surface must be black, must be heated directly by the sun, and must come in contact with the air flowing through the collector.

Metal absorber plates are also possibilities, of course. In fact, not only is metal durable and effective, but it is preferable for situations in which the sun cannot reach every surface in contact with the moving air. Metal can also alleviate the "hot spots" caused by an uneven flow of air on the surface by conducting the excess heat build-up to other surfaces and thence to the air. Investigations by J. D. Close (CLO.2) determined the relative merits of air duct locations with respect to opaque, metal-type absorber surfaces. The three basic configurations are shown in Figure V·B·30: Type I, in which the air duct is located between the plastic cover and the absorber surface; Type II, in which an additional air duct is located behind the absorber plate; and Type III, in which the upper air duct is eliminated and only the air duct behind the absorber plate is used. The Type II air heater has higher efficiencies than the other two types when its air temperatures are about the same as the outdoors. However, as the difference between collector temperature and outdoor temperature increases, Type III is dramatically better. An approximation of the air-type collector used in 1973 at the experimental solar house at the Institute of Energy Conversion at the University of Delaware is shown in Figure V·B·31. Basically, it is a Type III collector with solar photovoltaic cells added so that, in addition to heated air for house space heating, the collector assembly also converts sunlight into electricity.

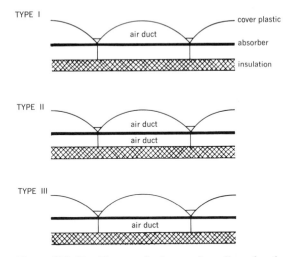

Figure V·B·30. Three collector configurations for the passage of air.

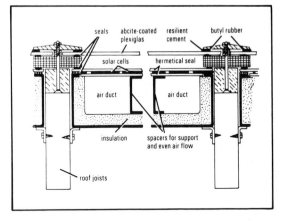

Figure V·B·31. Cross section through a solar electric-thermal flat-plate collector for the solar house at the University of Delaware. (Drawing courtesy of Institute of Energy Conversion, University of Delaware.)

Close also concluded that the higher the operating temperature of the collector, the more important is the heat transfer coefficient between the absorber surface and the air. For summer crop drying, for example, in which the collector temperature may be only 30° to 40°F higher than the outdoor temperature, the effectiveness of a single solid sheet of metal may be comparable to that of a finned absorber surface or a surface which has "vee" corrugations (see Figure V·B·32). However, in cool and cold climates, in which the temperature difference between the collector and the outdoors is as much as 100°F, the finned plated (see Figure V·B·33) adds 5 to 10% in efficiency to that of the flat plate, and the vee corrugated plate adds 10 to 15% beyond that.

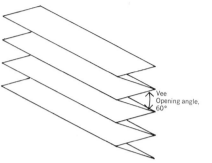

Figure V·B·32.
Vee-corrugated absorber plate.

The back side of an absorber surface should be painted black if air is in contact with it. The surface which separates that back air duct from the insulation should be lined with reflective foil; if a perforated absorber surface is used, however, the separating surface can be black and serve as another heat transfer surface.

For example, Whillier showed (WHI.3) that by replacing the conventional solid absorbing plate with a blackened screen of wire or plastic which allows about half of the radiation through to be absorbed at the lower surface of the air passage, the value of h (the effective coefficient of heat transfer between the absorber and the air stream) would double, resulting in increased heat collection of 10 to 15%. Good values of h are 6 to 12 Btu/hr/ft²/°F. Higher values of h are preferred, then, provided the air pumping costs do not become too large. Other methods of increasing the effective area of the absorber surface, and thereby increasing the heat transfer coefficient, are shown in Figures V·B·34, V·B·35, and V·B·36.

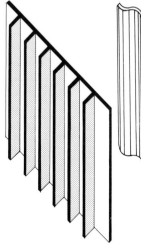
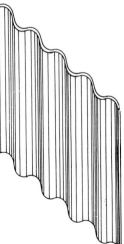
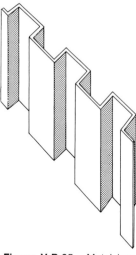

Figure V·B·33. Finned absorber plate.

Figure V·B·34. Corrugated absorber plate.

Figure V·B·35. Metal box duct absorber plate.

Figure V·B·36. Steel flooring type absorber plate.

Since it is possible for the absorber surface to be made of materials other than metal, it would seem that the cost of solar collectors might be considerably reduced, providing an incentive to manufacturers to explore this alternative to liquid systems as a means of producing a product which is more cost competitive with other energy systems. Unfortunately, however, there has been relatively little work done on the research and development of air collectors. This may be due in large part to the traditional emphasis on liquid systems. Manufacturers and others are bombarded by information on liquid systems and find them the easiest to pursue. Because of this availability of information, further research is done on liquid systems, and the imbalance is perpetuated.

Whether absorber surfaces are metal or not, it is important that the movement of air through the absorber plenum be turbulent. The usual case is for air flow to be laminar: the air next to a surface is relatively still while the air above the surface flows in smooth, undisturbed layers. Such flow results in poor heat transfer: the still air next to the absorber surface becomes hot, but the moving air above it does not come in contact with the absorber surface.

Turbulent flow on two scales can help to ameliorate this difficulty. On the macro scale, turbulent flow occurs when smoke blown through the space "tumbles" over itself on a scale visible to the eye. On the micro scale, the same effect should occur right at the absorber surface.

To help create turbulence at the macro scale, the absorber should not be flat, but should be broken up as much as possible, forcing the air to move in and out, back and forth, up and down. The finned plate and the vee corrugations perform this function well.

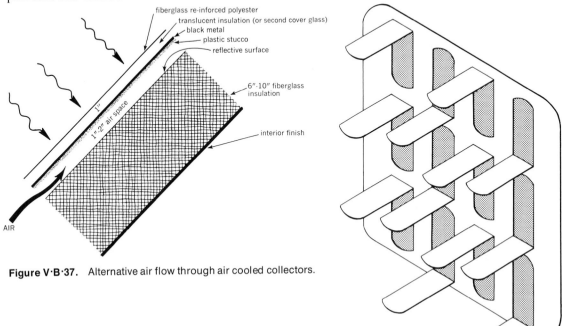

Figure V·B·37. Alternative air flow through air cooled collectors.

Figure V·B·38.
Pierced metal absorber plate.

To help create turbulence at the micro scale, the surface should also be rough, with as many sharp edges as possible. Examples of this are coarse surfaces such as gravel (see Figure V·B·37); air filters such as those for furnaces; cloth; mesh; and pierced metal plates (Figure V·B·38). If an aluminum sheet has thousands of tiny filaments attached to it (e.g., straight pins embedded in its surface), heat transfer would be much better because of the added surface area and because the filaments also cause micro scale turbulence. If the aluminum sheet with filaments is comprised of a series of discontinuous pieces which cause air to flow in a turbulent way at a macro scale, so much the better.

The tradeoff with turbulence is the pressure drop across the collector. Too many surfaces and too much restriction of air flow will require a large fan, and thus a large amount of energy, to push the air through. The energy required for this can cancel out the savings from using solar energy, particularly if the fan is electrical and if the amount of energy which is burned at the power plant to produce the electrical energy is included. Ray Bliss and Mary Donovan overcame this difficult tradeoff in their design for the Desert Grassland Station. Instead of creating turbulent air flow, the air was blown through four layers of black cotton mesh (see Figure V·B·39).

Figure V·B·39.
Black gauze solar collector by Donovan and Bliss.

For air type collectors, the considerations of absorber paints, selective surfaces, and cover plates are similar to those already discussed for liquid-type collectors. A few observations should be made, however. Perhaps one of the primary disadvantages of non-metal absorber surfaces for air-type collectors is the relative difficulty of applying selective surfaces. Until the technology for selective surfaces improves, metal absorbers for air-type collectors will be preferred. When selective surfaces can be easily applied to non-metal absorber surfaces, relatively low-cost but efficient air-type collectors will become available.

Close (CLO.2) demonstrated the importance of selective surfaces for air-type collectors. All other considerations being equal, the absorber with a selective surface will raise a 50% collection efficiency to 65% at low operating temperatures and a 15% efficiency to 35% at high operating temperatures. Although Close did not investigate the relative merits of one or two cover plates, it can be extrapolated from the work done with liquid-type collectors that a second cover plate or a selective surface affects collector efficiency to a similar extent.

Hollands (HOL.) showed that the performance of an air heating absorber plate can be significantly improved by the use of the vee corrugation already discussed (see Figure V·B·32). In addition to increasing heat transfer surface, this plate increases absorption since the direct radiation striking the vees is reflected several times, absorption occurring at each surface. When oriented properly, its absorption is much higher than that of a flat sheet and its increased emittance relatively small.

An important consideration in the design of the collector is the prevention of leakage. Since the solar-heated air is under pressure from the fan, it can escape through the tiniest crack, and cold outside air can also be drawn in. Many designers feel that leakage prevention is important for liquid systems, but, in fact, it is one of the most important elements in making air collectors efficient. Air tightness is crucial in the design of the entire air flow system, including the ducts and the dampers. Special care must be taken in the glazing frames of the cover plates to help prevent leakage; the use of large plastic sheets reduces the number of glazing joints considerably. Just as storm windows help reduce air leakage into buildings, so second and third cover plates help reduce air leakage in air-type collectors.

The various collector panels in George Löf's houses were designed and built as individual units and then assembled to form a large collector surface. This construction method can reduce the probability of leakage and can allow for mass production and possible reduction in cost.

If the collector is designed for on-site construction and assemblage, the construction members which separate the individual absorber spaces from each other can be perforated to allow air to flow from one space to the other, equalizing the pressure, and thus the air flow, through the various panels. Compared with liquids, the movement of gases such as air are relatively more difficult to predict and distribute evenly. In the case of premade panels which are assembled on the site, allowance must be made in the design for the measurement and adjustment of air flow. The difficulty in predicting and controlling air movement is one reason why engineers are more inclined to favor liquid solar systems.

Whillier (WHI.3) reported that the use of thin sheet plastic to cover a gauze type absorber surface had an unexpectedly high performance. He reports that random pressure fluctuations through the absorber space, "owing to wind blowing over the outside of the plastic cover, causes slight flapping of the cover, with a resulting small in-and-out pulsating flow of air through the gauze. This small pulsating flow considerably increases the heat transfer coefficient, with resulting large improvement in overall collector performance." He reported an increase of up to 10%.

Turbulent flow of air through the absorber space causes convection losses from an air-type collector to be somewhat larger than those from a similar water-type collector. Therefore, the long-wave reradiation losses from the absorber surface through the cover plates is a smaller component of the total collector heat loss than for water-type collectors. This means that although glass is nearly opaque to long-wave reradiation, plastic covers, such as Tedlar (polyvinyl fluoride), with a long-wave transmittance as high as 0.4, can be used with relatively minor negative effect.

One of the difficulties in the use of air rather than water or other liquids is its relatively low capacity to contain heat and its low density. The specific heat of air is 0.24; that of water is 1.0. Air's density is about 0.075 pounds per cubic foot under most conditions, water's over 62 pounds per cubic foot. Thus, significantly greater amounts of air than water, both in volume and in weight, have to be transferred through the collector. For example, if 64 pounds (eight gallons, or approximately one cubic foot) of water is circulated through a collector, 260 pounds (about 3500 cubic feet) of air would be required to transport the same number of Btu.

One implication of the low heat capacity of air is the need for large spaces through which to move the air, even in the collector itself. Air duct spaces within collectors (for example, between the absorber plate and the cover plate) have ranged from one-half to six inches. In general, the larger the space, the lower the pressure drop (the resistance to air movement), but the poorer the heat transfer from the absorber to the air. For flat sheet metal absorber surfaces, the duct-space usually ranges from one-half to one inch. Larger ducts can cost more, both in materials required and in space consumed. Duct sizes of from 1½ inches to 2½ inches are considered optimum for most large collectors used for natural circulation systems or for systems with unusually long ducts (fifteen feet or more).

The absorber in the Type I collector shown in Figure V·B·30 is relatively hot and loses heat directly through the back. In collectors such as Types II and III in the same illustration, the hot absorber plate is separated from back heat loss by the turbulent air movement which is somewhat cooler than the absorber plate. Less insulation is required for these types than for those represented by Type I.

COLLECTOR PERFORMANCE

Many extensive explanations of collector performance by others in the field of solar energy are available, among them an excellent one by Austin Whillier, "Design Factors Influencing Solar Collector Performance" in *Low Temperature Applications of Solar Energy* (JOR.); two others are the classic "Performance of Flat Plate Collectors," by Hottel and Woertz (HOT.1) and "The Derivations of Several 'Plate-Efficiency Factors' Useful in the Design of Flat Plate Solar-Heat Collectors," by Raymond Bliss (BLI.2). Few gains are made in overall efficiency without paying the penalty of increased collector cost. Higher collection efficiencies often require an increased expense not easily amortized over the life of the collector. The tendency of many solar engineers to develop collectors of ever-increasing efficiency without considering their increased cost is unfortunate.

Generalized Collector Performance

The following steady state equation of the thermal performance of collectors is basic to an understanding of the amount of energy which a solar collector can absorb and deliver to the fluid which comes in contact with the absorber plate:

$$
\begin{array}{cccccc}
H_c & = & H_a & - & H_l \\
\text{the useful heat} & = & \text{the total solar energy} & - & \text{heat loss from} \\
\text{collected} & & \text{received by the} & & \text{the absorber} \\
& & \text{absorber} & &
\end{array}
$$

The useful heat collected, H_c, is usually found by measuring the fluid flow rate and the difference between its temperature when it enters the collector and when it leaves. The flow rate, in pounds per square foot per hour, is multiplied by this outlet/inlet temperature difference; this number is in turn multiplied by the specific heat of the fluid, which is 1.0 Btu/lb/°F change for water and 0.24 for air. The resulting value is the heat energy collected in Btu per square foot of collector per hour.

<table>
<tr><td></td><td>the useful heat
collected</td><td>=</td><td>fluid flow
rate</td><td>×</td><td>specific
heat</td><td>×</td><td>outlet/inlet
temperature
difference</td></tr>
</table>

$$ H_c \frac{Btu}{sq\ ft} - hr = \frac{lb}{sq\ ft - hr} \times \frac{Btu}{lb - °F} \times t_{out} - t_{in})°F $$

Although H_c suggests that all of the heat collected is used, this is not necessarily the case, particularly if there are energy losses from the fluid transport ducts or pipes, or from the thermal energy storage.

The total solar energy incident, I, on a solar collector itself is usually expressed either in Btu per square foot per hour, in calories per square centimeter per minute (Langleys), or in kilowatts. This is such a complex subject in itself that it is treated in the Resource Section on "Insolation." Basically (see Figure V·B·40), the higher the insolation rate, the greater the overall efficiency of the collector. Obtaining the actual values of insolation for a particular site, however, is one of the most difficult tasks in predicting collector performance. The Resource Section, as well as a later section of this chapter, give some helpful hints on solving this problem. Figure V·B·41 shows the variation in the calculated monthly performance of identical collectors in two different locations: Boston, Massachusetts, and Madison, Wisconsin. Although temperatures at the two sites vary, the differences in insolation values are most important in explaining the variation in performance.

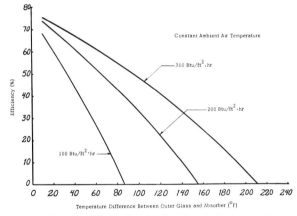

Figure V·B·40. Effect of solar flux (insolation) on collector efficiency. (The values shown are for a particular collector tested at the University of Pennsylvania and should not be applied for all types of collectors.) Source: (NSF.)

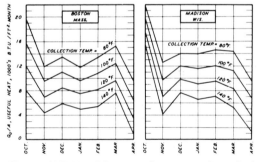

Figure V·B·41. Winter performance of vertical south-facing flat plate solar collectors in Boston and Madison. (Construction: 2 air-spaced glass plates over blackened water-cooled copper sheet; well insulated below.) Source: (MIT.)

When designing and working with natural energy systems, it is important to realize that since weather conditions vary greatly from hour to hour and day to day, monthly and yearly, a system designed for averages may not perform according to expectations at all times. In addition, historical solar data is very difficult to obtain, both because of the imprecise instruments which have been used for the last 20 or 30 years to measure solar radiation, and because of the rather small number of weather stations (approximately 75 in the United States) which measure solar radiation. As with temperature, insolation varies greatly from one location to another, even though the distances may be small between them.

As should be clear by now, the unpredictable availability of solar energy usually requires that it be looked upon as a bonus or as a supplement to other sources of energy. It is possible that in the future the development of long-term storage systems will solve the problems associated with wide fluctuations in available solar energy.

Precise calculations of solar weather, however, are less important than is usually assumed, since variations in solar irradiation of 10%, for instance, will affect the overall efficiency of a solar system as little as 3% out of a total overall efficiency of perhaps 40%. In addition, the collectors will not always be operating when the sun is shining for a number of reasons—when heat is simply not needed, for instance. Other times are during early morning and late evening, when the sun's intensity is not great enough to be collected and when the angle of the sun's rays to the collector is so low that most of the sun is reflected off the panel (see Figure V·B·42). Collectors may not operate during partly sunny periods when clouds break only for short times. During the course of a six month heating season, as much as 25% of the total solar irradiation which hits a well-designed collector will do so when the system is not in operation.

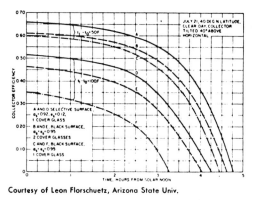

Courtesy of Leon Florschuetz, Arizona State Univ.

Figure V·B·42.
Variation of collection efficiency with time, expressed as hours from solar noon, for temperature differentials of 50° and 100°F for single- and double-glazed collectors with flat black absorber plates and for selective surface absorber (Tabor-type) and single glazing. (Reprinted by permission of ASHRAE.)

Of the total insolation, the amount which is actually received by the absorber, H_a, is reduced by the transmissivity, τ, of the cover plates, and by the absorptivity, α, of the absorber. Both of these are dependent on the angle of incidence. Although glass and plastic are about 90% transmittant to light at right angles, over the course of a day average transmittance may be as low as 70–75% for single glass and 62–67% for double glass. The absorptivity of the absorber surface is less affected by the angle of incidence and affected more by the characteristics of the surface itself. Also, the total figure may be reduced by 3–5% to account for losses due to dirt on the cover plates and to the shading effects of the collector's side walls and cover supports. The total solar energy received by the absorber, H_a, then, is

$$H_a = (.96) \text{ I } \tau\alpha$$

Heat loss from the absorber, H_1, is the sum of a large number of different heat losses, including the radiation losses by the absorber, the convection and conduction heat losses to the surrounding air, and the conduction heat losses from the collector to colder parts of the collecting system through the insulation and materials of construction.

These losses vary because of the following factors:

- The average temperature of the absorber plate (see Figure V·B·43).
- The outdoor air temperature (see Figure V·B·44).
- The effective sky temperature, which is usually 10 to 30 degrees cooler than the outdoor air and which affects the radiative heat losses.
- The wind speed, which has less effect with increasing numbers of cover plates and lower collector temperature.
- The number of cover plates and, to a lesser extent, their spacing; also their transmittance to long-wave infrared radiation.
- The amount and type of rear and side insulation.

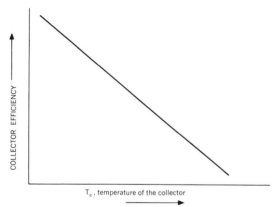

Figure V·B·43. Generalized relationship between collector temperature and its energy output (constant outdoor temperature).

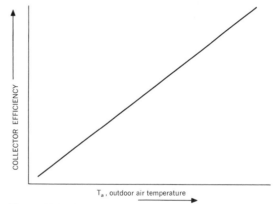

Figure V·B·44. Generalized relationship between collector output and outdoor air temperature (constant collector temperature).

The general effects of the last two factors, cover plates and back insulation, are shown in Figure V·B·45 by Jordan and Threlkeld. The first graphs indicate efficiencies for collectors using one, two, and three panes of glass, and for uncovered collectors with a U-value through the back of 0.15 Btu per square foot per hour per degree (Btu/hr/ft²/°F). The second set of curves represents the same collector exposed to heated indoor space and with theoretically no heat loss out the back. At low operating temperatures, there is almost no difference in performance, but at higher operating temperatures, there are significant increases in efficiency when back heat loss is reduced. As mentioned previously, the optimum number of cover plates for the corresponding temperature differences between the collector and the outdoor air must be tempered by considerations of the cost and difficulty of construction and maintenance.

The first item above, the average temperature of the absorber plate, is one of

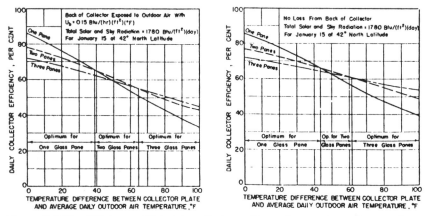

Figure V·B·45. Efficiency of flat plate collectors with varying numbers of cover glasses. (Courtesy of Jordan and Threlkeld.)

the crucial elements in optimizing solar system design. As Figure V·B·43 shows, for constant outdoor temperatures, collector efficiency drops off almost linearly as its collection temperature increases. This temperature depends on several factors:

- The fluid flow rate through the collector (Figure V·B·46).
- The type of collector fluid (gas or liquid).
- The temperature at which the fluid enters the collector.
- The tilt angle of the collector.
- The heat transfer coefficients between the heat-removal fluid and the absorber plate.

For liquid-type collectors with tubes and plates, these factors must be added:

- The conductance of the bond between the fluid tubes and the absorber plate.
- The fin efficiency of the flat plate, which is determined by the plate material, its thickness, and the spacing between the tubes.

The optimum fluid flow rate varies according to operating conditions. In general, of course, the higher the fluid rate, the lower the operating temperature and the higher the collection efficiency. Water remains the most efficient heat transfer fluid available for use with solar collectors. Its many disadvantages were discussed earlier, but its advantages are clear: low cost; a low viscosity resulting in low pumping energy requirements; a relatively high density and high specific heat; and a good thermal conductivity.

Other liquids under consideration include ethylene glycol, propylene glycol, oil, polyglycols, and silicone fluids. Most of these alternatives can be eliminated from further consideration because of their major unsatisfactory characteristics, e.g., high cost, high viscosity, corrosion problems, degradation at high temperatures, or low flash points.

Ethylene glycol (anti-freeze) is the most commonly used additive to water systems. The percentage of glycol to water is usually the same as that for automobiles in the same locale. Glycol is slightly denser than water but has a

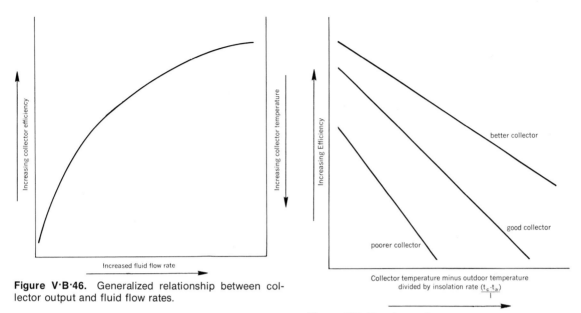

Figure V·B·46. Generalized relationship between collector output and fluid flow rates.

Figure V·B·47. Generalized relationship between collector efficiency and operating conditions.

lower specific heat. For a 25% glycol/water solution, the combined effect on efficiency is a net loss of about 5% for the same fluid flow rates in pounds per hour. For a 50% solution, the efficiency loss is about 17%. For space heating systems, optimum fluid flow rates are usually in the range of 4 to 10 pounds per square foot per hour for water, and 15 to 40 pounds (approximately 3 cfm to 9 cfm) for air.

John Minardi and Henry Chuang of the University of Dayton have proposed and experimented with a collector in which a highly absorbent black liquid flows in transparent channels and directly absorbs the solar energy. No metal is necessary, but the transparent tubes cover the entire surface of the flat plate collector. Tests showed that liquids such as India ink produced efficiencies comparable to those for metal flat plate collectors (MIN.).

Figure V·B·47 shows yet another relationship commonly used to determine collector efficiency: in order to eliminate the variable of solar radiation differences, the temperature difference between the collector and the outdoor air is divided by the insolation rate.

The actual performance of a solar collector made available by PPG Industries in 1974 at an insolation rate of 300 to 325 Btu per square foot per hour is shown in Figure V·B·48. Other experimental results of PPG's "Baseline" collector are shown in Figure V·B·49 for increasing rates of fluid flow. (The rate of 0.2 gpm corresponds to just over 5 pounds per square foot per hour for their 18 square foot collector.)

Figure V·B·49 also illustrates interesting relationships between insolation rates, time of day, outdoor temperature, and collector temperature. When defining a collector temperature, care must be taken to specify whether refer-

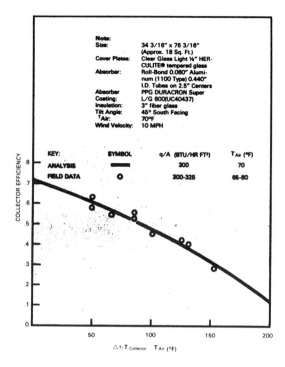

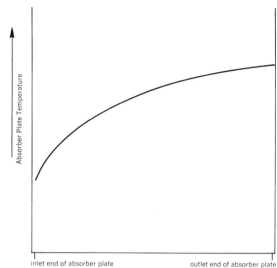

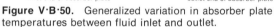

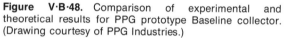

Figure V·B·50. Generalized variation in absorber plate temperatures between fluid inlet and outlet.

Figure V·B·48. Comparison of experimental and theoretical results for PPG prototype Baseline collector. (Drawing courtesy of PPG Industries.)

ence is being made to the outlet, inlet, or average temperature. A generalized temperature gradient from the inlet to the outlet is shown in Figure V·B·50; more detailed curves are shown in Figure V·B·51 for actual test conditions. The relationship between the collector's absorber temperature and the temperature of the water is also shown. (The "atmospheric" temperature refers to the temperature of the air between the absorber plate and the first cover plate. A generalized temperature gradient through a typical collector is shown in Figure V·B·52.)

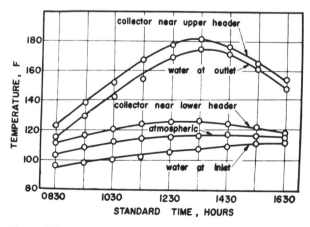

Figure V·B·51. Temperature variation at various points in the collector during a day. Source: (BHA.)

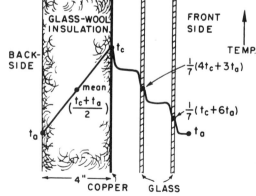

Figure V·B·52. Sketch of temperature distribution through a collector in equilibrium. Source: (MIT.)

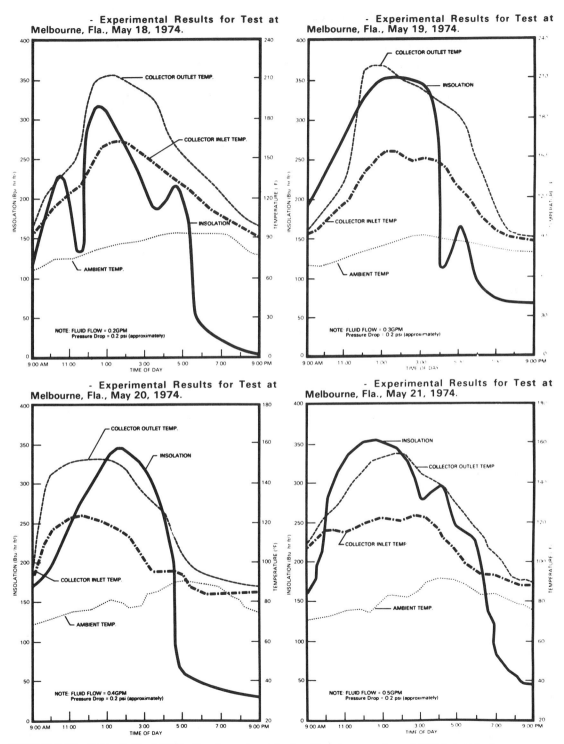

Figure V·B·49. Test results of PPG's Baseline solar collector. (Drawings courtesy of PPG Industries.)

Performance of Air-Type Collectors

Most of the general discussion of collector performance has been slanted towards liquid collectors rather than air collectors. Although the diagrams and curves can often be extrapolated to air systems, the performance of air-type collectors is less predictable than water-type. The difficulty of doing engineering analyses on air is one major reason for the reluctance of engineers to deal with it. *Solar Energy Thermal Processes* by John Duffie and William Beckman (DUF.) and *Principles of Heat Transfer* by Frank Kreith handle the subject well. *Solar Energy Thermal Processes* is particularly helpful in determining the efficiency of air collectors for given insolation rates, cover plate spacings, inlet temperatures, ambient air temperatures, air mass flow rates, the ratio of air duct width to length, and for given emmissivities and absorptivities of absorber surfaces.

Probably the most variable and relatively unpredictable aspect of the performance of air-type collectors is that of heat transfer between air and various absorber "plates" (metal lath, scrap metal, aluminum cans, gauze, and air filter materials). Also difficult is engineering the tradeoff between this heat transfer and the pressure drop through the panel, which increases as the air flow increases. Collector efficiency is also raised in the process, but the increased performance must be weighed against the increased fan horsepower and energy consumption required to move the extra air. This same tradeoff applies to the designing of heat storage compatible with air-type collectors. This is discussed in more detail in the section on heat storage.

One way to become informed about air collector performance is to review the results of the work done by others. Close (CLO.2) investigated various collector configurations for efficiency (see Figure V·B·30). In his calculations he assumed that the collectors were lying horizontally, that the dry bulb temperature was 74.5°F, that the wind was blowing 5 mph, and that the average hourly insolation was 160 Btu/sq ft on the horizontal surfaces. Figure V·B·53 shows the collection efficiencies for the three types of collectors as the air temperature within the collector varies. It also shows the effectiveness of the use of selective surfaces relative to the use of black paint. The assumed heat transfer coefficient for the flat plate was 2 Btu/hr/ft²/°F.

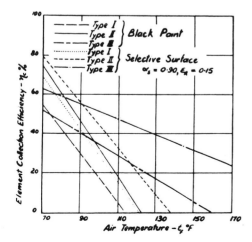

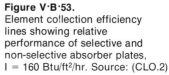

Figure V·B·53.
Element collection efficiency lines showing relative performance of selective and non-selective absorber plates, I = 160 Btu/ft²/hr. Source: (CLO.2)

For a range of models, including various absorber surface types and various selective surfaces, Close plotted Figure V·B·54 by using the best and poorest model of each type and shading the region between these two boundary lines according to heater type. The plots show the general superiority of the Type III heater, particularly at higher operating temperatures, from 120° upwards (for 74.5° outdoor temperatures); its superiority is marginal for lower temperatures. The best Type II heater has a somewhat better performance than the best Type III at temperatures below 100°. This may be satisfactory for some types of crop drying, as Close points out, as well as for low temperature or relatively warm climate space heating. Close summarized his findings in Figure V·B·55 which shows constant-efficiency curves as functions of insolation and air-stream temperature for the best and poorest models of each type.

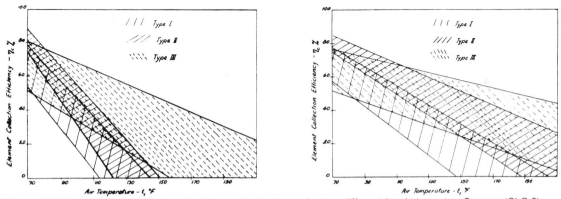

Figure V·B·54. Collection efficiency for three collector types for two different insolation rates. Source: (CLO.2)

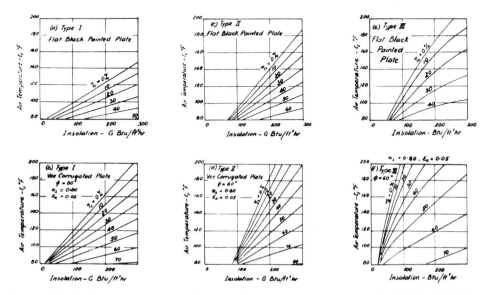

Figure V·B·55. Constant element-collection efficiency curves for varying insolations and air stream temperatures. Best and poorest models of each type are shown. Source: (CLO.2)

Buelow and Boyd (BUE.) showed how the efficiency of a particular air-type collector varies with air flow rate and with the overall coefficient of heat loss, U(Btu/hr/ft²/°F), from the air in the collector to outside. For a blackened flat metal absorber plate with one cover glass and with air moving on both sides of the plate (see Type II, Figure V·B·30), they found a U-value of 2.5. Figure V·B·56 shows, for an insolation rate of 350 Btu/ft²/hr, the efficiency of the collector as the air flow rate varies. Also shown are curves for collectors with U-values of 1 and of 0 (theoretically, no heat loss from the collector). No mention is made of a collector design to obtain a U-value of 1, but it probably has two or three cover plates and a large amount of insulation on the back side of the absorber.

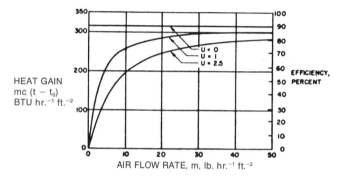

HEAT GAIN
mc (t − t₀)
BTU hr.⁻¹ ft.⁻²

AIR FLOW RATE, m, lb. hr.⁻¹ ft.⁻²

EFFICIENCY, PERCENT

Figure V·B·56. Relationship of efficiency and heat gain to air flow rate for three different over-all coefficients of heat transfer. Source: (BUE.)

Assumed value of coefficient. h	2	3	4	6	8	12
Collector heat-loss coefficient, $U_L = 1.2$ (single glass cover)						
Efficiency factor, F'	0.625	0.714	0.77	0.833	0.87	0.909
Collector heat-loss coefficient, $U_L = 0.8$ (two glass cover)						
Efficiency factor, F'	0.714	0.79	0.833	0.882	0.903	0.934

Figure V·B·57. Effect of heat transfer coefficient, h, on collection efficiency factor F'. Source: (WHI.3)

One of the best summaries of collector performance according to different absorber surfaces and cover plates is an article by Austin Whillier, "Black-Painted Solar Air Heaters of Conventional Design" (WHI.3). In particular, he provides detailed analytical equations which made collector comparisons possible without the necessity of experimentation.

Whillier stresses the importance of the heat transfer properties between absorber plate and air stream. In fact, as both he and Hottel had done earlier, he combined the overall U-value discussed by Buelow and Boyd with the heat transfer coefficient, h, and obtained an *efficiency factor,* usually designated in the literature as F':

$$F' = \frac{1}{1 + \dfrac{U}{h}}$$

The higher the value for F', the more efficient the collector is. Figure V·B·57 is Whillier's table showing the effect of heat transfer coefficients on the efficiency factor for a particular collector. It can be seen that a heat transfer coefficient (h) of 6 for an absorber with a single glass cover (U = 1.2) yields the same efficiency factor, $F' = 0.833$, as an absorber with a lower heat transfer coefficient, 4, in combination with an additional cover plate (U = 0.8).

NOTE; The U-values used by Whillier and Buelow are not comparable since they represent slightly differing assumptions and heat loss factors.

Whillier points out that a heat transfer coefficient of 3 is about average for flat metal plates; roughened and textured surfaces have greatly increased values, but these must be weighed against the corresponding increase in friction losses which require larger fans for pumping the air. A gauze screen approximately triples the heat transfer coefficient of the flat plate.

Figure V·B·58 is an adaptation of a graph by Whillier showing the efficiency of solar collection as solar irradiation varies and as the inlet air temperature above ambient temperature varies. The heat transfer coefficient is 4, the air flow rate is 30 lbs/sq ft/hr, the metal is painted black, the wind speed is 5 mph, and the sky temperature is assumed to be 10° colder than ambient. For air flow rates and heat transfer coefficients other than these values, the efficiencies derived from Figure V·B·58 must be multiplied by the appropriate correction factor given in Figure V·B·59 and V·B·60.

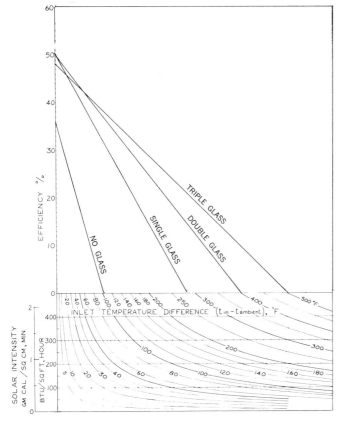

Air-flow Rate, G lb/hr, sq ft...	1	5	10	30	50	100
No cover (a)	0.14	0.57	0.78	1	1.06	1.10
One cover (b, c, d)	0.26	0.73	0.88	1	1.03	1.05
Two Covers (e, f, g, h)	0.34	0.79	0.91	1	1.02	1.03
Three Covers (i, j, k)	0.42	0.84	0.93	1	1.01	1.02

Figure V·B·59. Correction factors for different air flow rates, G. (Multiply the efficiency determined from Figure V·B·58 by the correction factor shown.) Source: (WHI.3)

Heat Transfer Coefficient, h, (Btu hr, sq ft °F)	2	4	6	8	12	16
No cover (a)	0.67	1	1.20	1.33	1.49	1.59
One cover (b, c, d)	0.80	1	1.09	1.14	1.20	1.23
Two covers (e, f, g, h)	0.85	1	1.06	1.10	1.13	1.15
Three covers (i, j, k)	0.88	1	1.04	1.07	1.10	1.11

Figure V·B·60. Correction factors for different heat transfer coefficients, h. (Multiply the efficiency determined from Figure V·B·58 by the correction factor shown.) Source: (WHI.3)

Figure V·B·58. Efficiency of conventional solar air heaters with different cover arrangements. Source: (WHI.3)

Donovan and Bliss obtained the data shown in Figure V·B·61 during their work at the Desert Grassland Station. Their collector consisted of four layers of black gauze spaced one-quarter inch apart (see Figure V·B·39). Air was drawn over and through the gauze and back to storage. The colder the outside air, the lower the collector efficiency; on the other hand, the lower the entering air temperature, the higher the efficiency.

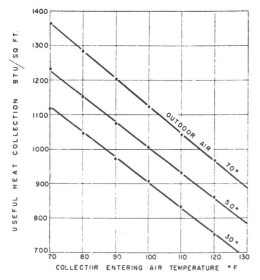

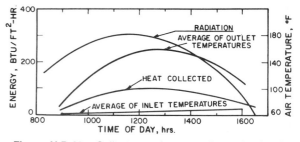

Figure V·B·62. Collector performance for a day for the Denver solar house, showing time dependence of collector output, i.e., storage unit input. Source: (GIL.)

Figure V·B·61. Variation in heat collected with ambient temperature and collector input-air temperature. Source: (BLI.1)

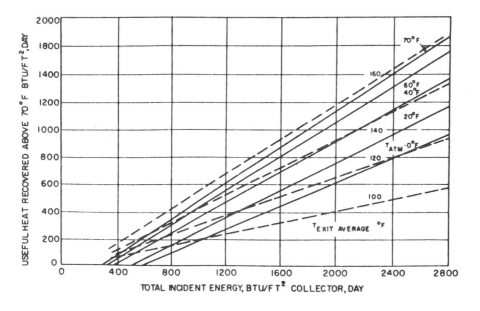

Figure V·B·63. Performance of overlapped-plate solar air heater—solid lines show heat recovered at various solar intensities (on sloping surface) and mean (24-hour) atmospheric temperatures, by a four-section collector of low reflectivity glass with single cover plates. Air rate is 1.5 standard cfm per square foot of collector. Dashed lines show heat recovered at various average exit air temperatures during the operating hours. Source: (LOF.2)

Some of the most complete performance data for air-type collectors comes from the operation of George Löf's second house, in Denver. Actual performance data for November 20, 1958, is shown in Figure V·B·62. More general data for his overlapped-plate collector (see Figure V·B·29) is shown in Figure V·B·63. The curves show the relationship between total incident solar radiation and the amount of useful heat collected as the outside temperature varies (solid lines) and as the inlet air temperature varies (broken lines). The air flow rate is 1 cfm per square foot of collector, well below the 4 to 6 cfm recommended by many other solar designers. The air velocity through the collector of one foot per second is also well below the 10 feet used by Close. To handle 1000 cfm, the system uses a fan of one horsepower. The actual speed of the air flowing through the collector is usually more critical to the determination of fan horsepower required than is the actual volume. Figure V·B·64 is a table showing the relationship for a particular model fan between air volume flow in cfm, pressure drop (static pressure in inches of water), and horsepower size (in BHP).

Buelow and Boyd (BUE.) showed that to achieve the best heat transfer characteristics between the collector and the moving air, the spacing between the absorber plate and the glass or insulation should be reduced until the pressure drop is the maximum that can be tolerated.

Fans
TYPE: Lau FGP10-6A

performance data

Performance is based on standard air (.075 lbs./cu. ft.) BHP does not include drive losses Performance is shown for FGP Blowers with outlet duct.

FGP 10 - 6A
Wheel Dia - 10⅝
Outlet Area 629 Sq. Ft
Tip Speed
(FPM) RPM · 2.78
Outlet Vel $\frac{CFM}{629}$

VOL. IN CFM	STATIC PRESSURE (IN. OF WATER)																	
	.500		.625		.750		.875		1.000		1.250		1.500		1.750		2.000	
	RPM	BHP	RPM	BHP	RPM	BHP	RPM	BHP	RPM	BHP	RPM	BHP	RPM	BHP	RPM	BHP	RPM	BHP
600	717	.08	792	.10	863	.12	932	.13										
700	742	.10	808	.12	873	.14	936	.16	997	.18	1115	.22						
800	774	.13	835	.15	892	.17	950	.19	1007	.21	1116	.26	1220	.30				
900	807	.16	867	.18	922	.20	974	.23	1025	.25	1126	.30	1225	.35	1318	.40	1410	.45
1000	842	.20	900	.22	954	.25	1005	.27	1052	.30	1145	.34	1236	.40	1325	.45	1411	.51
1100	879	.24	935	.27	987	.30	1037	.32	1084	.35	1171	.40	1255	.45	1338	.51	1420	.58
1200	917	.29	971	.32	1023	.35	1071	.38	1116	.41	1202	.47	1281	.52	1358	.58	1435	.65
1300	956	.35	1009	.38	1054	.41	1106	.45	1152	.48	1234	.54	1313	.60	1385	.66	1456	.72
1400	998	.42	1041	.45	1096	.48	1142	.52	1186	.55	1267	.62	1344	.69	1416	.75	1484	.82
1500	1041	.49	1088	.53	1135	.56	1179	.60	1222	.64	1305	.71	1376	.78	1448	.85	1515	.92
1600	1085	.58	1130	.62	1175	.65	1218	.69	1259	.73	1338	.81	1414	.88	1480	.96	1547	1.04
1700	1130	.68	1174	.71	1216	.75	1257	.79	1298	.83	1375	.91	1448	1.00	1516	1.08	1579	1.16
1800	1175	.78	1218	.82	1258	.86	1298	.90	1337	.94	1412	1.03	1483	1.12	1552	1.21		
1900	1221	.90	1263	.94	1302	.98	1340	1.03	1378	1.07	1451	1.16						
2000	1268	1.04	1308	1.07	1347	1.12	1384	1.16	1420	1.21								

Copyright 1974 Lau Industries. Used with permission of Lau Industries.

Figure V·B·64. A chart to aid in sizing fans. (Reprinted by permission of Lau Industries.)

COLLECTOR SIZE: PERFORMANCE, TILT ANGLE, AND ORIENTATION

It is difficult to predict the performance of solar collectors on a seasonal basis when they are tied into a heating (and/or cooling) system. The prediction of performance is, of course, a necessary prerequisite for the sizing of a collector. Although this delicate task is becoming simplified, particular site conditions, the heating (and/or cooling) demand of the building, the design of the space conditioning system, and particular design choices, such as collector operating temperature, tilt angle, and orientation, all affect seasonal performance of a particular collector.

Consider, for example, that doubling the collector size does not necessarily double the usable heat collection. This notion is illustrated in a general way in Figure V·B·65. The building heat load shown in the figure represents a house which has a heating demand of 12,000 Btu per degree day, or a total heating demand of 84 million Btu in a 7000 Degree Day climate.

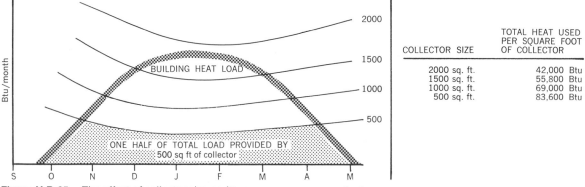

COLLECTOR SIZE	TOTAL HEAT USED PER SQUARE FOOT OF COLLECTOR
2000 sq. ft.	42,000 Btu
1500 sq. ft.	55,800 Btu
1000 sq. ft.	69,000 Btu
500 sq. ft.	83,600 Btu

Figure V·B·65. The effect of collector size on its per season energy output.

The average usable output of the collector is assumed to be 350 Btu per day per square foot of collector, or a total of about 83,600 Btu per seven month heating season. 500 sq ft, then, would provide about half of the total heating demand of 84 million Btu.

Figure V·B·65 shows that doubling the collector size (to 1000 sq ft) does not provide 100% of the heating load, as might be anticipated; instead, it provides about three-quarters. In addition, the per square foot usable heat collection drops from 83,600 to 69,000 Btu.

	July	Aug.	Sept.	Oct.	Nov.	Dec.	Jan.	Feb.	Mar.	Apr.	May	June	Annual
Assumed city water temperature, °F	65	65	60	55	50	45	40	40	45	50·	55	60	
Heat required for Domestic Water 10⁵ Btu/month	23.3	23.3	24.0	26.4	27.0	29.5	31.0	28.0	29.5	27.0	26.4	24.0	320
Degree Day, °F	8	26	122	391	693	1081	1195	1072	886	586	269	63	6392
Heat required for House 10⁵ Btu/month	1.5	5.0	23.4	45.0	73.2	103.8	114.8	103.0	93.6	67.5	51.7	12.1	695
Total heat required 10⁵ Btu/day	0.8	0.91	1.58	2.31	3.24	4.30	4.70	4.68	3.97	3.15	2.52	1.20	2.78
Mean outdoor daytime temperature, assumed, °F	75.0	70.9	64.7	53.6	42.8	34.5	30.2	31.1	36.5	47.7	58.1	67.7	
Mean storage temperature, assumed, °F	140	140	130	120	110	100	100	100	105	110	120	130	
Calculated solar energy collected, Btu/sq ft, day	468	567	567	567	477	477	414	567	630	567	468	468	
Required area of solar collector for complete solar heating, assuming a perfect, infinite storage system, sq ft.	171	161	279	408	680	901	1135	826	630	556	539	257	

Figure V·B·66. Calculated monthly heat requirement and solar collector output for MIT House IV. Source: (WHI.1)

Austin Whillier did a more complete analysis for the fourth solar house at MIT. He first determined the useful heat output of the solar collector for each month. The results, given in Figure V·B·66, assume that the collector was to be located in Boston at a latitude of 42°N, tilted at an angle of 55° from the horizontal, and oriented due south. Two glass covers were used, and the water flow rate was 6 pounds per hour per square foot of collector. The mean daytime outdoor temperature was computed by adding seven-tenths of the mean maximum daily temperature to three-tenths of the mean minimum daily temperature. (This yields about the same value as adding one-half the mean average daily temperature to one-half the mean maximum daily temperature.)

The area of the collector that would be required in a given month to satisfy completely the heating requirements of the house is shown in the table. Heat storage capacity is assumed to be infinite; heat loss from storage negligible. Actually, because of limitations in storage size and storage heat loss, even a collector of 1200 sq ft would need auxiliary assistance from December through February. Figure V·B·67 has taken these limitations into account. The table shows the fraction of the heating demand which would be provided by the auxiliary backup system. The limitation of thermal storage during long periods of cloudy weather is the major reason for providing backup heat.

Collector Area (57° tilt) sq ft	July	Aug.	Sept.	Oct.	Nov.	Dec.	Jan.	Feb.	Mar.	Apr.	May	June	Annual
1200						0.04	0.12	0.02					0.025
1100						0.069	0.16	0.041					0.037
1000					0.005	0.10	0.20	0.07					0.051
900					0.022	0.147	0.256	0.109	0.01				0.073
800					0.06	0.2	0.321	0.162	0.037				0.104
700					0.116	0.27	0.396	0.225	0.083	0.013	0.01		0.144
600					0.19	0.351	0.472	0.306	0.153	0.066	0.054		0.200
500				0.02	0.288	0.448	0.56	0.408	0.244	0.156	0.141		0.276
400				0.11	0.42	0.56	0.65	0.52	0.373	0.288	0.268		0.375
300			0.068	0.272	0.56	0.67	0.74	0.64	0.53	0.47	0.45	0.006	0.496
200	0.005	0.001	0.288	0.52	0.71	0.78	0.83	0.76	0.69	0.65	0.64	0.225	0.639
100	0.42	0.375	0.65	0.76	0.86	0.89	0.92	0.88	0.85	0.83	0.82	0.62	0.816

Figure V·B·67. Estimated fractional auxiliary heat requirement for MIT House IV. Source: (WHI.1)

Figure V·B·68 graphs the annual percentages of the heating load carried by auxiliary (see the last column of the previous table). A collector of 1200 sq ft would provide 97½% of the load. A collector of half the size, 600 sq ft, would take care of 80%, and one as small as 200 would carry 36%.

As the collector size increases, the amount of energy provided by each square foot drops because of the decreased *load factor* on each additional square foot. A modification of a graph by Whillier is shown in Figure V·B·69. As in Figure V·B·65, which showed that a 500 sq ft collector could provide 83,600 usable Btu per square foot per year while a 2000 sq ft collector would provide only 42,000, this figure shows that the annual return on a 300 square foot collector (for a different house) is 90 cents per square foot, while on the 1200 square foot collector it is only 45 cents per square foot.

One of the factors which most affects the performance of collectors is the tilt angle and orientation of collectors. Figure V·B·70 shows that the energy received on a surface is greatest if it faces the sun directly. In general, flat plate collectors are in a fixed position and do not follow (or "track") the sun to maintain full exposure year round. Although devices are being developed to aid this process, the extra energy gained is usually too small to warrant the increased expense.

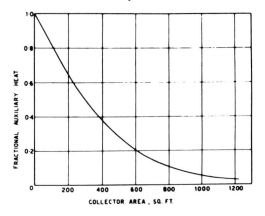

Figure V·B·68. Effect of collector area on auxiliary heat requirement. Source: (WHI.1)

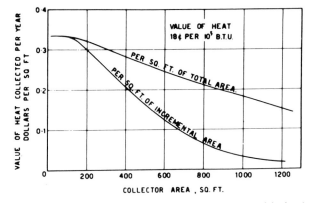

Figure V·B·69. Effect of collector area on the value of the heat collected. Source: (WHI.1)

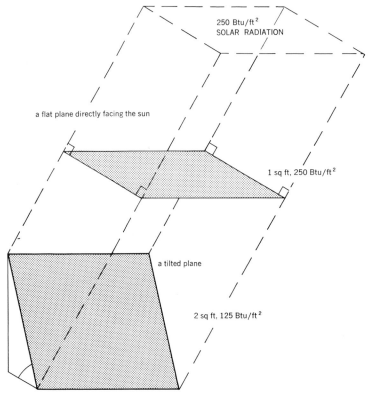

Figure V·B·70. The effect of angle of incidence on insolation rates.

Optimum orientations and tilt angles can be determined, but perhaps the most frequently asked questions by designers are: How much can I deviate from due south, and what are the penalties in efficiency?; and How much can I deviate from the optimum tilt angle, and what are the penalties in efficiency?

Deviations from due south are less frequent than those from the optimum tilt angle. However, in the first solar projects built many years ago, one of the major tasks was to show the extent to which solar energy can work; deviations from the optimum were rare. Although it is true that small deviations from the optimum may not be crucial, neither should they be ignored as a viable method of enhancing the compatibility of collectors with buildings.

Although most solar scientists believe due south is the optimum orientation, west of south is often chosen because of frequent early morning haze which reduces effective irradiation, and because of higher ambient temperatures in the afternoon, causing higher collection efficiencies. Fortunately, deviations from the optimum of 15 to 20 degrees cause relatively small reductions in collector output. Figure V·B·71 illustrates the variation of solar incidence on vertical south-facing surfaces as orientation varies. The time period represented on the graph is from November 21 to January 21 in latitudes 30 to 45°N.

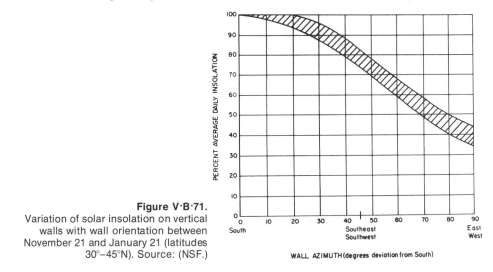

Figure V·B·71.
Variation of solar insolation on vertical walls with wall orientation between November 21 and January 21 (latitudes 30°–45°N). Source: (NSF.)

Figures V·B·72 and V·B·73 illustrate further the significance of orientation for particular situations and locations. Shown in both figures are vertical collector areas necessary for providing 50% of the heating demand in January for a house of approximately 1000 square feet which loses 9000 Btu per degree day in Boston (Figure V·B·72) and 13,500 Btu per degree day in Charleston. Note that deviations up to southwest (or southeast) require an extra collector area of only 10% in Boston and 30% in Charleston. Sizes of tilted surfaces are somewhat less affected by orientation.

Figure V·B·74 charts by latitude the effect on yearly insolation as the orientation of a surface deviates up to 45° from due south. The closer the location is to the equator, the less important the proper orientation.

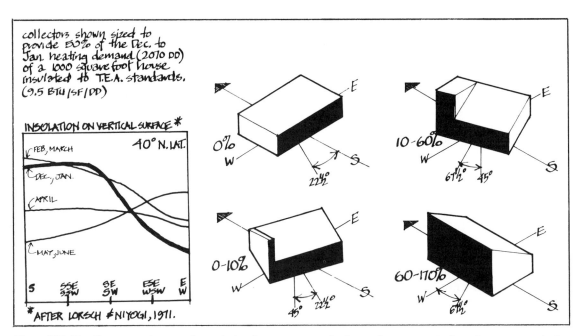

Figure V·B·72. Percentage increase in vertical collector area required as wall exposure deviates from true south: for Boston, Massachusetts. (Adapted from work done by Total Environmental Action, Inc., under subcontract with the AIA Research Corporation and published in *Solar Energy Home Design in Four Climates.*) Source: (TOT.2)

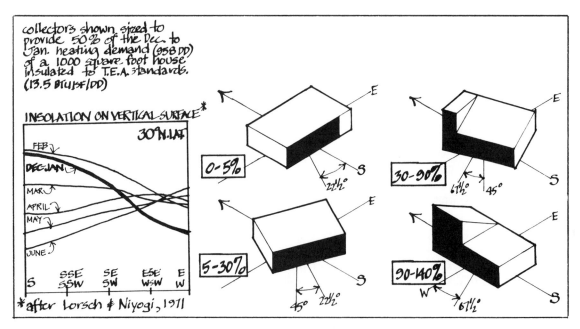

Figure V·B·73. Percentage increase in vertical collector area required as wall deviates from true south exposure: for Charleston, South Carolina. (Adapted from work done by Total Environmental Action, Inc., under subcontract with the AIA Research Corporation and published in *Solar Energy Home Design in Four Climates.*) Source: (TOT.2)

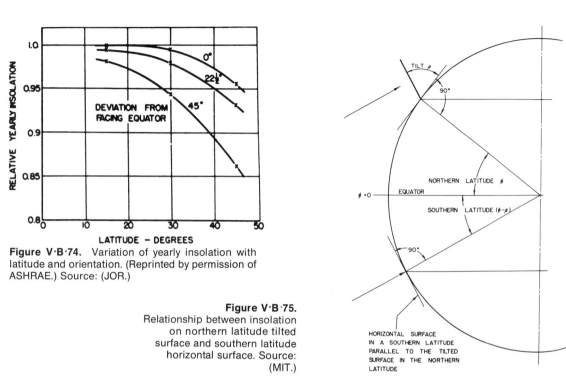

Figure V·B·74. Variation of yearly insolation with latitude and orientation. (Reprinted by permission of ASHRAE.) Source: (JOR.)

Figure V·B·75.
Relationship between insolation on northern latitude tilted surface and southern latitude horizontal surface. Source: (MIT.)

Optimum tilt angle for collectors varies according to the use for which the collector is built. This purpose defines the season of use and hence the angle of tilt which best matches the altitude and azimuth of the sun. Collectors used for winter heating will need a steeper tilt than those which are used for cooling during summer months. If a collector is to be used year round, e.g., for hot water heating, an angle must be chosen which is a compromise among the varying solar seasons. However, in locations where the heating and cooling demands are not evenly balanced, a more critical analysis must be done to make sure that the tilt angle chosen is properly biased toward the larger of the two demands.

Before exploring optimum tilt angles, examine Figure V·B·75, prepared by Lawrence Anderson, chairman of MIT's solar effort in the 1950's. He has represented the surface at a particular tilt by another surface tangent to the globe at the southern latitude of the northern latitude, ϕ, less the tilt angle of the collector, β. By finding the insolation on a horizontal surface in the southern latitudes of ϕ minus β, the insolation on a surface at a tilt angle of β at the northern latitude of ϕ can be found.

Figure V·B·76 shows the *approximate* optimum tilt angle according to latitude and climate. The line represented by the number 1 is the optimum angle for maximum summer insolation of approximately the angle of the latitude minus 15°. For year round maximum insolation, represented by line 2, an angle of the latitude is the optimum. An angle of latitude plus 15° is about the optimum for winter collection.

Total daily insolation varies for different tilt angles during the course of a year.

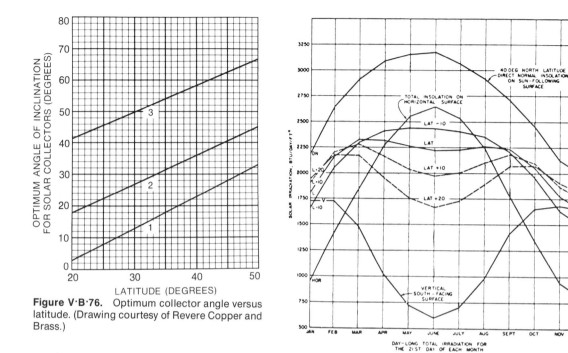

Figure V·B·76. Optimum collector angle versus latitude. (Drawing courtesy of Revere Copper and Brass.)

Surface temperature = t_a; surface emittance = 1.0.

Figure V·B·77. Total daily insolation at 40° N latitude. (Reprinted by permisssion of ASHRAE.) Source: (ASH.1)

In Figure V·B·77, the values are for clear days at 40°N and do not take cloud conditions into account. The tremendous difference between insolation on vertical and horizontal surfaces is evident: the insolation rate on vertical surfaces peak in the winter months and drops off sharply during the summer. A collector in this position follows the heating demand more closely and is easily shaded when not in use during the summer. A horizontal surface (at 40°N) receives almost no irradiation during winter months but very large amounts during the summer. Actually, when ground reflectance and hazy conditions are taken into account, the relative performance of a vertical collector is even better than what is shown here. In fact, its winter performance can surpass that of the optimum tilt angle of latitude plus 15°.

Vertical collectors are particularly amenable to ground reflection. Clean, fresh snow has been found to have the highest ground reflectance (0.87) of any natural surface (ASH.1) and can add 15 to 30 percent to the output of collectors. Other potential reflectors include asphalt, gravel, and concrete which have reflectances of 10%, 12 to 15%, and 21 to 33%, respectively. Bright green grass reflects 20% and 30% at incident angles of 30 degrees and 65 degrees respectively (ASH.).

Figures V·B·78 and V·B·79 show how the size of collectors supplying 50% of the January heating load in Minneapolis and Phoenix varies with tilt angle. The collector is shown sitting on the roofs of the same representative houses shown in Figures V·B·72 and V·B·73. The 25 by 40 foot house is oriented

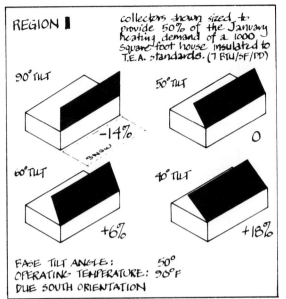

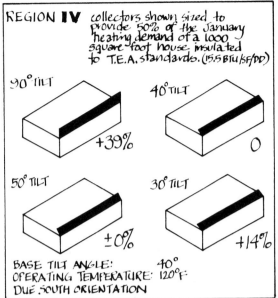

Figure V·B·78. Percentage change in collector area resulting from changes in tilt angle: for Minneapolis, Minnesota. (Adapted from work done by Total Environmental Action, Inc., under subcontract with the AIA Research Corporation and published in *Solar Energy Home Design in Four Climates.*) Source: (TOT.2)

Figure V·B·79. Percentage change in collector area resulting from changes in tilt angle: for Phoenix, Arizona. (Adapted from work done by Total Environmental Action, Inc., under subcontract with the AIA Research Corporation and published in *Solar Energy Home Design in Four Climates.*) Source: (TOT.2)

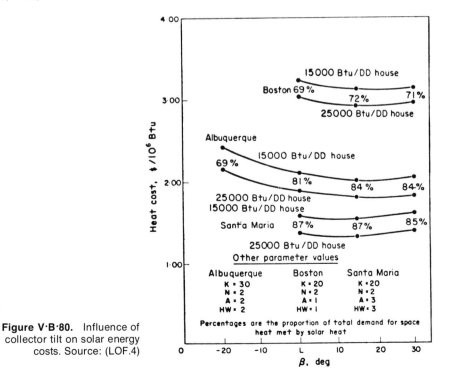

Figure V·B·80. Influence of collector tilt on solar energy costs. Source: (LOF.4)

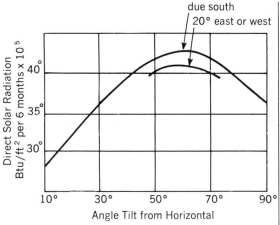

Figure V·B·81. Effect of orientation on the direct solar radiation incident upon a flat plate collector. (Values for Toronto, Canada, 43°.) Source: (ALL.)

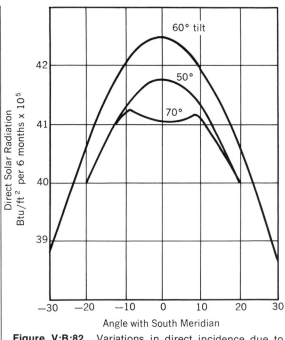

Figure V·B·82. Variations in direct incidence due to eastward or westward orientations. (Values for Toronto, Canada.) Source: (ALL.)

with its long axis in the east-west direction so that the collector sizes shown are all 40 feet long. The curves show clearly the relatively small differences in size as the tilt angle varies.

Tybout and Löf's economic studies (LOF.4) in Figure V·B·80 show curves for the optimum tilt angle in Boston, Albuquerque, and Santa Maria when the collectors are used for heating only during the winter. In all cases, the optimum is the latitude plus 15°, but only small deviations occur for other angles.

The combined effects of tilt angle and orientation are shown in Figures V·B·81 and V·B·82, prepared by Allcut and Hooper (ALL.) for Toronto, Canada, 43°N, during the six middle winter months. The first figure shows that for tilts of 50° to 70°, deviations in orientation of 20° east or west result in fairly minor losses in total direct solar radiation. Similar information is found in the second figure. Although the variations in direct solar radiation may seem rather large at first, the values chosen for coordinates of the curves exaggerate them; the differences are actually fairly small. For example, a collector facing due south and tilted 60° from the horizontal is shown to have the highest solar gain. A deviation of 20° east and a 50° tilt yields a solar incidence of only 5% or so less (4.25 versus 4.0 million Btu per square foot of collector for the six month period).

As should be apparent, solar energy design is hampered by the difficulty of quickly predicting the energy output of solar collectors in specific locations. The following is not a panacea for this problem, but a method which lies between simple guess work and detailed analytical engineering, such as proposed by Liu and Jordan in ASHRAE's *Low Temperature Engineering Applications of Solar Energy* (1957). This method of estimating heat collection is probably accurate to within 20%, resulting in somewhat conservative figures.

The basics of the method follow. The numbers in the margins refer to corresponding rows on the accompanying worksheet, Figure V·B·83, in which a Boston location is hypothesized.

The number of sunshine hours for the given locale are found by month (see *1* Resource Section B, "Insolation"). The day length for each month is obtained from almanacs or from Figure V·B·84. Reference is then made to the charts, *2* "Solar Insolation Values," prepared by Morrison and Farber for ASHRAE. These are reproduced in Resource Section B, "Insolation." The hourly values of insolation are located for the months (the months used to determine heating season energy output of collectors should be limited to those in which 100 or more degree days occur, on the average) and for the particular tilt or tilts to be analyzed. Only those hours of the day during which there is an average insolation rate of about 100 Btu/ft²hr, or more, are counted as collection hours. These usually occur during the middle two-thirds of the day. However, for insolation rates of 100 to 150 Btu/sq ft/hr, one-half hour of collection should be recorded; for rates over 150, a full hour should be used.

The "collection length of day" during which there are 100 Btu or more varies with the angle of tilt. The variation is least in mid-winter and greatest in the spring and fall. The number of hours of useful collection should be noted for *3* each month and for each angle of tilt being investigated.

The total daily insolation (during sunny conditions) is determined by adding the hourly rate in the ASHRAE tables corresponding to the daily hours which *4* provide useful amounts of sunshine. To obtain the number of hours during the month which are sunny during the period of collection, the hours of useful collection per day is divided by the normal length of the day and multiplied by the total hours of sunshine per month. *5*

The total daily insolation during sunny day conditions is divided by the number of hours of useful collection per day to get the hourly insolation rate *6* during the collection period. This number is the average rate of insolation while the collector is operating and can be used with performance curves which compare this number with the difference between the operating temperature of the collector and the average ambient temperature during the time of collection.

To obtain this differential temperature, the average ambient temperature during the operating time of the collector must be estimated. The normal daily *7* maximum temperature and the normal daily average temperature are available from many sources (e.g., the *Climatic Atlas of the United States,* pp. 2–24). The average ambient temperature during the hours when a collector operates is approximated by taking the average of these two temperatures.

The proposed operating temperature of the collector must then be determined. This is based on many parameters, such as fluid flow rate and storage size, which can, in turn, be controlled to some extent once the collector operating temperature has been chosen. In general, the faster the fluid flow rate and the larger the storage size, the lower the operating temperature. A range of temper-

DETERMINATION OF SOLAR COLLECTOR OUTPUT

LOCATION Boston
LATITUDE 42°
HEATING SEASONS Sept thru May

	TOTALS OF AVERAGES	SEPT	OCT	NOV	DEC	JAN	FEB	MAR	APR	MAY	
1 Total sunshine hours per month	1752	232	207	152	148	148	168	212	222	263	
2 Mean day length, hours	12	12	11	10	9½	10	11	12	13½	14½	
3 Hours of useful collection per day		8	8	7	6½	7	8	8½	9	9	30° angle of tilt
		8½	8	7	7	7	8	8½	8½	8½	40°
		8½	8	7	7	7	8(+)	8½	8½	8½	50°
		8	8	7	7	7	8(+)	8½	8	8	60°
		6½	8(−)	7	7	7	8	6½	4½	2	90°
		6½	8(−)	7	7	7	8	6½	4½	2	90° + 20% snow reflection
4 Daily solar radiation, Btu/ft²		2041	1828	1510	1334	1660	1908	2128	2083	2214	30° angle of tilt
		2059	1917	1634	1544	1662	1998	2250	2092	2010	40°
		2081	1947	1714	1640	1744	2093	2107	1966	1829	50°
		1918	1922	1744	1688	1774	2065	2007	1725	1542	60°
		1118	1622	1524	1534	1558	1579	1167	562	222	90°
		1342	1946	1828	1841	1870	1895	1400	674	266	90° + 20% snow reflection
5 Hours of useful collection per month: 3 ÷ 2 × 1	1196	155	151	106	101	104	122	150	144	163	30° angle of tilt
	1196	164	151	106	109	104	122	150	136	154	40°
	1196	164	151	106	109	104	122	150	136	154	50°
	1170	155	151	106	109	104	122	150	128	145	60°
	942	127	151	106	109	104	122	115	72	36	90°
	942	127	151	106	109	104	122	115	72	36	90° + 20% snow reflection
6 Hourly average insolation, Btu/ft²: 4 − 3		255	229	216	191	237	239	250	231	246	30° angle of tilt
		242	240	233	221	237	250	265	246	237	40°
		237	243	245	234	249	262	248	231	215	50°
		240	240	249	241	253	258	236	216	193	60°
		172	203	218	219	223	197	167	125	111	90°
		206	243	261	263	267	237	200	150	133	90° + 20% snow reflection

7 Air temperature

	TOTALS	SEPT	OCT	NOV	DEC	JAN	FEB	MAR	APR	MAY	
7a		73	63	52	40	37	37	45	56	68	normal daily maximum
7b		65	55	45	33	30	30	38	48	59	normal daily average
7c		69	59	49	37	34	34	42	52	64	7a + 7b/2

8 Average temperature difference between collector and air temp: (coll. temp.) − 7

	SEPT	OCT	NOV	DEC	JAN	FEB	MAR	APR	MAY	
	21	31	41	53	56	56	48	38	26	90° coll. temp.
	51	61	71	83	86	86	78	68	56	120° coll. temp.
	71	81	91	103	106	106	98	88	76	140° coll. temp.

9 Average hourly solar heat collected, Btu/ft²: (from graphs)

	SEPT	OCT	NOV	DEC	JAN	FEB	MAR	APR	MAY	
	140	130	130	120	120	120	120	115	105	90° coll. temp., 60° tilt
	115	131	135	130	125	125	95	70	70	90° coll. temp., 90° tilt with snow
	120	105	105	90	95	95	95	95	85	120° coll. temp., 60° tilt
	100	106	110	105	100	100	75	45	50	120° coll. temp., 90° tilt with snow
	105	85	85	75	75	75	75	80	65	140° coll. temp., 60° tilt
	80	86	90	85	80	80	50	30	30	140° coll. temp., 90° tilt with snow

10 Average monthly solar heat collected, Btu/ft²: 9 × 5

TOTALS	SEPT	OCT	NOV	DEC	JAN	FEB	MAR	APR	MAY	
143,255	21700	19630	13780	13080	12480	14640	18000	14720	15225	90° coll. temp., 60° tilt
109,611	14615	19781	14310	14170	13000	15250	10925	5040	2520	90° coll. temp., 90° tilt with snow
115,600	18600	15855	11130	9810	9880	11590	14250	12160	12325	120° coll. temp., 60° tilt
88,076	12700	16006	11660	11445	10400	12200	8625	3240	1800	120° coll. temp., 90° tilt with snow
94,160	16275	12835	9010	8175	7800	9150	11250	10240	9415	140° coll. temp., 60° tilt
69,021	10160	12986	9540	9265	8320	9760	5750	2160	1080	140° coll. temp., 90° tilt with snow

Figure V·B·83. Determination of solar collector output. Source: (TOT.2)

Month	Latitude									
	0°	10°	20°	30°	40°	50°	60°	70°	80°	90°
Jan	12.07	11.35	11.02	10.24	9.37	8.30	6.38	0.00	0.00	0.00
Feb	12.07	11.49	11.21	11.10	10.42	10.07	9.11	7.20	0.00	0.00
Mar	12.07	12.04	12.00	11.57	11.53	11.48	11.41	11.28	10.52	0.00
Apr	12.07	12.21	12.36	12.53	13.14	13.44	14.31	16.06	24.00	24.00
May	12.07	12.34	13.04	13.38	14.22	15.22	17.04	22.13	24.00	24.00
Jun	12.07	12.42	13.20	14.04	15.00	16.21	18.49	24.00	24.00	24.00
Jul	12.07	12.40	13.16	13.56	14.49	15.38	17.31	24.00	24.00	24.00
Aug	12.07	12.28	12.50	13.16	13.48	14.33	15.46	18.26	24.00	24.00
Sept	12.07	12.12	12.17	12.23	12.31	12.42	13.00	13.34	15.16	24.00
Oct	12.07	11.55	11.42	11.28	11.10	10.47	10.11	9.03	5.10	0.00
Nov	12.07	11.40	11.12	10.40	10.01	9.06	7.36	3.06	0.00	0.00
Dec	12.07	11.32	10.56	10.14	9.20	8.05	5.54	0.00	0.00	0.00

Figure V·B·84. Day length in northern latitudes (in hours and minutes for the 15th of each month).

atures can be chosen to show the effect which operating temperature has on relative overall performance.

The average outdoor temperature occurring during collector operation is subtracted from a given collector operating temperature to obtain the differential temperature. Reference is then made to a performance curve for the collector being used. Examples of such curves are shown in Figure V·B·85; these graphs were prepared by Revere Copper and Brass Company. The lower left graph can be used with reasonable accuracy for collectors of good to moderate performance.

The temperature difference between the collector and ambient temperature during a given month is found on the bottom left of the graph. From this point move vertically to the curve or an approximation of the curve which corresponds to the hourly average insolation determined above. Moving to the left gives the average efficiency; this value is for purposes of information only. Moving to the right, intersect again, on the right hand side of the graph, the curve representing the same insolation rate. Drop down vertically and read the average collector output in Btu/ft²/hr.

This average hourly collection rate per square foot is then multiplied by the number of hours during which the sun is shining during that month.

For any tilt, the total collection per square foot during the heating or cooling season can be found by adding up the values for the appropriate months. The optimum tilt is that which results in the largest sum. Other considerations, of course, affect collector tilt: the type of collector, the integration of the tilt with the building design, and the relative costs of building, owning, and operating the collectors (a vertical collector, for example, is usually cheaper to build and is more accessible for maintenance).

The accompanying worksheet (see Figure V·B·86) illustrates how this elaborate, but relatively simple, analysis works. This worksheet is for Boston (approximately 40°N latitude). Tilt angles chosen for analysis are 30°, 40°, 50°, 60°, 90°, (vertical), and 90° with 20% added for snow reflection. Collector efficiencies and energy output are compared for operating temperatures averaging 90°, 120°, and 140°F.

In general, the larger the percentage of heating load the collector is intended to supply, the more difficult it is to determine size using simplified techniques.

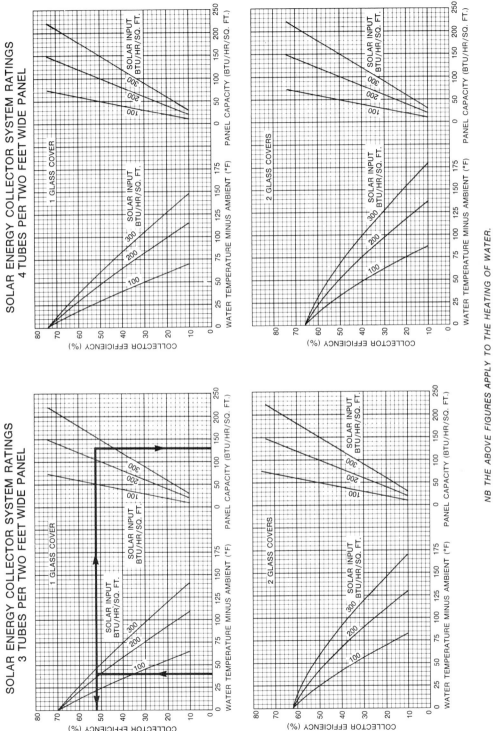

NB THE ABOVE FIGURES APPLY TO THE HEATING OF WATER.

Figure V·B·85. Generalized performance curves for four collector configurations. (Reprinted courtesy of Revere Copper and Brass.)

DETERMINATION OF COLLECTOR SIZE

LOCATION Boston
LATITUDE 42°
AVERAGE OPERATING COLLECTOR TEMPERATURE* 90

	TOTALS OR AVERAGES	SEPT	OCT	NOV	DEC	JAN	FEB	MAR	APR	MAY
11 Average monthly solar heat collected,* Btu/ft²	130,405	21700	17940	13130	11880	11880	13440	16920	17020	18375
12 Degree days per month	5,627	98	316	603	983	1088	972	846	513	208
13 Heat loss per month: $12 \times$ (Btu/DD)**	5346×10^4	93×10^4	300×10^4	573×10^4	934×10^4	1034×10^4	923×10^4	804×10^4	487×10^4	198×10^4
14 Solar heat collected: $(245\dagger$ ft²$) \times 11$	3486×10^4	532×10^4	440×10^4	322×10^4	291×10^4	291×10^4	329×10^4	415×10^4	417×10^4	450×10^4
15 Heat loss not provided by solar: $13 - 14$	2690×10^4	0×10^4	0×10^4	251×10^4	643×10^4	743×10^4	594×10^4	389×10^4	70×10^4	0×10^4
16 Heat loss provided by solar: $13 - 15$	2656×10^4	93×10^4	300×10^4	322×10^4	291×10^4	291×10^4	329×10^4	415×10^4	417×10^4	198×10^4
17 Percent heat loss provided by solar: $16 \div 13$	50%	100%	100%	56%	31%	28%	36%	52%	86%	100%

Repeat rows 14 through 17 until the desired percentage of heat loss by solar is achieved

*This information is obtained from row 10 in "Determination of Solar Collector Output."
**The heat loss of the house must be determined on the basis of Btu per degree day through the use of standard heat loss calculations.
†Use a square footage which is a first approximation of the necessary area for providing the desired percentage of heat loss by solar energy, in this case, 245.

Figure V·B·86. Determination of collector size. Source: (TOT.2)

Simple approximations in size, however, can usually be made for collectors which provide 60% or less of the heating demand.

Suppose that the collector must provide 50% of heating demand. It can be seen that if the collector is sized to provide half of the heating load during January, it will be much larger than if it is sized to provide half of the load during April. Sizing the collector based on an average demand for the year may cause the collector to supply too much heat during the early fall and late spring and too little during the middle of the winter.

Once the per square foot total of collected solar energy has been predicted and a tilt angle and operating temperature have been chosen, list the degree days per month. Then determine the heat loss of the house in Btu per degree day. To obtain the monthly heating demand, multiply the number of degree days in that month by the per degree day heat loss of the house. *11* *12*

Next, add the monthly heat losses to obtain a total seasonal heat loss (row 13). *13* Divide this total by the sum of the monthly solar heat collected (row 11). Fifty to sixty percent of the resulting number gives a first approximation of a collector size to provide 50% of the seasonal heat loss.

Multiply this approximate collector size by each of the figures for monthly *14* solar heat collection. Subtract each of these results from the corresponding monthly heat loss (negative numbers should be recorded as 0). The resulting *15* numbers represent that portion of each monthly heat loss which is NOT provided for by solar energy, but rather by the back up system. Obtain the number of Btu provided monthly by the collector by subtracting the figure in 15 from the monthly heat loss in 13. *16*

Divide the total heat loss provided by solar energy by the monthly heat loss of the house in 13 to get the percentage of heat loss provided by solar. If this *17* percentage is not sufficiently close to 50%, or whatever other percentage is being sought, a refinement in the original approximation can be made through observation, and rows 14 through 17 can be repeated. Row 16, the heat loss provided by solar, is the "useful" energy output of the collector and can be used to determine the value of the energy which it provides.

The final collector size will be tempered by factors other than heating demand, including the preheating of domestic hot water, solar heat gain through windows in combination with proper thermal storage mass of the house, building program and design constraints, economic return, other choices of energy supply, and the final supply efficiency of the overall solar energy system.

Two important factors are considered here in addition to the issue of heating. First of all, collector surface area can be increased to obtain the energy required for domestic hot water heating. Roughly speaking, the increment in Btu requirement is found by multiplying daily hot water consumption in gallons times eight pounds of water per gallon, times the specific heat of 1 Btu per pound per degree increase in temperature, times the average temperature increase of the water to be provided by the solar collector. If the water is normally increased in temperature from 55° (the temperature of water from the well, perhaps) to 125° (the useable temperature), the collector, on a yearly average, can be easily sized to provide heat for about half of that temperature difference for annual average sunshine conditions of 50%. Solar energy will easily provide more than half for sunnier locations such as Miami or Phoenix. Warm annual average outdoor temperatures will also provide higher percentages of the total heating load.

If the preheat tank for the system is located inside the heat storage mass of the solar heating system, the house heating collector is, in effect, the domestic hot water solar collector as well; the storage heat is transferred directly to the pre-heat tank 24 hours per day. If a heat exchanger for the domestic hot water is located inside the heat storage mass, it transfers heat to the domestic water heater. If the pre-heat tank is not located in the heat storage mass of the solar heating system, and if the collector is of the air type, the solar hot water system will probably need its own collector, different from and built independently of the space heating system.

CHAPTER V·C

Other Collector Types

As the prototype of solar heat collectors, the flat-plate collector has been the subject of much research effort. Because of the limitations which its design places on mass production, however, other types of collectors may rise in importance in future years. Some other collector types are variations of the flat plate, while others, such as concentrating collectors, are in altogether different categories.

One variation is itself flat but does not have a "plate," in the traditional sense, for solar collection. A solar pond, shown in Figure V·C·1, is a body of water, often several meters deep, which both collects and stores heat year round. The required size varies with climate, building type, and system design, but may need to be the same volume (in cubic feet) as the area of the space (in square feet) which heats it. Concepts similar to this are discussed in a later section on long-term storage.

Increasing the amount of insolation which strikes a collector is a major method of increasing its performance. An adaptation of the flat plate collector designed by Gerald Falbel concentrates solar energy by using what he calls *reflective pyramidal optics*. The result, illustrated in Figure V·C·2, allows a reduction in the necessary size of the usually costly absorber plate. At the mouth of the pyramid is

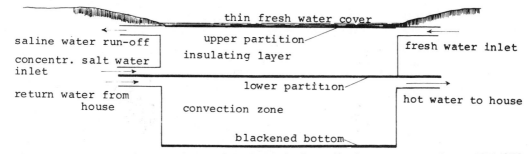

Figure V·C·1. Solar Pond Collector. Source: Solar Ponds for Space Heating, presented at August 1974 ISES Conference, Ft. Collins, Colorado. Source of drawing: (DAV.)

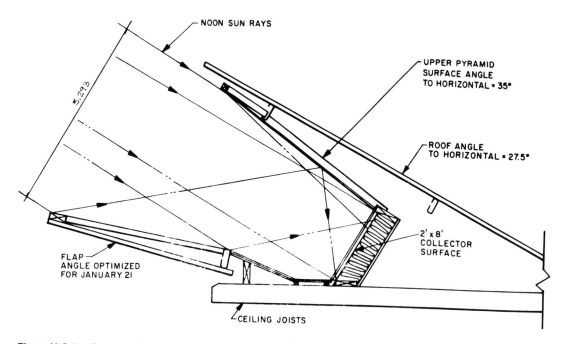

Figure V·C·2. Pyramidal Optics System. (Reprinted courtesy of Wormser Scientific Corporation.)

a hinged, reflector door which can be closed during cloudy weather. When the sun is shining, its tilt can be adjusted to obtain maximum reflection through the mouth and onto the absorber surface. A concentration of two to four times more than the ambient insolation rate can be obtained. Although the structure and container for the smaller absorber may be expensive, perhaps one of the best features of the design is the potential for eliminating the freezing problem of

Figure V·C·3. An installation of the pyramidal optics system showing the door in the open position. (Photo courtesy of Wormser Scientific Corporation.)

Figure V·C·4. An installation of the pyramidal optics system showing the door in the closed position. (Photo courtesy of Wormser Scientific Corporation.)

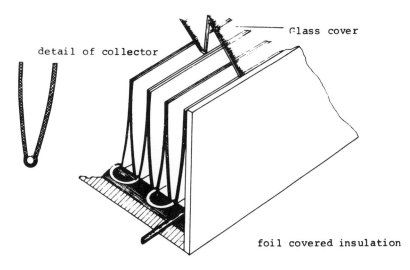

detail of collector

Glass cover

foil covered insulation

Figure V·C·5. Winston Collector. Source: Argonne National Laboratory. Source of drawing: (DAV.)

water-type collectors by closing the hinged panel when the sun does not shine; this can reduce collector and system costs. Additionally, higher fluid temperatures may be obtained without a sacrifice of collector efficiency. Figures V·C·3 and V·C·4 are photographs of the system installed at the home of the inventor, showing the hinged door in both open and closed positions.

Another kind of concentrating collector was invented by Roland Winston (see Figure V·C·5). His compound parabolic troughs concentrate sunlight on a relatively small percentage of the total absorber surface. Not only does the increased insolation raise efficiency, but it also uses a smaller absorber surface, a fact which is particularly important, for example, if the absorber consists of expensive photovoltaic cells for the conversion of solar energy into electricity.

James Eibling (NSF.) classified solar collectors according to their ability to concentrate sunlight. His results, shown in Figure V·C·6, include operating temperature range and collection efficiency. Although collectors of medium and high levels of concentration produce quite high temperatures (up to 1200°F) at relatively high efficiencies (up to 75%), the primary reason for not considering them for space heating or cooling at this time is their high cost and the complexity which they introduce into heating or cooling system.

Category	Example	Application Temp. Range, F	Probable Range* of Collection Efficiency, %
High concentration	Paraboloid	500–1200	60–75
Medium concentration	Parabolic cylinder	300–800	50–70
Low concentration	Flat plate	150–300	30–50

*Percent of solar energy received at the aperture that is converted to heat in the working medium.

Figure V·C·6. Classification of solar collectors. Source: (NSF.)

The evacuated glass cylinders of about four inches in diameter were designed by Owens-Illinois to be mass produced at a level similar to that for flourescent light bulbs. Figure V·C·7 shows how the cylinders can be located on a flat surface, either vertically or horizontally. With sufficient spacing to prevent shadowing, they can be rotated to achieve an optimal tilt angle. Such versatility could aid immeasurably in making solar collectors compatible with building design.

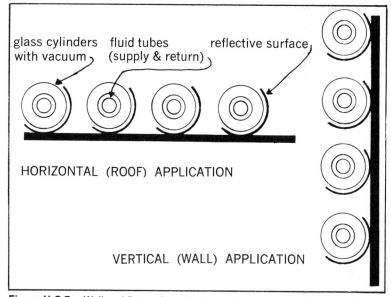

Figure V·C·7.   Wall and flat roof applications of vacuum tube solar collectors.

CHAPTER V·D

Solar Heat Storage

The most important reason for providing heat storage in a solar energy system is that the sun does not always shine, but the need for energy continues. Also, more energy than necessary is usually available when the sun is shining, and by storing it for later use, energy can be provided at times when the sun is not shining.

Solar heat storage design must weigh cost against performance. Some of the most pressing determinants of cost include the choice of storage medium, such as rocks, water, or eutectic salts; the amount, or size, of the storage medium required, measured by weight or volume; the location, whether part of the space to be heated or separate from it; the type and size of the container for the medium; heat exchangers, if required, for moving the heat in and out of the medium; and the mechanical device to move heat transfer fluid through the storage or heat exchangers.

In addition to these factors, performance is also affected by the average operating temperature, the pressure drop of the heat transfer fluid moving through the medium, and heat loss by the container to the surroundings.

There are basically three choices of storage medium: rock, water, and phase-changing (eutectic) salts.

The capacity of different materials to store heat varies according to their specific heat. As defined in a previous part, the specific heat of a material is the number of Btu required to raise one pound of the material 1°F. The energy, often referred to as sensible heat, can be reclaimed as the material drops in temperature. This is the basic principle of most forms of solar heat storage. Figure V·D·1 lists the heat storage capabilities of several common materials.

The choice of storage medium and solar collector must be made simultaneously. Almost without exception, liquid-type systems, whether the open type (e.g., Thomason's trickle system) or the closed tube-in-plate, require a liquid heat storage medium. In most air-type systems, the heat storage medium consists of small pieces—rocks, small (one gallon or less) containers of water, or containerized eutectic salts are most common—which allow the air to travel around and

Sensible-Heat Storage	Sp. Ht., Btu/lb F	True Density, lb/ft³	Heat Capacity, Btu/ft³ F	
			No Voids	30% Voids
Water	1.00˜	62	62	43
Scrap Iron	0.12*	490	59	41
Magnetite (Fe_3O_4)	0.18*	320	57	40
Scrap Aluminum	0.23*	170	39	27
Concrete	0.27	140±	38	26
Stone	0.21	170±	36	25
Brick	0.20	140	28	20
Sodium (to 208 F)	0.23	59	14	—

Figure V·D·1. Heat storage materials. Source: (HOT.)

between, transferring its heat to the medium. Those systems which combine the solar collector with heat storage, discussed in Part III, are also alternatives.

STORAGE FOR LIQUID SYSTEMS

A significant advantage of liquid systems, including the water storage tank component, is their compatibility with solar cooling. The water can be used in virtually all types of solar cooling, including nocturnal radiation, off-peak cooling using small compressors, and Rankine and absorption refrigeration cycles. The greatest advantage of water as a heat storage medium is its relatively low cost, except in areas of the world where water is scarce. There are, however, some difficulties with water, the solutions for which can cause considerable expense.

Containing a large volume of water (anywhere from two to ten gallons per square foot of collector) has been somewhat simplified in recent years by the advent of waterproofing products and large plastic sheets. Previously, the only available container was the steel galvanized tank, which eventually leaked. The replacement of large tanks, which are usually located in basements or underground because of their size and weight, is difficult and expensive. The introduction of glass linings and fiberglass tanks alleviated the corrosion problems but increased the initial cost. The use of poured concrete tanks has, until recently, been hampered by the difficulty and cost of insuring their long-term water tightness; concrete is permeable and develops cracks. Large plastic sheets or bags

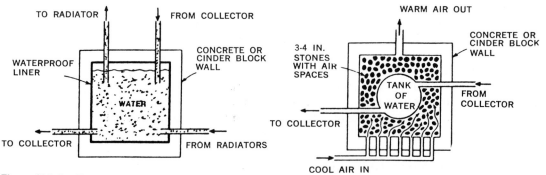

Figure V·D·2. Two methods of water-type heat storage. At the left is a concrete tank filled with water; at the right, a water tank encircled by rocks.

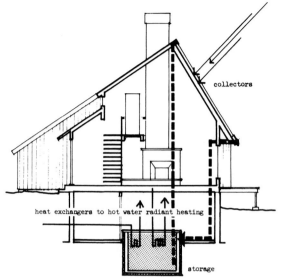

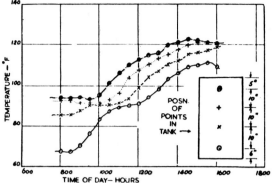

Figure V·D·4. Daily temperature variation of water at points inside tank. Vertical distances between curves show linearity or otherwise of temperature distributions. Source: (CLO.)

Figure V·D·3. External flow water collectors with tank storage in Vermont house. (Design by architect Sue Burton Tenner, consulting by Total Environmental Action, Inc.)

used as liners, however, can eliminate the need for concrete; lightweight wood or metal frames can be used to support the plastic.

Figure V·D·2 shows two methods of storing water: the first is a water-filled concrete (or cinder block) container; and the second is Dr. Harry Thomason's system, a water tank encircled by rocks. In the first method, the warm water from the tank is circulated to the building either directly through radiators or radiant heat panels or indirectly through heat transfer coils over which cool building air is blown and heated. This latter method was used in House IV by MIT in 1959. Figure V·D·3 is a cross section of a house in midwestern Vermont, designed by Sue Burton Tenner. The solar system, designed by Total Environmental Action, includes an open trickle type collector. A heat exchanger takes heat from the storage and delivers it to the house via large wall and ceiling radiant panels, allowing for the use of relatively low temperature storage water. A second heat exchanger preheats the domestic hot water as it travels to the conventional hot water heater. The second storage configuration in Figure V·D·2 transfers heat slowly but continuously from the water tank to the stones. Cool air from the house is circulated slowly in large volumes through the warmed rocks and back into the house. In both configurations, the coolest water, at the bottom of the tank, is delivered to the collector for heating and then returned to the top of the tank. This warm water is used for heating the house.

Temperature stratification within water tanks is demonstrated in Figure V·D·4 by Close (CLO.1). The 41 inch high tank shows temperatures at the beginning of the day ranging from less than 70° six inches from the bottom to almost 95° five inches from the top. At the end of the day, the difference is somewhat smaller, about 15°.

The large size and high cost of heat exchangers can be major issues in the use of water storage tanks. Thomason's 25 to 50 tons of rock, although serving as additional heat storage, is in some sense a tremendously large heat exchanger. For typical metal heat exchangers which lie submerged in the water, the total heat exchanging surface area can be as large as one-third the surface area of the solar collector.

Heat exchangers are necessary when the water in the tank cannot be used directly for purposes other than storage. For example, when an anti-freeze/water solution is used in the solar collector, it must travel through a heat exchanger to prevent it from mixing with the water in the tank. Also, in providing buildings with heat, heating engineers are likely to require that the water from the tank not be used in the heating system. This is true particularly when the water in the tank is circulated through the collector.

The limitation in the choice of location for large water containers may prove advantageous for building designers who wish to be relieved of the burden of deciding where to place such a large object. However, for the designer who wishes to use the heat storage as an integral part of the total design, placement of the large, unwieldy tank can be difficult. Self-draining liquid-type systems, of course, require that the storage be located below the bottom of the collector; thermosiphoning systems require that it be above the top of the collector. If the storage system is tied into other mechanical equipment, such as the furnace, pumps, heat exchanger, and domestic water heaters, close proximity to them may be required.

STORAGE FOR AIR SYSTEMS

Of the several heat storage mediums for air-type systems, rocks are perhaps the best known and most widely used. Although this medium appears to present a relatively cheap and easy solution, it is not always a clear cut choice. The most significant advantage of rocks is their low cost, if they are in fact available. Through much of New England, for example, the only available rock is 1 to 1½ inch diameter gravel. Depending on the design and dimensions of the rock bin, rock diameters up to four inches may be required. Even if the proper size is available, a supplier may be unable or unwilling to deliver uniform rock sized to the specified diameter. Collecting rocks by hand sounds romantic but becomes drudgery after the first hundred pounds; thousands of pounds, from 80 to 400 pounds per square foot of collector, are required because of rock's low specific heat. The huge quantities involved further compound the problem of handling the rocks and of providing a bin sufficient in size to contain them. With 30% voids, the volume of rocks needed to store the same amount of heat as a tank of water is about two and one-half times greater.

The large perimeter area of these storage containers result in both higher construction costs and in greater heat loss. The potential for greater heat loss from large rock bins than from smaller water tanks, however, is offset by the relatively slow natural movement of heat through rocks compared to the constant movement of water within a large container as it changes temperature (e.g., because of heat loss).

One of the greatest limitations of rocks is their relative lack of versatility for purposes other than containing heat, such as preheating domestic hot water, cooling, and even heating the living space. One of the few and most obvious methods of preheating domestic hot water is to place a small (30 to 100 gallons), uninsulated water tank within the rocks. The heat transfer is slow but continues 24 hours per day.

Methods of solar cooling are limited to those in which the rocks store coolness for later use. This coolness can be obtained by the circulation of cool night air, of air cooled by nocturnal radiation, or of air cooled by off-peak refrigeration compressors. Air-type collectors which will efficiently produce temperatures high enough to operate refrigeration cycles (180 to 300°F) are unlikely to be developed. Air conditioning equipment which is compatible with hot air rather than hot liquid as a source of heat is not now available.

Air systems limit the manner by which heat can be delivered to the space. Almost without exception, heating systems must be of the forced warm air-type, in contrast to the water tank type of storage in which either forced warm water or forced warm air can be used. However, as discussed in Part III, air can circulate naturally through rocks to heat the space without the need for fans.

Figure V·D·5 shows a domehouse designed by Total Environmental Action, Inc., in which the rock storage is located within the confines of the space. Heat is transferred by slow loss from the tank to the space, by natural convection from the room to the bottom of the tank and out through the top, and by small auxiliary fans when necessary. (The dome was the client's choice, and the separate structure necessary for the collector indicates some of its limitations for human habitation.)

The storage location of rocks can be an important limitation to their use. If the location is a crawl space under the building, the cost of constructing that space may not necessarily be included in the overall cost of the solar system. However, if otherwise useable basement or living space is used, the cost of building that

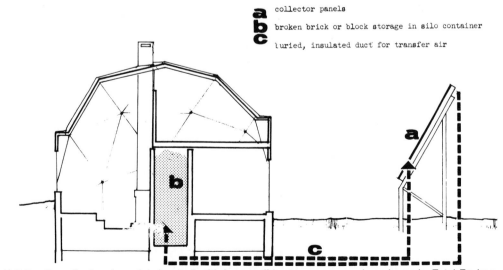

a collector panels

b broken brick or block storage in silo container

c buried, insulated duct for transfer air

Figure V·D·5. Air collectors (remotely located) with loose-solids storage use on dome home by Total Environmental Action, Inc.

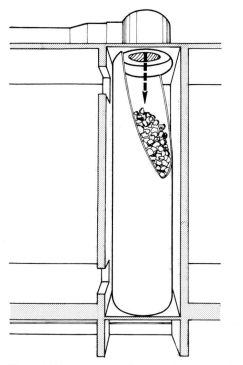

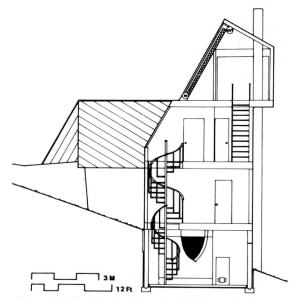

Figure V·D·7. Section through solar house design concept for Boston. (By Total Environmental Action, Inc., under subcontract with the AIA Research Corporation and published in *Solar Energy Home Design in Four Climates.*) Source: (TOT.2)

Figure V·D·6. Loose solids contained in vertical fiberboard cylinder.

space may be applied to the system cost. Figure V·D·6 shows how rock can be part of an architectural element of the building. Dr. George Löf's house in Denver used this method in a rather elegant manner. Because of the massive weight, however, strong foundations must be provided under the containers.

Figure V·D·7 is a section of a house design concept for the Boston area by Total Environmental Action, Inc., under a grant from the AIA Research Corporation of the American Institute of Architects (TOT.2). The urban site is a steep north slope with tall buildings to the south. The collector is as high in the air as possible because of potential shading from neighboring structures. Because of its large size and weight, the rock heat storage container is on the lowest floor of the house.

The design illustrates a rather straightforward method for transferring heat to and from the bin. Figure V·D·8, a solar system diagram for the house, shows the warm air from the collector being delivered to the top of the tank. It is drawn down and out through the bottom and back to the collector. To heat the house, cool air is brought to the bottom of the tank and heated as it rises through the rocks. The warmest rocks at the top warm the house air to the greatest extent possible. The figure also shows the furnace heating cycle, which allows the house air to bypass the rock storage. Usually, the storage container should not be heated by the furnace, except when it is located within the living space.

One of the important reasons for bringing the warm collector air in through

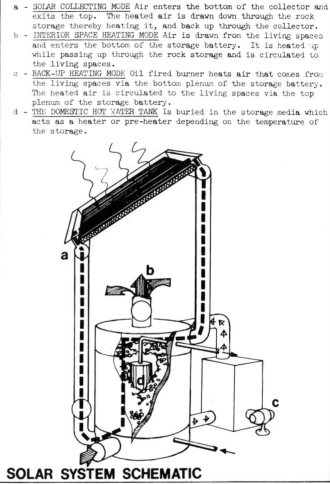

a - SOLAR COLLECTING MODE Air enters the bottom of the collector and exits the top. The heated air is drawn down through the rock storage thereby heating it, and back up through the collector.

b - INTERIOR SPACE HEATING MODE Air is drawn from the living spaces and enters the bottom of the storage battery. It is heated up while passing up through the rock storage and is circulated to the living spaces.

c - BACK-UP HEATING MODE Oil fired burner heats air that comes from the living spaces via the bottom plenum of the storage battery. The heated air is circulated to the living spaces via the top plenum of the storage battery.

d - THE DOMESTIC HOT WATER TANK Is buried in the storage media which acts as a heater or pre-heater depending on the temperature of the storage.

SOLAR SYSTEM SCHEMATIC

Figure V·D·8. For solar house design concept in Boston. (Adapted from work done by Total Environmental Action, Inc., under subcontract with the AIA Research Corporation and published in *Solar Energy Home Design in Four Climates.*) Source: (TOT.2)

the top of the tank is to achieve temperature stratification. This allows the house air to be heated to the highest temperature possible by the warmest rocks at the top. If the warm air is brought through the bottom of the tank, even without air movement within the tank, the heat from the bottom tends to distribute itself evenly through the entire tank, resulting in low temperatures throughout. Bringing building air in at the same point as the warm collector air tends to promote this distribution of heat through the storage rather than to achieve the heating of the air for the purpose of heating the building.

Figure V·D·9 shows temperature profiles for one of the heat storage cylinders during one day at the home of Dr. Löf. The stratification would be more pronounced if the air were brought in through the top and out through the

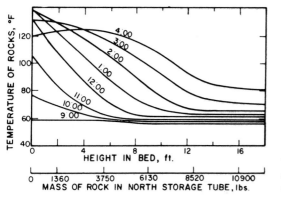

Figure V·D·9.
Temperature profiles in the packed bed storage unit of George Löf. Source (GIL.)

bottom. As it is, a range of as high as 70° occurs during the course of the day.

The shape of the storage bin is particularly important when rocks are being used as the storage medium. In general, the longer the distance which air has to move through rocks, the larger the rock diameter required to reduce the pressure drop and the necessary fan size. For example, if the storage is of the tall cylindrical type in Figure V·D·6, large diameter rocks are required. If the cylinder is more than eight feet tall, the rocks should be at least two inches in diameter, larger for taller containers. For squat, horizontal bins, such as might be used in a crawl space, one to two inch diameter gravel can be used (see Figure V·D·10).

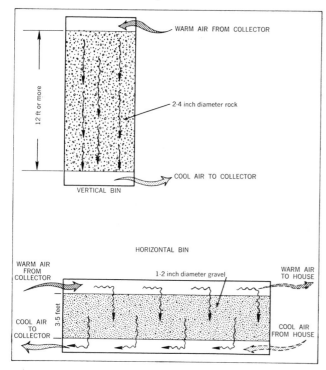

Figure V·D·10.
Rock diameters for solar heat storage vary with the shape of the bin and the length of gravel through which the air must flow.

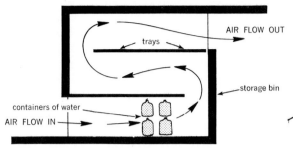

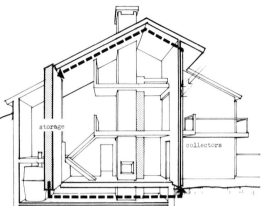

Figure V·D·11. Heat storage bin for air systems using small containers of water and large trays.

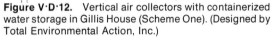

Figure V·D·12. Vertical air collectors with containerized water storage in Gillis House (Scheme One). (Designed by Total Environmental Action, Inc.)

The numbers suggested above are greatly affected by the velocity of the air moving through the rocks. The slower the air speed, the smaller the rocks which can be used or the deeper (or taller) the bed of rocks can be. In fact, increased pressure drop through rocks is directly proportional to increased air speed. Of course, the smaller the rock diameter, the greater the amount of surface area of rocks to which the heat from the air can transfer. For example, a cubic foot of one inch diameter rock has about 40 square feet of surface area while the same volume of three inch rock has about one-third that much.

Löf and Hawley (LOF.1), Close (CLO.3), and Bird, *et al.* (BIR.) are among the best references for determining relationships among rock diameter, air velocity, and pressure drop. In general, the rocks or pebbles should be large enough to keep the pressure drop low but small enough to obtain good heat transfer.

Air-type systems can also utilize small (one gallon or less) containers of water (or other material for that matter) which can be arranged in racks, on shelves, or in some other way as to allow unobstructed air to flow around them. Such containers include plastic, glass, or aluminum jars, bottles, or cans. The labor intensiveness of handling thousands of small containers—gathering, filling, and stacking them—is potentially costly and cumbersome. The stacking or arranging is solved in numerous ways but perhaps most successfully by sandwiching them between large trays and then blowing the air horizontally between the trays (see Figure V·D·11). Part III shows small containers located between floor joists, which in turn act as air plenums. Vertical heat storage cavities serving also as room partitions or exterior walls can also be used. Again, when the heat storage is located within, or contiguous to, the heated space, all of the heat loss from the storage goes to the space being heated. Figure V·D·12 shows a section through a house designed by Total Environmental Action, Inc., for a location in Massachusetts. In this preliminary design, the air circulates up through a vertical, south-facing collector and then down through a vertical partition filled with small containers of water.

Such a wall configuration is not easily possible with rocks and reveals one of the principal advantages of containerized water. Another advantage is that some-what smaller volumes of space are needed in order to contain enough water to store the same amount of heat as rock. Assuming 50% voids between the containers, water stores 32 Btu per cubic foot per degree temperature differ-ence. Rock, with about 30% voids, stores 25. If the water containers can be spaced with only 30% voids, 43 Btu will be stored under the same conditions.

Leakage is unlikely to be a problem since no more than one gallon of water can escape from any single leak. Should this type of heat collection and storage prove to have widespread applicability, it is likely that special containers, e.g., of rigid molded plastic, will become available. Such containers could be self-supporting as they stack one above the other. Similar containers of softer vinyl could be shipped easily and filled on location. The proper design would incorporate integral air spaces for the flow of air.

Also under contract with the AIA Research Corporation, Total Environmental Action, Inc., (TOT.2) used the structure of a house to store the heat. The system, shown in Figure V·D·13, was a design concept for Minneapolis.

Alternatives to rocks and water for air-type solar collector systems are phase-changing (eutectic) salts. The heat storage principle of salts is that a material stores significant amounts of heat as it changes phase from solid to liquid (melts) and gives that stored heat up again as it solidifies (freezes). For example, only one Btu is required to change the temperature of one pound of water one degree in temperature. However, in order to melt ice, 144 Btu are required to change its temperature only 1°F.

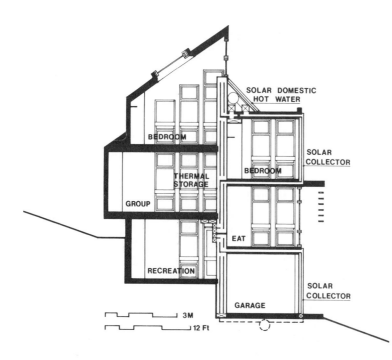

Figure V·D·13.
Solar house design concept for Minneapolis. (Adapted from work done by Total Environmental Action, Inc., under subcontract with the AIA Research Corporation and published in *Solar Energy Home Design in Four Climates.*) Source: (TOT.2)

Since water melts at a temperature too low for useful heat storage, eutectic salts, with higher melting temperatures, have been investigated for more than 30 years. The salt most widely studied is commonly known as Glauber's salt, which melts at 90°F and, in the process of melting, stores 105 Btu per pound. The use of salts, therefore, can result in significantly smaller storage volumes. Some of the other salts which have been investigated are listed in Figure V·D·14. The various melting temperatures make possible the selection of the salt which provides a system with its most effective average storage temperature.

The small-size advantage of salt storage decreases as the temperature range of other storage mediums increases. For example, in Figure V·D·15, the volume of salt required to store the same amount of heat as a given volume of water increases as the temperature range of the storage increases. The figure assumes no voids in the water. Some of the requirements of a successful salt for solar heat storage include the following (TEL.):

- it must be of relatively low cost;
- its phase change must have a high latent heat effect, that is, it must store large quantities of heat;
- it must be available in large quantities;

	Chemical Compound	Melting Point, °F	Heat of Fusion Btu per pound	Density lb/ft
1. Calcium chloride hexahydrate	$CaCl_2 \cdot 6H_2O$	84–102	75	102
2. Sodium carbonate decahydrate	$Na_2CO_3 \cdot 10H_2O$	90–97	106	90
3. Disodium phosphate dodecahydrate	$Na_2HPO_4 \cdot 12H_2O$	97	114	95
4. Calcium nitrate tetrahydrate	$Ca(NO_3)_2 \cdot 4H_2O$	102–108	60	114
5. Sodium sulfate decahydrate (Glauber's salt)	$Na_2SO_4 \cdot 10H_2O$	88–90	108	97
6. Sodium thiosulfate pentahydrate	$Na_2S_2O_3 \cdot 5H_2O$	118–120	90	104

Figure V·D·14. Salt hydrates for solar heat storage. (Reprinted by permission of ASHRAE.) Source: (TEL.)

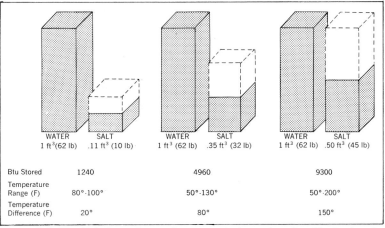

	WATER 1 ft³ (62 lb)	SALT .11 ft³ (10 lb)	WATER 1 ft³ (62 lb)	SALT .35 ft³ (32 lb)	WATER 1 ft³ (62 lb)	SALT .50 ft³ (45 lb)
Btu Stored	1240		4960		9300	
Temperature Range (F)	80°-100°		50°-130°		50°-200°	
Temperature Difference (F)	20°		80°		150°	

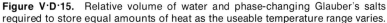

Figure V·D·15. Relative volume of water and phase-changing Glauber's salts required to store equal amounts of heat as the useable temperature range varies.

- it must be reversible over a large number of phase-changing cycles without serious degradation of the latent heat effect;
- its phase change must occur close to its actual melting temperature;
- its preparation for use must be relatively simple;
- it must be harmless (non-toxic, non-inflammable, non-combustible, non-corrosive);
- it must be contained in such a way and in materials which allow heat to be transferred to and from it.

The present cost of Glauber's salt is over a penny a pound, but other salts can cost significantly more. The preparation and containerization of salts adds greatly to their price; it is unlikely, for instance, that Glauber's salt can be installed in a solar system for less than 20 cents per pound, not including the large bin in which smaller containers of salt are stored (NSF.). Through a temperature range of 30°F, each pound will store about 105 Btu as it melts and an additional 20 or so as it changes 30°F in temperature. To store the same 125 Btu requires about four pounds of water and about 20 pounds of rock. This relationship can be determined for other conditions by determining the temperature range of a particular heat storage system. Then the cost for construction, insulation, and so forth can be compared for each choice of storage medium.

Glauber's salt is presently available for wide application in most parts of the country, but this may not be true of other salts. The preparation of salts for use in solar heat storage is somewhat complicated by several difficulties. Considerable effort has been made, particularly by Dr. Maria Telkes, first at MIT in the 1950's and most recently at the University of Delaware, to resolve some of these difficulties. Included are the loss of the salts' high latent heat effect because of the stratification of the various chemicals during the liquid state. Thickening agents, up to 7% by weight, have been added to prevent this separation of chemicals. The salts also tend to supercool before solidifying, dropping to a temperature well below their melting temperature. Nucleating agents, such as Borax (up to 3% by weight) have been added with good results; supercooling is virtually eliminated with crystallization beginning at between 83 and 85°F. More than a thousand cycles have been tallied for some salts, and further experimentation is producing even more.

Because of the corrosive nature of most salts, care must be taken to containerize them properly even though inhibitors can be added to reduce corrosiveness. Solar I, the experimental solar house built by the University of Delaware in 1973, has ABS type plastic tubes of 1¼ inch diameter and a 0.030 inch wall thickness. The salts have also been placed in horizontal pan-shaped containers about 21 inches square and one inch deep. The pans are spaced one-quarter inch apart for the passage of air. Leakage and vapor losses must be eliminated in the containerization (TEL.). The shape of the container assists in reducing stratification and supercooling. Perhaps the most significant paper on the subject is one by Dr. Telkes (TEL.) in which she reports on salts in addition to Glauber's salt which melt at 55°, 75°, and 120°, and which are used by Solar I.

The relatively small volumes required by salts made possible unusual versatility in storage location. For example, closets, thin partition walls, structural voids, and otherwise unuseable small spaces within buildings become possible locations.

Paraffin, or wax, lies somewhere between water and eutectic salts in its ability to store heat. Its specific heat is only 0.7, but its density is almost that of water, 55 pounds per cubic foot. When it melts, however, it stores about 65 Btu per pound, as contrasted to over 100 for Glauber's salt. Its cost may be significantly lower than salts, particularly if its sales and shipping container can be used in the solar system; not only does paraffin oxidize, but it also corrodes plastics and some metals such as copper. In the oxidation process, of course, it slowly disappears into the atmosphere unless tightly sealed. Its coefficient of expansion is particularly large, expanding up to 20% when it melts. Conversely, as it solidifies and gives up its heat, it contracts, pulling away from the container walls and significantly reducing the transfer of heat out of the container.

Paraffin's greatest disadvantage is its combustibility. Few building codes permit its use inside a building, in spite of the acceptance of small oil tanks for house heating. Heat storage using paraffin would probably have to be located outside the building, or the storage bin would have to have a significantly high fire rating.

STORAGE SIZE

Countless methods and inputs have been devised to refine the process of determining storage capacity. Most are more complicated than they need to be and can be summarized in relatively simple fashion since some inexactness can be tolerated. Most methods, however, require rather precise information about the performance of the collector as well as detailed weather data. Since such information is usually unavailable, approximations must usually be made, and, if possible, provisions for altering the final storage capacity after actual performance is determined during operation. An oversized concrete water tank, for example, can be filled to various levels during actual operation until the best system performance is obtained.

In general, storage should be oversized rather than undersized to keep the average temperature as low as possible. The limit in size is usually determined by available space and by the ability of the living space to use the low temperature heat in the storage. A heating system which requires 130° air, for example, cannot readily use a storage system with an average temperature of less than 130°. Studies by Total Environmental Action, Inc., (TOT.1), show that a collector which operates at an average temperature of 90° collects about twice as much energy per square foot during a heating season in Boston as one which averages 140°. The corresponding average storage temperatures are 85° to 135°, respectively. The useable temperature range for the lower temperature storage may be from a low of 80° to a high of 120°, or a total temperature difference of 40°. The higher temperature storage, however, has a potentially wider temperature range, assuming the heating system is designed to use low storage temperatures. If it is, the useable temperature range may be from 80° to 170°, or a total overall temperature difference of 90°, about double that of the low temperature storage. The high temperature storage, therefore, can be half the size of the low temperature storage and still store the same amount of heat.

As an example, 1000 pounds of water (about 125 gallons) stores 20,000 Btu as it increases in temperature from 70° to 90°, and 500 gallons of water stores the same amount as it increases in temperature from 80° to 120°.

In determining storage size, total heat flow must be determined. Except in unusual designs, the sole source of heat to the storage is from the collector (other sources include heat pumps, wells, and electric resistance heating). This heat is lost in two main ways: first, and continuously, 24 hours per day, is the conduction (and occasionally convection) heat loss from the storage tank; second, and in quantities that vary tremendously from hour to hour and from month to month, is the loss of energy which is supplied as heat to the building. Often, too, energy which is used to heat (or to preheat) the domestic hot water is lost, but this consumption is somewhat more regular and predictable because of the relatively constant daily demand.

Every type of storage system requires massive amounts of insulation to reduce its heat loss to the surrounding environment. The higher its average temperature and the colder the surrounding environment, the more insulation should be used. If the storage is located within the confines of the space to be heated, of course, the insulation need not be as great as if it is located outside or in a basement. The ground *can* provide insulation for heat storage but should rarely be relied upon; even a small amount of moisture movement through soil will virtually destroy its insulating value.

Insulation equivalent to at least 6 inches of fiberglass insulation (an R-value of about 20) should be used for low temperature storage located within the confines of heated space. Up to three feet of high performance urethane-type insulation should be used when temperatures of over 200° are being stored for several months, for example, when heat is collected during the summer and stored for winter use. The MIT House IV, built in Massachusetts in 1959, collected about 41 million Btu during a six month heating season. However, about 6.5 million were lost from the warm tank of water located in the unheated basement and only 34.5 million were actually used.

All ducting for plumbing should be insulated according to the same high standards as for the storage tank or bin. Of course, close proximity of the storage tank to the collector reduces the heat losses from ducting and plumbing runs; energy transport costs are also slightly reduced.

One of the most inaccurately communicated concepts in solar heating concerns the number of days of carry-through of the solar heat storage. A system designed to provide heat during two sunless days in April is considerably smaller than one which does so for two sunless days in January. Also, a system which has two days of carry-through during average temperatures in January is about half as large as one which provides for the two coldest days in January. Providing heat storage for two January days in Charleston requires a considerably smaller system than one which does the same thing in Minneapolis. Even in Minneapolis, a system for a well-insulated structure can be significantly smaller than one for a building which is conventionally insulated. Additionally, a house which has large quantities of solar heat gain through windows and which has considerable thermal mass in which to store the heat will not need energy from the solar collector system until long after the sun has disappeared.

Probably the best method of describing the heat storage capacity of a system is to define the number of useful Btu which it can store through a given temperature range. Further qualification would include defining the heat loss of the building on a Btu per degree day basis. Dividing this number into the heat storage capacity yields the number of degree days of heating demand which the heat storage is able to provide the building. For example, a well-insulated house of 1200 sq ft in Minneapolis may lose 10,000 Btu per degree day. Its 8000 pounds (1000 gallons) of water solar heat storage holds 320,000 Btu through a temperature range of 40°, providing sufficient heat to carry the house through 32 degree days (this would occur when the temperature is 33° for 24 hours or 49° for 48 hours).

The sequence of sunny and cloudy days is also important to storage sizing. The extreme optimum is a regular alternation of sunny and cloudy days. The solar system (including the solar heat gain and the thermal mass of the house) could be sized to collect heat for one sunny day and for the following cloudy day. As much as 100% of the heating demand could be provided by solar energy if the system were sized for the coldest two day period.

If the weather sequence were two sunny days followed by two cloudy ones, however, the storage would have to be twice as large to accommodate two cloudy days of heating demand, but the solar collector would have been increased in size only slightly or not at all. If the sequence were one sunny day followed by two cloudy ones, the solar collector would have to be increased in size so as to collect sufficient heat in one day to provide for several days, but the storage would be about the same size as the two-day sunny, two-day cloudy sequence. The wide variation in the sequence of sunny and cloudy days for particular locations and for the country as a whole makes it impossible to generalize about this input. In addition, this weather data is not widely available.

At the Blue Hills weather station near Boston, half of the sunless days are sandwiched between two days during which solar energy can be collected. About 80% of the sunless day periods are two days long or less, 90% are less than three days, and a sequence of four sunless days is rare. A solar system which will carry a building in Blue Hills through three cloudy days during the coldest weather will provide nearly 90% of the load.

As long as the goal of the solar heating design is to provide 60% or less of the building's heating needs, there are two basic approaches to determining final storage size in areas of the country with about 50% possible sunshine. For the first method, begin by sizing the collector according to the needs of the building in the particular location, a process described earlier. Then determine the average storage temperature range. (Earlier, in order to determine the energy output of collectors and the collector size, the average operating temperature of the collector was determined. This temperature can be assumed to be about five degrees higher than the average storage temperature.)

When the temperature range of the storage is determined, calculate the amount of heat which can be stored per pound (or cubic foot) of storage medium. Divide this number into the number of Btu which the solar collector will collect during an average sunny day. This gives an approximate total weight, or volume, of storage required.

For example, if the average collector temperature is 100°, the average storage temperature is about 95°. The useable temperature range during a 24 hour period may be 85° to 105°, or 20° total. For a cubic foot of rock, which weighs about 120 pounds, about 500 Btu will be stored through the 20° temperature range. If the solar system collects 1000 Btu per square foot per average sunny day, two cubic feet of rocks should be used for each square foot of collector surface.

The larger the temperature increase in the heat storage during a single day of sunshine, the higher the average operating temperature of the collector and the lower the overall efficiency. In addition, during relatively warm winter weather, a second day of sunshine will raise the storage temperature even higher, particularly if little of the heat was needed to maintain the temperature of the building. A third day will raise the temperature even more, resulting in an even higher average collector operating temperature.

The second method of sizing heat storage is based on the number of degree days of heating load through which the system must carry the building. A temperature range and average temperature are determined, and the corresponding collector size is chosen.

Perhaps one of the most important issues in solar energy is the need for a full-sized backup heating system for periods of cold cloudy weather and depleted solar heat storage. If the source of energy for the backup system is gas or electricity, utilities may have to construct full-sized facilities to provide for the backup heating, which may be needed when the utility experiences its peak demand. At this time there is increasing public pressure to restrict the construction of power plants, and the use of solar energy may not alleviate this pressure unless it is designed to reduce peak loads on the utilities. If gas and electricity are used for backup, then, they should be used during off-peak hours. A solar house in Norwich, Vermont, designed by architect Stu White, uses electricity to heat an extremely large rock pile during off-peak hours, 9 p.m. to 7 a.m., to a temperature of 85°. This temperature was determined by Total Environmental Action, Inc., the solar engineers, to be high enough for the rocks to be able to carry most of the house through all but the very coldest 24 hour period. The 85° temperature, however, is low enough to allow the addition of considerable solar heat should the following day be warm and sunny.

A more sophisticated version of this method is designed by Burt, Hill & Associates for the Alumni Conference Center in Albany, New York. A simple logic circuit senses the heat storage temperature and compares it with the projected heating demands of the building based on daily weather forecasts of temperatures, wind, and sunshine. When no sunshine is forecast, off-peak electric resistance heating warms the heat storage to a temperature sufficient to carry the building through the following day. If sunshine is predicted, the electric heating is triggered only long enough to supplement the projected solar collection should it be insufficient to provide for the total heating demand.

Long-Term Storage. The economics of solar energy are complicated and made more difficult by the fact that there are occasional long periods of cold, sunless days. In addition, the various components of solar heating systems are sized and detailed to squeeze the last bit of possible efficiency out of their interface with the conventional auxiliary system. This design process is expensive, and the resultant solutions are complicated in operation.

The use of long-term storage can reduce the complications of and possibly solve some problems of economics. Long-term storage is best described as the means by which solar energy is stored for use long after it is collected, for example, from one season to the next, in the manner to which Nature is accustomed. The primary difference between a long-term storage system and a conventional solar system is the elimination in the first of the auxiliary back-up system (the furnace) and the attendant components which are required as they interface. Compare the flow diagram of such a system (see Figure V·D·16) with those for some other systems. A heat pump might use this long-term storage as a heat source (see the next chapter); or, if the heat in the large storage "tank" is of sufficiently high temperature, it can be used by buildings directly, e.g., with radiant panels or forced hot air.

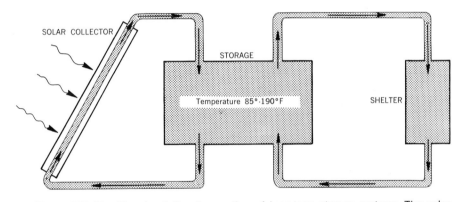

Figure V·D·16. The simplicity of operation of long-term storage systems. The solar system collects and stores heat year round when the sun shines. The building uses the heat when required. No auxiliary backup (fossil fuel) is used.

The money which is saved by the elimination of the back-up system can be utilized in the construction of the long-term storage container, and since 100% of the heating needs will be met by solar energy (except for electricity required to operate fans and pumps), a higher initial cost can be justified. For example, a solar system with 500 square feet of collector and 1500 gallons of storage may cost $5,000 to $7,500, including the conventional furnace (for backup heat when there is no sun). Such a design might save the equivalent of 500 gallons of fuel oil each season, or 50% of a total heating load of 1000 gallons. The furnace system itself might cost $1,500. By putting this money into a long-term storage system (eliminating the furnace), and perhaps spending an additional $1,000–$2,000, 1000 gallons, or 100% of the total heating load, might be saved. The total expenditure for the system might then be $6,000 to $9,500, but the only heating expense would be in operating the pumps for the collector.

H. C. Hottel reported that an analysis of MIT's first solar house in 1939 showed the collection of summer heat for use during the winter to be uneconomical (see Figure V·D·17). The house had a 17,000 gallon water tank for storing the heat. About two feet of insulation was used around the tank. The temperature of the tank ranged from 125 to 195°F, and no auxiliary heat was needed for the two seasons of operation. The cost, however, was excessive. It is possible that these economics have changed and that they may be more favorable for larger projects than for single-family residences.

E. A. Allcut and F. C. Hooper used a thermal computer model to show that for a house in Toronto which had a heating load of 775 Btu/°F/hr. (18,600 Btu per degree day), a heat storage tank of 50,000 imperial gallons would vary in temperature from 140° down to 80°F over the course of a heating season (see Figure V·D·18). The storage volume is equivalent to the volume of the space being heated.

The Solterra Home, invented by William B. Edmundson in 1966, uses a roof-mounted collector through which air moves and is heated (see Figure V·D·19). The heated air is circulated through 4 inch diameter tubes (e.g., glazed soil pipe or asphalt impregnated paper), which are buried in a heat storage bin under the house. The bin is built with concrete walls, floor, and cover, and is filled with water-saturated loam, sand, gravel, or even crushed rock. Heat might be stored in sufficient quantities so that no additional heat, solar energy or fossil fuel, would be needed for many weeks. If this were the case, solar collectors could be sized to provide all the heating for a house, and an auxiliary heating system would not be needed.

In his design, Edmundson assumed that wet soil weighs 100 lbs/cu ft and has a specific heat of .44 giving it a heat storage capacity of about 44 Btu/cu ft/°F. If the soil were heated from 80° to 130°F, it would store 50 × 44 = 2200 Btu/cu ft. Edmundson's bin is about 8,800 cubic feet; he uses about 2,000 linear feet of tubing, resulting in a total heat transfer surface between the tubes and the soil of about 2,800 sq ft. Under the above conditions, the bin would store about 19 million Btu. If a house had a heating load of 15,000 Btu per degree day, the

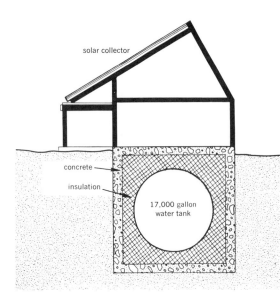

Figure V·D·17. First solar house, MIT, Cambridge, Mass., 1939. The heat storage tank temperature ranged from 125°–195° and no auxiliary heat was needed for two seasons.

Figure V·D·18. Cross-section of a 100% solar heated house, Toronto, Canada. A thermal model showed storage temperature varied from 80° F at the end of March to 140° F on September 30. House heated load was 775 Btu/°F/hr.

outdoor temperature could average 30°F for about 40 days before the 19 million Btu would be depleted (assuming no heat loss from the bin).

During the summer, heat is collected and stored in the bin. The heat is then "mined" by a heat pump as required to raise the temperature from perhaps 140°F to about 250°F, enough to operate an absorption-type air conditioner.

Ernst Schönholzer, an engineer in Switzerland, looked at the problem of long-term storage in 1969, but on a large scale and without economic analysis. His particular interest was in reducing city pollution during the winter, but his recommendations would also reduce fossil fuel consumption. Below is a summary of his findings (SCH.).

For A Single Dwelling

- Assume that a 195°F temperature can be attained in the storage; begin the solar heat collection on April 1 at an 85° storage temperature and continue until October 1, reaching 195°.
- Assume that the storage temperature drops back to 85° at the end of the heating season (4,380 hours).
- Assume a seasonal heating demand of 40,000,000 Btu (MIT House IV used twice this amount, including domestic hot water).
- The required heat storage is a cylindrical water container with a diameter equal to its height, 20 feet (6000 ft³ and about 48,000 gallons of water).

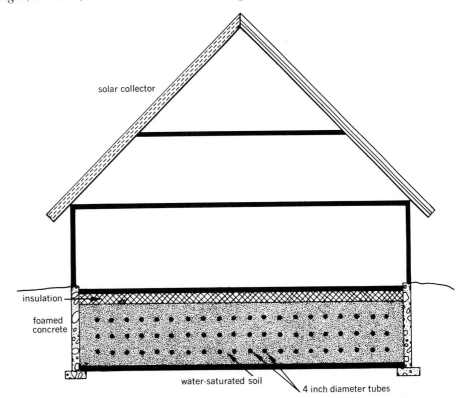

Figure V·D·19. The Solterra home by William B. Edmundson.

For 100 Apartments

- Make the same assumptions as above:
  40,000,000 Btu/apartment/season, 4×10^9 Btu total for 100 apartments;
  the storage temperature is 85° April 1 and 195° October 1.
- The required heat storage is a cylindrical water container with a diameter
  equal to its height, 95 feet (600,000 ft³ and about 4,800,000 gallons of water).

Results

Schönholzer computed that if the heat were not used during the heating
season, the temperature of the small tank would drop to 170° and that of the
large tank to only 185° (for the period October 1 to April 1). He made no
economic analysis but suggested that the larger installation would probably be
more economical.

Additional Ideas on Collecting Heat for Long-Term Storage

- The collectors might have to be more expensive (e.g., the focusing type) to
  attain temperatures of 195°. However, each square foot of collector will be
  saving larger amounts of fuel than if it operated only in the winter.
- The collectors could operate year round, not just in the summer, to furnish
  heat to the storage tank even when it is being used to heat the building. This
  might result in the use of a smaller tank that Schönoholzer suggested, or,
  instead, domestic hot water could be added as part of the total demand.
- MIT House IV collected 350 Btu/sq ft per day during the winter. It is not
  unreasonable to assume that this figure could double during the summer
  collection period. If heat were to be collected only in summers, collection
  might average 126,000 Btu/sq ft per summer. To collect 40,000,000 Btu (an
  apartment's heating load), the collector would have to be about 400 sq ft, less
  than normally required. Collectors might be able to be horizontal on flat
  roofs, instead of tilted.
- Projects might have to be completed in the spring or early summer so that
  the large storage tank would have time to be heated to capacity before
  winter.

CHAPTER V·E

Auxiliary Heating

Perhaps one of the most frequently asked questions among people who wish to understand the use of solar energy for heating (or for any other use) is: What do I do when the sun isn't shining? When they understand the concept of energy storage, the next question is: But what do I do when there is no longer any heat energy left in storage? The question is not a trivial one, and the need for a back-up, often conventional, system is a major stumbling block to widespread acceptance of solar energy as an alternative to present energy sources.

The relative infrequency of unusually cold and cloudy weather does not really ameliorate the problem. If the solar system were insufficient to carry the building through a period of cold, sunless weather, the consequences, even once during a winter, could be serious enough for a full-sized conventional heating system to be necessary as a back-up. In many instances, such as small homes in low-population areas with healthy people who can take care of themselves, those few times could probably be handled by wood stoves or small plug-in electric heaters. However, most solar-heated buildings need a full-sized backup system. At this time, solar energy must be viewed as a means of decreasing the consumption of oil, coal, gas, or electricity rather than as a complete substitute.

Full-sized furnaces are appropriate backups, but so are many other alternatives. Instead of using electricity, oil, or gas, backup furnaces might use methane gas generated from organic wastes or hydrogen gas which could be generated by alternative energy-producing methods such as windmills or ocean thermal gradients. Windmills could also be used to generate electricity through a resistance coil to produce heat inside a heat storage container (similar to what a standard electric immersion heater coil would do in the same situation). Electric space heaters could be used if the need for them were occasional rather than continuous. Other "space heaters" include fireplaces, wood furnaces, and the wide variety of wood stoves, which range in efficiency from 15 to 75%. Heat pumps, another possible backup, are discussed in the next section.

Suppose, however, that we wished to make the solar system large enough to provide for all extremes of weather. Since the combination of very cold days and long periods of cloudy weather is an occasional occurrence, the extra solar energy system (collector and storage) size which would be required for handling these occasions would be expensive for the relatively small amount of fuel saved. In addition, the system would be working well below capacity most of the time.

The example diagram (Figure V·B·65) and table (Figure V·B·67) in the section on "Collector Sizing" show that a solar system designed to provide for 50% of the heating load of a house may provide enough heat for only one day of intensely cold weather. If the solar system were doubled in size (up to 1000 sq ft in the example), it would provide for two intensely cold days without sun. During periods longer than two days, the second increment would be just as inadequate as the first. In addition, there will be times of mild weather when this second increment will not be needed.

Now, if the system were enlarged to 1500 sq ft to provide for three cold and cloudy days, it would theoretically be large enough to collect enough heat over the course of the winter to provide for half again as much as the entire heating demand of the house. But of course, this would not occur in actual practice since there would occasionally be four (and sometimes more) consecutive days of cold cloudy weather. To provide for this fourth day, we would need a solar heating system which could theoretically collect twice as much heat as would be needed by the building over the course of the heating season. It is clear then that no matter how large the solar system is, there might be periods of time when the weather is colder and cloudier than the most extreme conditions which had been planned for in the design. The larger the collector, the less intensely each additional increment in size is used, the less energy is saved per square foot of collector surface, and the less is the return on the investment in each additional square foot.

Valiant efforts have been made, however, in trying to provide enough solar energy to supply the total heating demand and to eliminate the auxiliary heating system. MIT achieved this goal with their first solar house back in 1939, but they did it for other reasons, and they did it by collecting and storing heat year round in a huge water tank. With the possible exception of systems such as Harold Hay's flat-roofed solar house in California, long-term storage is probably the only real alternative to an auxiliary system (see section in previous chapter on "Long-term Storage"). In their 1954 Desert Grassland Station, Donovan and Bliss took an existing building and provided a solar system which carried the house through five days of cold cloudy weather. The system was expensive, and the climate is relatively mild in Amado, Arizona. Harry Thomason came close to 100% solar heating with his first house in Washington, D.C., in 1959; only 5% of the heating load was provided by his full-sized oil furnace.

If the auxiliary provides only a small percentage of the total load, electric heating might be considered in spite of the fact that it requires considerable amounts of energy at the power plant to produce the heating energy for the building. (10,000 to 13,000 Btu are burned at the power plant to produce 1 kwh (3400 Btu) at the building. An oil furnace at 65% efficiency burns about 5200 Btu to achieve the same result.) An electric resistance heater would be cheaper than an oil or gas furnace in most cases, and the relatively small amount of electricity required to heat the building might justify its installation. It might also be a less energy-intensive installation because of the relatively small amount of material (compared with a furnace) which is needed to manufacture the electric coils. Unfortunately, such a system may not relieve the problem of peak demand, since, during the winter, this will usually occur during cold cloudy days and nights when solar heat has been depleted.

Since the efficiency of a solar collector is greatly increased by operating it at low temperatures, the heating system should be designed to use temperatures as low as possible, even 75° or 80°F. One of the virtues of the Thomason warm air system is that it continues to extract useable heat from the heat storage at temperatures almost as low as the temperature of the house. The MIT House IV had a water-to-air heat exchanger which used heat at water storage temperatures as low as 84°F. Steam, at more than 212°, is a relatively high-temperature heat delivery mode, compared to hot water heating (ranging from 90° to 160°) or warm air heating (ranging from 80° to 130°). Trying to tie into steam heating systems with solar heat, therefore, is usually impractical. If steam or high-temperature hot water systems are used, the heat from the solar system should be provided to the spaces through an independent system, such as forced warm air or lower-temperature hot water radiation. This is particularly true for existing buildings, and most large buildings have high-temperature systems.

In new construction, heating systems can be designed to use lower temperatures, for example, by using more linear feet of hot water finned-tube radiators, larger radiant panels, or larger volumes of lower-temperature air. The choices most often made by designers are forced warm air or larger-than-usual radiant water panels. A system of forced warm air makes best use of low-temperature storage heat. Radiant heating panels have a long time lag (between switching on the system and the warming of the air space) and usually need higher operating temperatures than hot air systems. Heat from the storage unit, therefore, is not useful at the lower temperatures that are useable by warm air systems, and the overall efficiency of the system is less. The necessary oversizing of a radiant panel system to produce similar results with forced air can result in a considerable additional expense.

In order to increase overall system efficiency (the solar system and the auxiliary backup) and simultaneously to reduce the total cost by eliminating redundancy of components, many designers have chosen to integrate the solar collector and storage with the auxiliary. Blowers, pumps, heat exchangers, controls, piping, and ducts are among the components which are shared. The figures in Chapter V·G ("Systems Design") represent various arrangements for such systems. The pitfall in designing the interface between the systems is adding more controls and movable parts, thereby increasing the likelihood of a mechanical failure. The temptation to try to achieve a one or two percent greater efficiency by adding another control and system interface is almost irresistible and may be the most common cause of solar heating system failures. (This is also one of the main reasons for this book's emphasis on low technology solutions.) Usually, the auxiliary heater should not heat the solar heat storage compartment. If it does, the storage will always be up to or close to useable temperature, and the solar collection phase of the cycle will be less efficient because it will almost always be operating at the higher temperatures of useable heat. In other systems, the drop in storage temperature resulting from the use of the heat by the building actually adds to the overall system efficiency.

Additional inefficiencies of this scheme result from the higher heat loss from the storage due to its continuously high useable temperatures. With systems in which the auxiliary is not heating the storage, it will lose considerably less heat when the sun has not shone for several days and the tank is relatively cool. Even in systems which have been designed in this way, heat losses from the container have ranged from 5 to 20% of the heat collected by the solar system. With an auxiliary-heated storage, the heat loss would be considerably higher and might be justified only if the storage container were within the heated building space.

Heat Pumps. The heat pump is used as a combination back-up and booster for some solar energy systems. (Its operation is explained in detail in the Resource Section on "Heat Pump Principles.") Although it is also used for cooling, in the heating mode the heat pump is basically a refrigeration device working in reverse: it takes heat from one location and delivers it at a higher temperature to another location. The first location is cooled in the process, and the second is warmed.

On the average, for every two Btu which it takes from a location, one Btu is required to operate the machine; the pump delivers the sum (three Btu) to the space to be heated. The performance ratio is commonly defined as the ratio of the heat delivered to the heat required for operation. This is called the Coefficient of Performance (COP) and would be 3 in this case. COP's usually range between 2.5 and 3 for heat pumps. Straight electric resistance heating, on the other hand, delivers only one Btu of heat for one Btu expended, or COP = 1.

The primary advantage of the use of heat pumps for space heating in combination with solar energy is that the stored heat is useful over a much wider range in temperature. A forced warm air system without a heat pump may use storage temperatures from 75° to 80°F (Thomason's system), up to perhaps 125° to 140°F; the range for a hot water system would be slightly higher, 90° to 140°F. With the addition of a heat pump, however, heat storage temperatures as low as 40° are boosted to higher temperatures (90° to 140°F). This results in an effectively larger heat storage container, and since the heat storage is often at low temperatures, the overall efficiency of the solar system is increased. In addition, the system will operate to collect heat when the collector temperature is as low as 50° or 60°F. This increases the amount of energy which each square foot of collector can save each year, more easily justifying its high initial cost.

When there is insufficient heat in solar storage (temperatures below 40°F), the source of the heat pump can become the outside air (in the case of an "air-to-air" or "air-to-water" heat pump). When outside air temperatures drop below 40°F, electric resistance coils switch on to provide heat, and the system becomes, in a sense, an electric furnace.

The heat pump can be used in reverse during the summer for conventional cooling. For designs in combination with solar systems, it can operate during the night, cooling the heat storage for the next day. This makes it possible to use electricity at night during low demand periods instead of in the daytime during peak demand periods. This can ameliorate the peak demand load on the power plant but does not reduce the overall energy consumption of the cooling process.

Cooling can also be accomplished by using the heat pump to cool the building during the day, warming the heat storage in the process (using the storage as the heat sink). The heat is discharged at night through the collector, which then acts as a "radiator" instead of as a collector, radiating heat to the cold night sky. Collectors which are used for this purpose, *nocturnal radiation,* work best if they slope northward. More critical than their orientation, however, is that they have no transparent cover plate. All such covers obstruct some of the outward radiation of heat; glass performs this task so well that it blocks radiation almost 100%. Absorber plates acting as night radiators should be coated with a flat black of high radiational capability rather than with a selective coating which reduces emittance.

Although the heat pump will usually move two or three times more heat than is required for input into its operation, the difficulty in justifying its use is that, in generating electricity, three times more energy is expended at the power plant than is expended by the heat pump. When 11,300 Btu are burned at the power plant, 3,400 Btu (1 kwh) are produced for use by the heat pump to move 7,000 Btu from the heat source to the heat sink. Total heat into the sink, then, is 10,400 Btu (3,400 and 7,000), while 11,300 Btu were used at the power plant. Therefore, the real overall Coefficient of Performance of the heat pump is closer to one, instead of three. When the electricity used by the fans and pumps of the solar system is added to the energy which is required by the heat pump, the total overall energy consumption is often equivalent to that of an oil- or gas-fired system without solar energy.

Solar projects which use heat pumps vary in the amount of electricity used to supplement solar heat. For example, in a laboratory at Nagoya, Japan (by Government Industrial Research Institute, 1958), the heat pump used 1,370 kwh (4.7 million Btu). This required 15.5 million Btu at the power plant, but the solar system itself delivered only 14.5 million Btu. The total heating demand of the building for the season was the sum of the solar and the heat pump energy (4.7 + 14.5 million = 19.2 million Btu).

Masanosuke Yanagimachi's second solar house (1958) required 64 million Btu over the course of the season. The solar system furnished 47 million, but since the heat pump required 5,580 kwh, 63 million Btu were expended at the power plant, almost as many Btu as were required by the house.

Donovan and Bliss did somewhat better than this in their laboratory in Phoenix (1959). The total heating demand of the building was about 35.5 million Btu. More than 30 million of this was supplied by the solar system, and approximately 17 million Btu were burned at the power plant to furnish the heat pump with the 1,470 kwh which it used.

The solar heated and cooled office building by Bridgers and Paxton in Albuquerque (1956) probably showed the best performance in this regard. The total heating load was almost 170 million Btu, of which 145.6 million were supplied by the solar system; only 44 million were expended at the power plant to produce the 3,900 kwh used by the heat pump. The use of heat pumps, then, is not a straightforward design choice for use with solar energy. A complete energy consumption analysis for each month of the heating season is necessary to determine the electrical energy which is needed by the heat pump to supplement the solar energy. Included in the analysis must be the amount of energy required at the power plant to produce the electrical energy used by the heat pump.

CHAPTER V·F

Solar Cooling

Although the main emphasis of this book is on heating, solar cooling deserves some discussion. The cooling of buildings represents only a small percentage of our national energy consumption. In fact, energy used for air conditioning of residential structures is about one-tenth that for space heating. However, the demand for cooling energy is increasing at a much higher rate than that for heating. Energy demand for cooling is also somewhat regional, and in some areas, such as the Southwest, more energy is used for cooling than for heating.

Solar cooling has some inherent advantages over solar heating. Cooling demand is more in phase with the shining of the sun, both annually and daily. Since solar radiation is the most important factor in the determination of outdoor temperature, the hottest seasons of the year usually occur during periods of greatest solar intensity. Similarly, on a daily basis, the hours of highest temperatures correspond closely with highest solar radiation levels.

Other factors, such as wind and humidity (and internal building heat loads), affect cooling requirements and can alter the correlation between solar intensity and cooling demand. Solar cooling systems are usually designed with the assumption that peak cooling loads may not necessarily correspond to peak solar intensities.

Summer peak electrical demand often corresponds closely to peak solar intensity, which causes peak cooling demands; thus peak electrical loads can be drastically reduced when solar cooling is widely utilized. Even though the energy used for summer cooling in residential and commercial buildings is less than 3% of the total national annual consumption, it is 42% of the summer total for these buildings.

The two principal methods for lowering air temperature for space cooling are refrigeration (actually removing energy from the air) and evaporation cooling (the vaporization of moisture into it). Dehumidification is also used. Evaporative cooling occurs when water vaporizes; the evaporation of perspiration from the skin is one of the mechanisms which our bodies use to remain cool. Evaporative cooling for buildings works best in dry climates, e.g., the Southwest.

Removing moisture from the air, dehumidification, is a cooling method used in humid climates. Often referred to as absorption dehumidification, this method uses hygroscopic materials called desiccants to remove most of the moisture from the air with which they come in contact. Solids, such as silicon gel, or liquids, such as triethylene glycol, are used as desiccants. In order to reuse the same desiccant material, heat is used to evaporate the absorbed liquid. Solar energy is being investigated as a source of heat for this regeneration (drying) process. If the dehumidification cooling effect is insufficient, the dried air can be partially reduced in temperature in an evaporative cooler. Figure V·F·1 is a diagram of the use of solar energy for absorption dehumidification.

Electrically driven vapor compression is the most commonly used cooling system. It can be employed for conditioning air which enters under almost any condition; in most parts of the country, it can also provide a level of comfort higher than that of other systems. Most through-the-wall air conditioners are small versions of this type. Investigations are being made to employ a solar driven Rankine cycle engine directly coupled to the compressor shaft. Concentrating collectors have been used to obtain 600°F temperatures to produce steam, which drives a steam turbine, which in turn powers a conventional compressor of a compression refrigeration cooling system.

The other form of vapor compression can work without electricity and is employed in the gas-actuated absorption refrigeration cycle. This is the cycle used by gas refrigerators and air conditioners, and is described in detail in the Resource Section on "Absorption Cooling Principles."

The lowest temperature generally used in the operation of absorption refrigeration systems for space cooling is about 185°F. Even this seemingly high temperature is actually well below the standard temperatures of 250° to 300°F. Figure V·F·2 diagrams a system in which solar energy is used as the source of heat in the distillation process. The solar collectors need to operate 15° to 20°F warmer than the operating temperature of the absorption unit. Collector efficiency, of course, drops off sharply at these high temperatures. Extensive technological development is being done on the design of cooling equipment

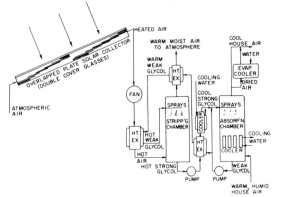

Figure V·F·1. Solar cooling by absorption dehumidification. Source: (LOF.3)

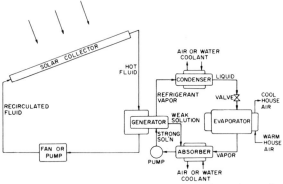

Figure V·F·2. A diagrammatic illustration of a solar operated absorption refrigeration cooling system. Source: (LOF.3)

which can operate at lower temperatures and on solar collectors which maintain high efficiencies at high operating temperatures.

Dr. George Löf has shown (LOF.3) that if water at 210°F were pumped from the collector to the generator, it would transfer heat at about 180° and return to the collector for reheating at 200°. On a 90° day, a square foot of collector might deliver up to 900 Btu, corresponding to about 40% overall collector efficiency. The resultant cooling effect would be about 450 Btu. A collector of 600 sq ft could provide a cooling system with a day-time heat removal capacity of about 30,000 Btu/hr, equivalent to about two and one-half tons of refrigeration.

Because of the high performance required, solar collectors designed for such cooling systems will undoubtedly be more expensive than those which are used for winter heating. However, if the same collector can be used for both purposes, the higher cost may be amortized within a reasonable period of time.

Collectors which are being considered for summer cooling include concentrating types and flat plate types. The concentrating collectors have the limitations described earlier but are able to provide high temperatures at good efficiencies, given the proper climatic conditions (a high percentage of direct solar radiation).

Flat plate collectors with selective absorber surfaces must use at least one transparent cover and up to three or more. If glass is used, it must have high transmittance and low reflectance. Most absorption equipment requires that the collectors be of the liquid, rather than the air, type. It is less likely that air-type collectors will be developed for use by absorption cooling equipment.

Collectors which operate only during the summer should have a much shallower tilt angle than that needed by collectors which operate only during the winter, or, for that matter, which operate year round. The shallower tilt is probably more compatible with conventional roof slopes, and in fact, may even be accommodated by flat roofs.

Night (nocturnal) sky radiation, discussed further in the previous Chapter and in Resource Section A on "Heat Theory," is a cooling process which occurs when objects radiate their heat to a cooler night sky. Dry climates, particularly those with warm days and cool nights, are the most suitable locations for this cooling method because of their low sky temperatures and clear atmospheres. Air or water can be cooled as they circulate past a surface exposed to the night sky.

One of the first examples of nocturnal cooling was devised by Donovan and Bliss at the Desert Grassland Station in Amado, Arizona in 1954. This project utilized a black gauze cloth stretched across an opening in the ground. Cool night air became cooler as it was drawn through the cloth radiating to the night sky; the cooler air was blown through the rock pile used for the storage of solar heat during the winter.

Dr. Harry Thomason used the same principle on his first solar house in Washington, D.C., in 1959. Water from his winter solar heat storage tank was pumped to the ridge of the north facing roof and cooled by night sky radiation as it trickled down over the surface of black shingles. The coolness was stored in his large water tank for use during hot days.

Harold Hay's house in Atascadero, California, is cooled without the use of moving parts, except for the sliding of the movable horizontal insulating shutters which open at night, allowing the water-filled bags to cool as they radiate to the sky. The shutters are moved back into place during the day to trap the coolness underneath and to ward off the intense heat of the sun.

Solar cooling can also be accomplished, in a sense, through the proper use of the storage designed for winter solar heating. In some instances, cool outdoor night air can be used to cool the storage for use during the day. It can be cooled further by using a small refrigeration compressor. If the compressor operates only at night during off-peak, its size may be as small as one-half that of the compressor which operates during the peak cooling times. Compressor size can be reduced even further if it can operate continuously, cooling the solar heat storage when cooling is not needed by the space. Both of these methods are unlikely to decrease operating costs but may reduce initial cost.

Dual storage systems can be used in combination with heat pumps for summer cooling. Heat pumps can transfer heat from one storage to another, cooling the first and heating the second. The coolness is used for the building, and the heat from the second is discharged to the outside environment, either mechanically or naturally by nocturnal radiation.

CHAPTER V·G

Systems Design

Up to this point, the components of solar space conditioning have been discussed individually; the next logical step is putting them all together into a system. The many preceding allusions to systems integration will allow this discussion to be concise; in any case, the more intricate details of heating, ventilating, and air conditioning will be left to the mechanical engineers who can make design decisions according to the specific requirements of a project.

The complexity (but not necessarily the sophistication) of solar space conditioning systems increases with the following inputs:
- The scientific and technologic training of the designer
- The development and promotion of new products
- The level and precision of comfort which the user requires of his/her environment (the setting of a thermostat in contrast with the stoking of a wood stove)
- The lack of participation which the user exercises in the control of his/her environment (if a person cannot control his/her environment, then a complex central control system may need to)
- The increasing number of tasks which the system must perform
- The increasing size of the project
- The increasing overall efficiency which must be achieved.

The simplest uses of solar energy have been discussed earlier. A south-facing window combined with the building's thermal mass and an insulating shutter is potentially the simplest and most sophisticated solar heating system. Thermosiphoning air collectors or solar hot water heaters are also relatively straightforward. Harold Hay's Skytherm, Steve Baer's drumwall, and the Trombe-Michel concrete wall fall into this same category. Simple systems may not necessarily be the most efficient (although they often are), but over the long term they are more likely to use fewer natural resources and less energy in their construction, operation, and maintenance.

Apart from the examples above, the simplest solar heating system makes use of collectors which operate only when the sun is shining and the building needs heat. Such collectors can be set up in a yard during the winter and dismantled during the summer. They can also be attached to the walls and roofs of existing buildings. In any case, air from the building is ducted from the building to the collector, heated by the sun, and blown back into the space. The fan operates on the basis of a comparison between two temperatures. A sensor determines whether the sun is shining and whether the collector is sufficiently hot to heat air; a second device senses whether or not the space needs heat. This sensing device should operate on an upper-limit thermostat, that is, solar heat should be delivered to the space until it reaches a higher-than-usual temperature, in order to take advantage of the sun's energy when it is shining. (This process, of course, can also be done manually with a simple on/off switch for the fan.) Since there is, in this mode of operation, no provision for storage of the heat for later use, the building must act as the heat storage container. It should thus be heated to as high a temperature as can be tolerated by the occupants. The more massive the building, the more heat it can store, the longer the building can go without heat after the sun stops shining, and the better the overall efficiency of this simple system. Earth-covered and underground buildings with insulation located between the concrete and the earth are very compatible with these simple systems since the large amount of concrete stores the heat.

The system at the next level of complexity delivers solar heat to solar heat storage. When the space needs heat while the sun is shining, the back-up heating

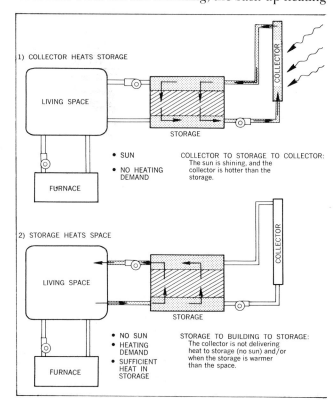

system comes on. Ideally, however, solar heat gain through windows should satisfy the heating demand of the building while the collector is in operation. The back-up heating system is completely separate from the solar heat collection and distribution system in order to reduce the overall complexity. When the sun is not shining and when there is heat in storage, the call from the house for heat goes first to solar storage. If that is insufficient, the back-up heating system comes on. A summary of the four modes of operation are shown for an air system in Figure V·G·1. The component shown as the "furnace" is independent of the solar system and can use oil, gas, electric, wood, or any other energy.

Systems increase in complexity as an attempt is made to integrate all of the components. For example, in large installations it is generally undesirable to have two separate heat delivery systems, one for the solar and the other for the back-up. Integrating these two into one ducting system (for forced warm air systems, for example), causes dampering and other control complications but can reduce the initial cost of heat delivery.

Figure V·G·2 is an example of a single heat delivery system using air as the heat transport fluid. If, after a period of about 10 minutes, the heat from the collector storage cannot satisfy the thermostat, the furnace turns on. When the collectors are operating, heated air goes to storage unless the building needs it.

The various dampers which serve the different modes of operation are often in close proximity to each other and are subject to strong fan pressures. They should be designed so that air pressure forces them to close tightly in order to

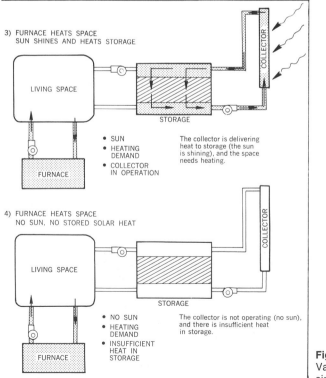

Figure V·G·1.
Various operating modes of a simple air system.

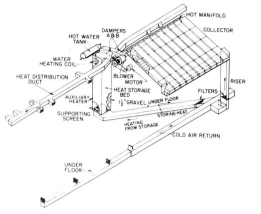

Figure V·G·2.
Solar heating system for
"Denver Design" house. Source:
(LOF.2)

prevent leakage.

Primarily because of their cost, and notwithstanding the need for simplicity of operation, the amount of ducting and the numbers of dampers and damper motors should be reduced as much as possible. Efforts should be made to design each damper so that when a particular mode of operation calls for it to be closed or opened, the movement of the air caused by the triggering of the fan for that mode will operate it accordingly. However, if motors are required to operate dampers, they should be juxtaposed so that one motor operates several dampers. This usually means that some of the ducting must be side by side at the damper locations and that the dampers must be arranged in a line to pivot along the same axis. Damper motors can open one duct and simultaneously close another.

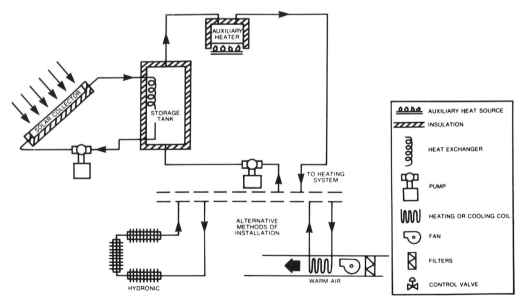

Figure V·G·3. Piping system design for solar space heating. (Drawing courtesy of Revere Copper and Brass.)

As discussed in Chapter V·B on "Flat Plate Solar Collectors," liquid systems are usually not as cost effective as air systems for heating single family homes. However, as the size of the project increases and as the need for solar absorption cooling increases, the choice of liquid systems becomes more probable. It is also somewhat easier to combine domestic hot water heating with a liquid system than with an air system.

Figure V·G·3 is a simplified diagram of such a liquid system. Two alternative methods are shown for the heating of the space. The hydronic system circulates the hot liquid directly through the building; the warm air heating system blows cool air from the space over heat transfer coils through which the solar heated liquid circulates.

The next system, Figure V·G·4, also preheats domestic water before it travels to the conventional hot water heater. Figure V·G·5 adds solar absorption air conditioning. The cool liquid circulates through coils, and warm building air is cooled as it blows across them. Excess heat from the equipment is exhausted through a cooling tower.

If solar space conditioning were as easy as it appears in these relatively simple diagrams, it would have been adopted long ago and would now be gaining favor even more quickly than it is. Unfortunately, the valves, control systems, heat exchangers, pumps, fans, pipes, ducts, and expansion tanks required for accomplishing the various modes of operation do not appear clearly in diagrams such as these. Figure V·G·6, for example, is from an excellent and elaborate article (WHI.2) about the plans for MIT's House IV in 1959. Note the numerous fluid loops, valves, heat exchangers, and controls. All of these components are discussed in detail in the article.

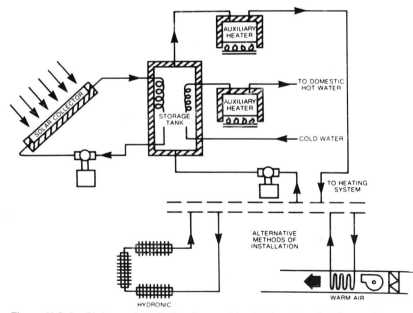

Figure V·G·4. Piping system design for combined solar space heating and domestic water heating. (Drawing courtesy of Revere Copper and Brass.)

HEATING, AIR CONDITIONING & DOMESTIC WATER

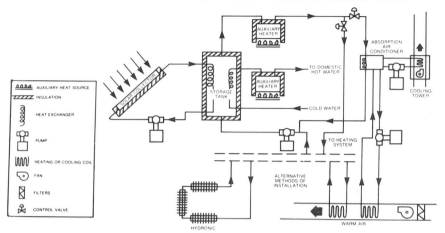

Figure V·G·5. Piping system design for combined solar space heating, air conditioning and domestic water heating. (Drawing courtesy of Revere Copper and Brass.)

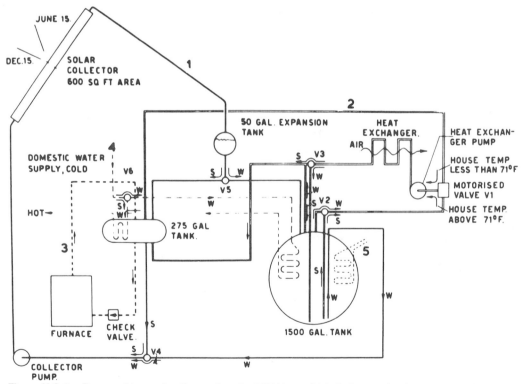

Figure V·G·6. Proposed house heating system for MIT House IV:1. Collector circuit, pump on when sun shines; 2. Heat-exchanger circuit, pump and motorized valve controlled by house thermostat; 3. Furnace circuit, flow by natural circulation whenever furnace is on—in winter furnace keeps 275-gallon tank between 135° and 150° F, in summer it is turned off; 4. Domestic hot-water circuit, the heating coil in the large tank is by-passed during the summer; 5. Evaporator (Freon) coil of the 3/4-ton refrigerator, used in summer only. Source: (WHI.)

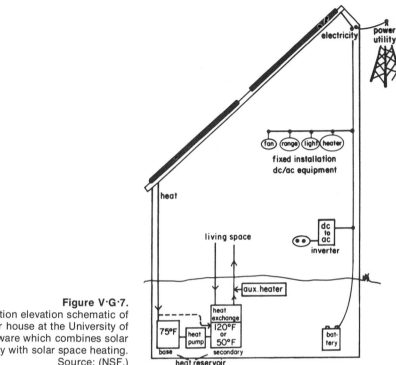

Figure V·G·7.
Cross section elevation schematic of
the solar house at the University of
Delaware which combines solar
electricity with solar space heating.
Source: (NSF.)

When solar photovoltaic cells become incorporated into building design, the system diagram may look something like Figure V·G·7, the solar house at the University of Delaware's Institute of Energy Conversion. Wind energy may also be incorporated into many such designs.

A word about heat exchangers and control systems is in order. According to Duffie and Beckman (DUF.), a heat exchanger increases the operating temperature of the collector by an amount corresponding to the temperature drop across the exchanger. For approximately each degree Celsius of temperature drop across the exchanger, the useful gain of the collector decreases by 1–2%. The fewer the number of heat exchangers in the system, the higher the performance of the collector (or the smaller the required surface area).

The control system required for delivering heat or cool to a building from the collector, heat storage, and backup heating (or cooling) system is determined by the final requirements of the building. In general, the thermostat calling for energy from storage can operate at a different temperature level than the conventional one which calls for heat from the backup system. For example, the solar thermostat might be set at 70°F and the standard thermostat at 65°. If the heat storage cannot maintain the 70° temperature, the backup system comes on when the temperature drops to 65°. Alternatively, a single thermostat can trigger the backup system if the storage system cannot maintain the temperature.

The controls to operate the collector are relatively simple and are readily available. For both air and liquid systems, the effect of solar radiation on collector temperature can be measured directly on the absorber or on a surface similar in

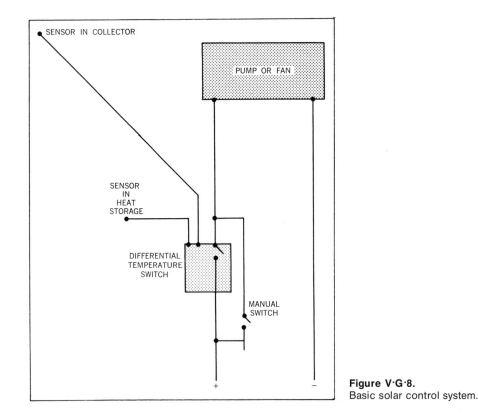

Figure V·G·8.
Basic solar control system.

thermal performance to the absorber but without metallic bond to it. Another temperature sensor is located near the exit of the storage unit(s). Customarily, the collector pump comes on when the collector registers 5°F warmer than the storage; for air systems, the differential may be as much as 20° before the fan is triggered. A time delay of about five minutes should be included in the control to prevent the system from turning on and off during intermittent sunshine. Once the system is on, another time delay insures that the system does not shut off as soon as the collection fluid from storage cools the absorber down to a temperature less than 5° above storage. Figure V·G·8 is a basic diagram of a solar differential temperature switch. For liquid systems, the control may need to have an upper limit temperature cutoff to prevent the pressure of the liquid within the piping system from increasing to uncontrollable levels.

A major consideration in the difficult task of optimizing solar system design is the comparison of performance and cost. Performance is measured by the amount of energy which can be saved by the system each year; cost is a comparison of the value of the energy saved with the extra costs incurred to save that energy.

The performance of several past solar systems are given in Figures V·G·9, V·G·10, V·G·11, and V·G·12. The first figure is for Dr. Löf's first house, built in 1945 in Boulder, Colorado. The total collector area of 463 sq ft saved approximately 20 million Btu during the first season. It is likely that refinement

Month	Degree days	Btu req. $\times 10^6$	Btu from gas $\times 10^6$	Cost of gas, $	Cost of gas without solar, $	Heating saving, %	Fuel cost saving, %
Sept.	55	0.83	0	0	1.03	100	100
Oct.	111	1.68	0	0	1.50	100	100
Nov.	945	14.20	10.65	7.71	9.91	25.0[1]	22.0[1]
Dec.	899	13.52	12.60	8.63	9.47	10.0[2]	8.9[2]
Jan.	772	11.60	9.10	6.66	8.38	21.6	19.5
Feb.	1143	17.23	15.34	10.79	12.63	11.0[3]	14.7[3]
Mar.	1027	15.38	11.40	7.96	10.03	25.9	21.0
Apr.	302	4.56	2.05	2.19	4.02	55.0	45.5
May	248	3.73	0.81	1.79	2.90	78.1	38.2
Total	5502	82.72	61.51	45.73	59.87	25.6	23.4

Winter cost of electricity for heating system operation is $14.29; computed cost of electricity without solar unit is $9.59.
[1] Blower operating improperly for one week.
[2] Controls not working properly—no stored heat was used during month.
[3] Solar system turned off for two-weeks test.

Figure V·G·9. Performance of solar heated house, Boulder, Colorado. Source: (LOF.2)

and smooth operation would save considerably more, perhaps as much as 75,000 Btu per sq ft of collector per season. Dr. Löf estimated his costs at that time at about $1.00 per sq ft of collector.

Figure V·G·10 shows the performance and operating costs of the solar space heating system at the Desert Grassland Station of Donovan and Bliss in 1954. They estimated that the 315 sq ft of air-type collector would save more than 14 million Btu during the season, or slightly less than 50,000 Btu per sq ft of collector. The system, however, was greatly over-sized, requiring no backup heat.

In 1959, Donovan and Bliss built a liquid system in Phoenix with 1623 sq ft of surface; it was designed both to heat and cool. During 1960, the system delivered 35.3 million Btu for heating and required about 5 million Btu for the operation of pumps, compressor, controls, and ventilating fans. Figure V·G·11 lists some performance data.

Also during the 1959–1960 heating season, the 640 sq ft of liquid-type collector at MIT's House IV collected just over 40 million Btu and delivered to the space as heat and to the domestic hot water about 34.5 million. Ten years earlier, their third house delivered between 60,000 and 70,000 Btu per sq ft of collector each season as compared with less than 60,000 for this house. Figure V·G·12 shows the cumulative values of the performance of the collectors.

	Jan. 21-Feb. 20	Feb. 21-Mar. 20	Mar. 21-Apr. 20	3-month Total	Full Winter Season (estimated)
Degree-Days.....................................	513	288	162	963	1800
Heat Supplied to House by Space-Heating System: (2.3 kwhr per Degree-Day)					
Expressed in Btu x 10⁻³....................	4030	2260	1270	7560	14,150
Expressed in kwhr........................	1182	663	372	2217	4140
Electrical Energy Consumption of Space-Heating System:					
Fans and Controls, kwhr.....................	139	93	84	316	600
Auxiliary Heating, kwhr....................	0	0	0	0	0
Operating Cost (2¢ per kwhr)........................	$2.78	$1.86	$1.68	$6.32	$12.00
Estimated Operating Cost of Comparable Butane System: Tank Rental (for 6 months—Summer Rental Assumed Charged to Other Appliances)... $12.00 Furnace Fan and Controls (2 kwhr per day).. 7.20 Butane (300 Gallons at 20¢ per gallon)...... 60.00					$79.20

Figure V·G·10. Performance and operating costs—space-heating system Desert Grassland Station. (Initial 1955 operation.) Source: (BLI.1)

	Average temperature of upper tank, F.	Average temperature of lower tank, F.	Electrical energy used by compressor, kWh	Total electrical energy used by heating-cooling system, kWh	Electrical energy used in building (instruments, shop, lights, etc.) kWh	Approx. energy removed from lower tank by compressor, therms	Approx. energy added to upper tank by compressor, therms
January. . .	92	89	264	517	332	28.7	37.4
February . .	102	102	8	192	279	0.9	.1.2
March . . .	94	59	19	304	293	2.0	2.6
April	48	49	0	164	270	0	0
May	58	58	113	363	280	12.9	16.8
June	82	65	1 035	1 451	317	113.0	148.3
July	88	65	1 261	1 707	318	133.0	176.0
August . . .	86	66	1 141	1 597	287	121.0	160.0
September. .	80	65	882	1 306	271	93.3	123.4
October. . .	73	57	103	408	313	11.2	14.7
November. .	90	54	35	320	341	4.0	5.2
December . .	94	93	93	346	294	10.5	13.7

Figure V·G·11. Miscellaneous performance data, Arizona Solar Laboratory—1960. Source: (BLI.3)

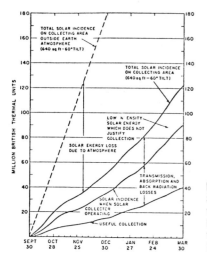

Figure V·G·12.
Solar collector performance during the winter season 1959–1960 for MIT House IV. Cumulative values for every week are plotted in million Btu. Source: (ENG.)

A WORD ABOUT SOLAR SYSTEM ECONOMICS

From the actual amounts of energy which past systems saved, it is clear that particular caution should be exercised in sizing a system so that the initial cost per square foot of collector is not inordinately high. When a conventional heating system is designed, the engineer does not need to be nearly as precise about sizing and initial cost since the cost of fuel represents the most significant portion of the fixed annual charges. A solar heating system, unfortunately, does not provide free heat; the initial capital investment is quite high and varies greatly because of the large number of components involved and the large number of design variations for each component.

There are numerous methods of evaluating cost. Just when the economic theoretician comes to a conclusion about a particular solar design, another variable, such as new tax incentives, rising fuel prices, or changing interest rates, alters the results. Also the inability to obtain fuel for a backup system may be far more important than the strict economic return on investment. A Federal call for decreasing the jobless rate can result in solar construction which may not necessarily be economically justifiable in the strictest sense. The drive for energy independence also alters economic evaluations.

Actually, the real questions to be answered by someone seeking an analysis of the economic justification for using solar energy are: How much energy will I save by using solar energy? What is the value of that energy to me? How much am I willing to spend now in order to save that amount of energy every year?

The above discussion of system performance attempts to answer the first question. A solar energy system will save from 50,000 to 250,000 Btu per sq ft of collector per year. The monetary value of that amount of energy varies greatly, depending on whether the energy saved is gas, oil, electricity, wood, or something else, and of course the value increases each year as the cost of energy rises. The last question is answered best by the person who has to spend the money.

The numerous issues of cost have been touched upon throughout the many sections of this chapter. The following remarks attempt to put some of these issues into perspective.

In spite of soaring fuel prices, the cost of building a complex sophisticated solar energy system is one of the factors inhibiting its wide adoption. In general, the more complex the system, the more expensive it is. Usually, but not always, higher overall seasonal efficiency accompanies the increased cost. This efficiency is here defined to mean the ratio of solar energy collected and used to the total insolation hitting the collector.

A seasonal efficiency of 60% is considered EXCEPTIONAL and requires engineering design and equipment which will result in an installed cost of $25 to $40 per sq ft of collector (costs are for the entire solar system, above the cost of the conventional system). Such a design might save up to 250,000 Btu per sq ft of collector per year, assuming twelve months of use and proper system design for the climate.

A 50% overall efficiency is EXCELLENT and would cost $5 to $10 per sq ft less than the EXCEPTIONAL system. Forty percent efficiency is ATTAINA-BLE, and will probably be achieved by numerous projects built in the 1970's. Many of these projects cost $15 to $25 per sq ft of collector. An efficiency of 30%, somewhat less than the 1959 MIT House IV, is relatively EASILY ATTAINA-BLE. Such systems cost $10 to $20 per sq ft.

A vertical, combination collector/storage performs with an overall seasonal efficiency of 25–50%. The cost per square foot is $5–$12 more than that of the wall it replaces. Windows vary considerably in performance, but can save up to 100,000 Btu per sq ft of surface if accompanied by a properly designed (and used) insulating shutter.

No discussion of the costs of solar energy can be complete without mentioning the intense computer work by Dr. George Löf and Dr. Richard Tybout. Their findings have been well-reported (see TYB., LOF.4, and LOF.5) and deserve the attention of anyone seriously interested in cost comparisons. Figure V·G·13 summarizes studies in which they compared the cost of solar heating and cooling with those for alternate fuels for eight different U.S. cities. The results are based on the least costly solar system design; the heating of domestic hot water is included. Fuel and electric rates in 1970 were used and adjusted for 56% oil operating efficiency and 67% gas operating efficiency. Solar costs assumed were

	Collector area (m²)	Storage (kg m⁻²)	% Load by solar				Cost combined ($/10⁶ Btu)	Cost combined ($/100 kW hr*)
			Covers	Cooling	Heating	Combined		
Albuquerque	48.4	48.8	2	56	73	63	1.73	0.59
Miami	96.7	48.8	3	58	100	60	2.13	0.73
Charleston	96.7	48.8	3	63	92	68	2.47	0.84
Phoenix	96.7	48.8	3	29†	100	33†	1.71	0.58
Omaha	96.7	48.8	2	57	60	59	2.48	0.85
Boston	96.7	73.2	2	66	64	65	3.07	1.05
Santa Maria‡	24.2	48.8	2	27	64	52	2.45	0.84
Seattle‡	48.4	73.2	2	39	44	43	3.79	1.29

Collector tilt = latitude (except Miami = Latitude − 10°).
Optimum criterion—least cost solar heat for *combined* use.
Hot water heating included.
Collector cost $21:50/m² ($2/ft²).
Storage cost 11¢ per kilogram (5¢/lb) of water.
Other constant costs $375 per system.
Air conditioner cost $1000 above conventional.
C.O.P. of absorption cooler 0.6.
Amortization 20 yr 8% interest.
Large house, heating demand = 0.55 $(t_a − 18)$ kW + 0.305 kW water heating.
*To convert $/100 kW hr to $/10⁹ cal, multiply by 11.63.
To convert $/100 kW hr to $/10⁶ Btu, multiply by 2.93.
To convert $/100 kW hr to $/10⁹ J, multiply by 2.78.
†Cooling demand for Phoenix based on building having walls with overall heat transfer rate approximately ten times the value used in all other locations.
‡One computation only.

Figure V·G·13.   Solar heating and cooling design for economic optima. Source: (LOF.5)

$2/sq ft for the collector, a 20 year amortization, 8% interest, and a $1000 surcharge for cooling. Other capital costs are about $375, plus about $1 per sq ft of collector.

Based on these assumptions, solar heating is cheaper than electric resistance heating in nearly all eight cities investigated. In addition, results of the studies are that the amount of heat storage above 1 gallon of water per sq ft of collector has almost no effect on the optimization of cost. Other cost optimization curves were also relatively flat, showing that tilt angle and system size are not as important for the determination of overall cost as is widely supposed. Of particular interest are their conclusions for collector size versus percent of heating load supplied by solar. For houses with identical thermal performance (a heat loss of 25,000 Btu per degree day), collectors of similar size provided 70% of the demand in Phoenix, Arizona, and 55% in Charleston, South Carolina. An increase in size of 65% provided only 40% of the heating load in Boston, however.

Computer Modeling. The intense interest in energy conservation and solar energy has been accompanied by the more extensive use of computers for purposes of modeling the effects of various design parameters. The complication of ever-changing weather combined with the increased complexity of heating, ventilating, and air conditioning systems for buildings has nurtured the development of computer programs for simulation work.

The cost analysis work by Tybout and Löf (TYB.) was among the first and most intensive uses of computer modeling in solar energy. Since then, the use of computers has greatly increased: work is being done by numerous groups, including the University of Wisconsin (their TRNSYS program); Colorado State University (their SIMSHAC program); the National Bureau of Standards; the University of Pennsylvania; Honeywell; and NASA. New and improved programs are being continually developed. For one of the most complete discussions of the use of computers in solar energy in 1974, see an article by J. L. Wilson in the *ASHRAE Journal* for January, 1975 (WIL.).

Nearly all major solar assisted buildings now being designed or under construction are being computer modeled. Several homes are also being modeled to obtain actual performance data for comparison with the performance projected by computer.

CHAPTER V·H

Separated Collectors

A separated collector is one which is not attached to the building which it serves. In the broadest sense, the term includes those collectors which are attached to buildings but not integrated into the exterior skin (roof or wall). Solar hot water heaters are commonly of this type. However, *separated collectors* will refer primarily to those which are used to heat and cool buildings and which are mounted on structures not attached to the building they serve.

Separated collectors are often used when an existing building is "refitted" for solar energy. Unfortunately, most buildings cannot accommodate a solar collector because they lack an appropriate roof or wall to which to attach it; many are shaded or face the wrong direction. Donovan and Bliss used a separated collector in their first solar house, the Desert Grassland Station in Amado, Arizona. The use of a separated collector allowed their design to be of sufficient size to provide all of the heat for this house.

There are several recent examples of separated structures. The National Science Foundation sponsored the refitting of five schools. Two of these projects—Fauquier County Public High School in Warrenton, Virginia, and North View Junior High School in Osseo, Minnesota—used separated structures. The structure at Warrenton is large enough to provide storage space and athletic locker rooms.

Under a U.S. Navy contract, Jerry Plunkett, of Materials Consultants, Inc., Denver, Colorado, designed and built a $15,000 solar heating system for two existing homes in Hawthorne, Nevada (see Figure V·H·1). The two collectors are each mounted on their own structures in the yard, away from the houses. The shallow collectors, each 15 x 45 ft are tilted at an angle of 55° from the horizontal. Two mylar (plastic) transparent sheets cover the blackened wire mesh heat-absorbing surface; air blows across the mesh and is warmed. The heat is transferred to water through heat exchangers and stored in two 2100 gallon tanks, from which it is transferred back to air in the house, with the help of a forced warm air furnace.

Separated collectors are also used in new construction. The 4500 sq ft collector for the Grassy Brook Village housing project in Vermont serves ten condominiums. The collector was detached from the houses for several reasons: this allowed greater architectural freedom; it made possible the use of windows in the south walls of the building for the direct utilization of solar energy; and it allowed flexibility in the sizing and design of the solar system.

The primary disadvantage of using a separate structure is the cost of building that structure, especially if it has no other purpose than to support the collector. The structure can be particularly expensive if it must be raised above the ground to prevent snow from covering it, or if it is thrust high into the air, acting as a giant sail, subject to large wind loads. The collector should be as close to the ground as possible, consistent with snow conditions, and inaccessible to vandals (but accessible for maintenance). The total surface area of individual collectors should be kept small to reduce the sail effect.

Figure V·H·1. Separated solar collectors for heating homes at the U.S. Naval Ammunition Depot, Hawthorne, Nevada, 1974. (Solar design by Jerry Plunkett, President, Materials Consultants. Photo courtesy U.S. Navy.)

Land area may be another major expense in some instances, especially in urban and suburban areas. However, since the collector does not serve as part of the exterior skin of the building, the connection between the individual panels does not have to be as water-tight, and this may reduce the cost of construction. For example, the glazing details of the transparent cover may be less critical, resulting in possible savings.

The overall efficiency of a separated collector will probably be somewhat less than a comparable collector attached to a roof or wall. There will be transmission line heat losses in the pipes or ducts between the collector and the building. These transmission lines are expensive, so the collector should be as close to the building as possible. The heat loss out the back of the collector (which is insulated but which still loses significant amounts of heat) will be greater because it will be exposed to cool outdoor temperatures instead of relatively warmer indoor temperatures. Instead of this heat being lost to the living space, it will be lost to the outdoors. In addition, the detached collector adds nothing to the overall insulating value of the facade of a building.

There are, however, compensations. The separated structure allows the construction of a collector which is larger than might be applied to a building. There is also greater freedom in the architectural design of the building. In fact, the south facade of the building can be used as a heat collector itself (through the use of windows and thermosiphoning panels) if it is not called upon to provide a substantial percentage of the total heat load. Many design programs do not easily permit south facades to be used as collectors or even as windows. In fact, many building sites may not even have sufficient sun to allow collectors to be integral with the building. For example, the site may be shaded by large trees, or by neighboring buildings or other structures. Or the site may be on a disadvantageous north, east, or west slope.

A separated collector also allows greater freedom in designing the collector. It permits innovative designs, such as inflatables and focusing devices. A separated collector makes possible the alternative of a collector which "tracks" the sun, that is, which varies in orientation and tilt angle: the collector may be mounted on a structure which swivels either automatically or manually as the sun moves from east to west. The collector itself may be mounted on the structure in such a way that it can vary its tilt angle, either automatically or manually, as the sun rises and sets in the sky, or from season to season, as the sun travels lower (in the winter) or higher (in the summer) across the sky. This would also allow the collector to be tilted vertically or even forward in such a way as to shade itself, thereby reducing the thermal stress on the collector when not in use. Movable collectors and automatic tracking devices are, of course, expensive.

REFERENCES FOR PART V

(ALL.) Allcut, E. A. and Hooper, F. C. "Solar Energy in Canada." *Proceedings of the United Nations Conference on New Sources of Energy*. Rome, 1961.

(ASH.) American Society of Heating, Refrigerating, and Air Conditioning Engineers (ASHRAE). *ASHRAE Handbook and Product Directory*. New York, N.Y., 1975.

(BHA.) Bhardwag, R. K.; Gupta, B. K.; and Prakash, R. "Performance of a Flat-Plate Solar Collector." *Solar Energy* 11(1967):160–161.

(BIR.) Bird, R. B.; Stewart, W. E.; and Lightfoot, E. N. *Transport Phenomena*. New York City: John Wiley & Sons, 1960.

(BLI.1) Bliss, R. W., Jr. "Solar House Heating—A Panel." *Proceedings of the World Symposium on Applied Solar Energy*, Phoenix, Arizona, 1955.

(BLI.2) Bliss, R. W., Jr. "The Derivations of Several 'Plate-Efficiency Factors' Useful in the Design of Flat-Plate Solar Heat Collectors." *Solar Energy* 3(1959):55–64.

(BLI.3) Bliss, R. W., Jr. "The Performance of An Experimental System Using Solar Energy for Heating and Night Radiation for Cooling a Building." *Proceedings of the United Nations Conference on New*

Sources of Energy. Rome, 1961.

(BUE.) Buelow, R. H. and Boyd, J. S. "Heating Air by Solar Energy." *Agricultural Engineering,* January 1957, pp. 28–30.

(CLO.1) Close, D. J. "The Performance of Solar Hot Water Heaters with Natural Circulation." *Solar Energy* 6(1962):33–40.

(CLO.2) Close, D. J. "Solar Air Heaters for Low and Moderate Temperature Applications." *The Journal of Solar Energy Science and Engineering* 7(1963):117–124.

(CLO.3) Close, D. J. "Rock Pile Thermal Storage for Comfort Air Conditioning." *Mechanical and Chemical Engineering Transactions of the Institution of Engineers,* Australia, May 1965, pp. 11–12.

(DAV.) Davis, Albert J. and Shubert, Robert P. *Natural Energy Sources in Building Design.* Blacksburg, Va.: Virginia Polytechnic Institute, Department of Architecture, 1974.

(DEW.) deWinter, Francis. "How to Design and Build a Solar Swimming Pool Heater." New York: Copper Development Association, 1974.

(DUF.) Duffie, John A. and Beckman, William A. *Solar Energy Thermal Processes.* New York City: John Wiley & Sons, 1974.

(DUN.) Dunkle, R. V. and Davey, E. T. "Flow Distribution in Absorber Banks." Paper presented at the Melbourne International Solar Energy Society Conference, 1970.

(ENG.) Engebretson, C. D. and Ashar, N. G. "Progress in Space Heating with Solar Energy." ASME Paper No. 60-WA-88, December 1960.

(GIL.) Gillette, R. B. "Analysis of the Performance of a Solar Heated House." M.S. Thesis, University of Wisconsin, 1959.

(HOL.) Hollands, K. G. T. "Directional Selectivity: Emittance and Absorption Properties of Three Corrugated Specular Surfaces." *Solar Energy* 7, 108–116.

(HOT.1) Hottel, Hoyt C. and Woertz, B. B. "The Performance of Flat-Plate Solar Heat Collectors." Publication No. 3. The Solar Energy Conversion Research Project, Massachusetts Institute of Technology, Cambridge, Mass., 1940.

(HOT.2) Hottel, Hoyt C. "Residential Uses of Solar Energy." *Proceedings of the World Symposium on Applied Solar Energy,* Phoenix, Arizona, 1955.

(HOT.3) Hottel, Hoyt C. and Howard, J. B. *New Energy Technology, Some Facts and Assessments.* Cambridge, Mass.: The MIT Press, 1971.

(JOR.) Jordan, Richard C., ed. *Low Temperature Engineering Applications of Solar Energy.* Technical Committee on Solar Energy Utilization of the American Society of Heating, Refrigeration & Air Conditioning Engineers, Inc. (ASHRAE), New York, N.Y., 1967.

(KLE.) Klein, S. A.; Duffie, J. A.; Beckman, W. A. "Transient Considerations of Flat-Plate Collectors." Paper presented at the Paris Solar Conference, 1973.

(KOB.) Kobayashi, Takao and Sargent, Stephen. "A Survey of Breakage-Resistant Materials for Flat-Plate Solar Collector Covers." Paper presented at the International Solar Energy Society, US Section Meeting, Ft. Collins, Colorado, August 1974.

(LOF.1) Löf, G. O. G. and Hawley, R. W. "Unstead-State Heat Transfer Between Air and Loose Solids." *Industrial and Engineering Chemistry* 40(1948):1061–1070.

(LOF.2) Löf, G. O. G. "Solar House Heating—A Panel." *Proceedings of the World Symposium on Applied Solar Energy,* Phoenix, Arizona, 1955.

(LOF.3) Löf, G. O. G. "Cooling with Solar Energy." *Proceedings of the World Symposium on Applied Solar Energy,* Phoenix, Arizona, 1955.

(LOF.4) Löf, G. O. G. and Tybout, R. A. "Cost of House Heating with Solar Energy." *Solar Energy* 14(1973):253–278.

(LOF.5) Löf, G. O. G. and Tybout, R. A. "The Design and Cost of Optimized Systems for Residential Heating and Coolins by Solar Energy." *Solar Energy* 16(1974)9–18.

(MIN.) Minardi, John E. and Chuang, Henry N. "Performance of a 'Black' Liquid Flat-Plate Solar Collector." Paper presented at the International Solar Energy Society, US Section Meeting, Ft. Collins, Colorado, August 1974.

(MIT.) Hamilton, Richard, ed. "Space Heating with Solar Energy." Proceedings of a course-symposium at MIT, August 20–26, 1950. Published by MIT, 1954.

(MOO.) Moore, S. W.; Balcomb, J. D.; and Hedstrom, J. C. "Design and Testing of a Structurally Integrated Steel Solar Collector Unit Based on Expanded Flat-Metal Plates." Los Alamos Scientific Laboratory, Los Alamos, N.M., 1973.

(NSF.) University of Maryland, Department of Mechanical Engineering. "Proceedings of the Solar Heating and Cooling for Buildings Workshop." National Science Foundation and Research Applied to National Needs, 1973.

(SCH.) Schönholzer, Ernest. "Hygienic Clean Winter Space Heating with Solar and Hydroelectric Energy Accumulated During the Summer and Stored in Insulated Reservoirs." *Solar Energy* 12(1969):379–385.

SPE.) "Spectrum." *Environment* 14(1972):25.

(TEL.) Telkes, Maria. "Storage of Heating and Cooling." Paper presented at the annual meeting of ASHRAE, Montreal, June 23, 1974.

(THO.) Thomason, Harry E. and Thomason, Harry Jack Lee, Jr. "Solar House Plans." Edmund Scientific Co., Barrington, N.J.

(TOT.1) Total Environmental Action, Inc. "Solar Energy Housing Design." A report written for the AIA Research Corporation under their contract with HUD and NBS, January 1975.

(TOT.2) Total Environmental Action, Inc. *Solar Energy Home Design in Four Climates.* Harrisville, N.H.: Total Environmental Action, Inc., May 1975.

(TYB.) Tybout, Richard A. and Löf, G. O. G. "Solar House Heating." *National Resources Journal* 10 (1970):268–326.

(WHI.1) Whillier, Austin. "Solar House Heating—A Panel." *Proceedings of the World Symposium on Applied Solar Energy,* Phoenix, Arizona, 1955.

(WHI.2) Whillier, Austin. "Principles of Solar House Design." *Progressive Architecture,* May 1955, pp. 122–126.

(WHI.3) Whillier, Austin. "Black-Painted Solar Air Heaters of Conventional Design." *Solar Energy* 8(1964):31–37.

(WIL.) Wilson, J. L. "Analysis of Solar Heating and Cooling of Buildings." *ASHRAE Journal* 17(1975):72.

RESOURCE SECTION

Energy And Solar Energy Phenomena

RESOURCE SECTION R·A

Heat Theory

It is difficult to consider the thermal behavior of buildings and to design for solar heating without understanding the nature of heat and how it flows from one place to another. There are two basic types of measurement used to describe heat: quantity and intensity. The measurement we are most familiar with, that of temperature, refers to intensity: if we refer to a swimming pool at a temperature of 70°, we are not saying anything about the quantity of heat in the pool. Intuitively, we know it will take a large quantity of heat to raise the pool temperature to 75°, while a much smaller quantity of heat would be required to effect the same increase for a kettle of water. The intensity of heat in the 75° pool is the same as the intensity in the 75° kettle, but the quantity of heat depends on the amount of material that is heated. Thus, the units for heat quantity are based on the amount of heat required to raise a unit mass of a reference material by a unit measure of temperature. The reference material used is water, because it is a universally available, standard material. Within the English system of measurement, the amount of heat required to raise the temperature of one pound of water one degree Fahrenheit is called a British Thermal Unit, or a Btu. In the metric system, the unit is a calorie, or the amount of heat required to heat one gram of water one degree Celsius. It takes the same quantity of heat, i.e., the same number of Btu, to heat 100 pounds of water one degree F as it does to heat 10 pounds of water 10°F, or one pound of water 100°F.

Another measure of heat, closely related to temperature and heat quantity, is heat capacity or *specific heat*. Not all materials absorb the same amount of heat in undergoing a particular temperature rise. While it will take 100 Btu to heat 100 pounds of water one degree F, it will only take 22.5 Btu to raise 100 pounds of aluminum one degree. The specific heat is the ratio of the number of heat units required to raise a certain mass of a given material a certain number of degrees, to the number of heat units required to raise the same mass of water the same number of degrees. This ratio is the same for any system of measurement—metric, English, or other. Thus, the specific heat of aluminum can be calculated in this way: 22.5 Btu/100 Btu = 0.225. The units can be Btu/lb/°F, or calories/gm/°C, or kcal/kg/°C. Specific heats of various materials can be found in the Appendix by that name.

The importance of all this, at least as far as buildings are concerned, is that the production of heat costs money and uses resources. The cost depends on the rate at which heat is used, which in turn depends on the rate of heat flow from inside of the building to outside (winter), or from outside to inside (summer). The rate of heat flow is proportional to the difference in temperature between the source of heat and the object or space to which it is flowing. Thus, heat will flow out of a building at a faster rate on a cold day than it will on a mild day. This assumes, of course, that some means is being employed within the building to keep the temperature constant, e.g., a furnace, boiler, or wood stove. While the rate is proportionate to the temperature difference, the quantity of heat actually flowing depends on how much resistance there is to the flow. Since little can be done about the temperature difference between inside and outside except by lowering the inside temperature, most effort is expended on increasing the resistance to heat flow. The actual mechanisms of heat flow and the methods of resisting such flow are numerous. Therefore, before discussing heat resistance, it is necessary to review the basic methods by which heat flows from a warm object to a cooler object: conduction, convection, and radiation.

We all learn about conduction at an early age in an intuitive but direct way. When an iron skillet sits on a hot stove for a period of time, the handle becomes hot. This is because heat is conducted through the metal from the burner to the handle. The heat flows to the handle because it is much cooler than the burner. The rate of flow to the handle of an iron skillet is much slower than that for a copper skillet because iron has less conductance (more resistance to heat flow) than copper, and also because it has a higher specific heat than copper. This means it will take less heat to warm the copper and less time to heat all the metal between the burner and the handle. These principles are basic to heat transfer by conduction.

Convection is heat transfer by the movement of fluids, i.e., liquids or gases. A heated fluid can move or be moved to a cooler area where it will transfer its heat to warm that area. The heated water at the bottom of a kettle on a stove rises and mixes with the cooler water above, spreading the heat and warming the mass much quicker than it would be warmed by simple conduction. A house with a warm air furnace is heated in the same way. Air is heated in the firebox and blown up to the living spaces. Since the house is cooler than the hot air, the heat is transferred from the air to the living spaces.

Heated fluids will move themselves by natural convection. As a fluid is warmed, it expands, becomes buoyant in the surrounding cooler fluid, and rises. Cooler fluid flows in to take its place and is, in turn, heated. The warmed fluid, meanwhile, has moved to a cooler place where the heat is absorbed, cooling down the fluid. The cooler fluid, which is heavier, tends to sink, and the cycle continues. When we want more control over the heat flow, or when we want heat to flow better than it would by natural convection (for instance, into a room distant from the furnace), we use a pump or a blower to move the heated fluid. It should be noted that convection works hand-in-hand with conduction. Heat

CONDUCTION - Heat will flow through any material, at a rate determined by the material's physical characteristics. Copper is an excellent conductor of heat; insulating materials are poor conductors.

CONVECTION - When two surfaces - one hot, the other cold - are separated by a thick layer of air, moving air currents (called convection currents) are established that carry heat from the hot to the cold surface. The process works like a thermal bucket brigade.

RADIATION - Any object that is warmer than its surroundings radiates heat waves (similar to light waves, but invisible) and, thus, emits heat energy.

BRITISH THERMAL UNIT (Btu) - A familiar measure of heat energy that is defined as the quantity of heat required to raise the temperature of one pound of water $1°F$. Btu per hour are designated Btuh.

k, or THERMAL CONDUCTIVITY - A measure of the ability of a material to permit the flow of heat. It expresses the quantity of heat per hour that will pass through a one-square-foot chunk of inch-thick material when a $1°F$ temperature difference is maintained between its two surfaces; k is measured in Btuh.

C is similar, but measures the heat flow through a given thickness of material. If you know a material's k, to find its C just divide by the thickness. E.g!, 3"-thick insulation with a k of 0.03 has a C of 0.10. The lower the k or C, the higher the insulating value.

U, or OVERALL COEFFICIENT OF HEAT TRANSMISSION - A measure of the ability of a complete building section (such as a wall) to permit the flow of heat. U is the combined thermal value of all the materials in a building section, plus air spaces and air films. The lower the U, the higher the insulating value.

R, or THERMAL RESISTANCE - A measure of ability to resist the flow of heat. R is simply the mathematical reciprocal of either C or U. Thus,

$$R = 1/C \text{ or } R = 1/U$$

depending on whether you're talking about the thermal resistance of a piece of insulation or a complete building section.

Insulation products are typically characterized by their R-values. Thus, a specification of R-11 means the insulation displays 11 resistance units. Clearly, the higher the R-value, the better the insulating ability.

R is a simple common denominator for describing all types of insulation and all kinds of dwelling construction. All insulation, for example, that is rated R-11 has the same insulation ability no matter what its material or thickness.

Figure R·A·1.
Glossary of heat transfer terms.

from the warm surface is transferred to the fluid by conduction before it is carried away by fluid flow; it is also transferred from the warmed fluid to the cool surface by conduction. The greater the temperature difference between the warm and cool surfaces, the greater the heat flow between them. The specific heat of the transfer fluid, its conductance, and the resistance to fluid flow are other factors affecting convection heat transfer.

Radiation is the transfer of heat through a space by electro-magnetic waves; most objects which stop the flow of visible light also stop the radiation of heat. The earth receives its heat from the sun by radiation, as we all know. We also experience radiation heat transfer when we stand in front of a fireplace or a hot stove. Radiation of heat is accomplished primarily by long-wave radiation, which is invisible. We will feel heat radiating from the hot stove, even if the stove is not hot enough to give off light. Heat is constantly radiated from warmer objects to cooler objects (as long as they can "see" each other) in proportion to their temperature differences and the distance between them. The same effect, though more subtle and more difficult to perceive, is what makes us feel cold sitting next to a window on a winter night: as heat sources, our bodies radiate to the cold night sky and become chilled in the process. Of the three basic methods

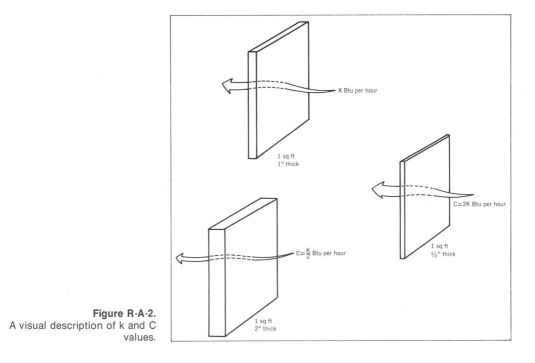

Figure R·A·2.
A visual description of k and C values.

of heat transfer, radiation is the most difficult to quantify for buildings.

CONDUCTION

Of the factors affecting the rate of heat conduction, the most important to seasonal heat loss is the resistance of the building materials. All materials have some finite resistance to heat flow; those with a particularly high resistance are called insulation. The inverse of resistance is conductance, or how much heat will be conducted by the building *to* the outside in the winter and *from* the outside in the summer. Thermal conductivity (k) is a measure of the capability of a given material to conduct heat; it expresses the quantity of heat per hour that will pass through 1 sq ft of inch-thick material when a 1°F temperature difference is maintained between its two surfaces; k is measured in Btu/hour (Btuh). C is similar, but it expresses the Btuh heat flow through a given thickness of material. Dividing k by the thickness of the given material (in inches) gives the value of C for the material (see Figure R·A·2); the lower the k or C, the higher the insulating value.

The overall coefficient of heat transmission (U) is a measure of the ability of a complete building section (such as a wall) to permit the flow of heat. It is the combined thermal value of all the materials in a building section, plus air spaces and air films. The lower the U-value, the higher the insulating value. U is expressed in units of Btu/hr/ft²/°F. To find overall heat loss, multiply the U-value by the number of hours, by the total square footage of surface, and the temperature difference between surfaces inside and outside. To find the loss through a 50 sq ft wall with a U-value of 0.12 over an eight-hour period when inside

temperatures are 65°F and outside temperatures are 40°F, multiply (0.12)(50)
(8)(65 − 40) = 1,200 Btu lost.

The U-value of any building section (wall, roof, window, etc.) can be computed
from the conductances of the various components of the section. This computation involves resistance. The individual resistance of each component of a
building section is the inverse of its conductance: R = 1/C [or R = (1/k)
(thickness)]. The higher the R-value of a material, the greater its insulating
ability. R_t is the sum of the individual resistances of the components. Therefore,

$$U = \frac{1}{R_1 + R_2 + R_3 \ldots R_x} \text{ or } U = 1/R_t.$$

The computation, then, involves adding up the R-values across the building
section, including the inside still air film, any air spaces greater than ¾ inch, all
the building materials, and the outside air film. Values of these resistances are
given in Appendix C, "Insulating Values of Materials." Figure R·A·3 computes
the U-values of two typical wall sections. Note that the uninsulated wall conducts
more than three times as much heat as the insulated one.

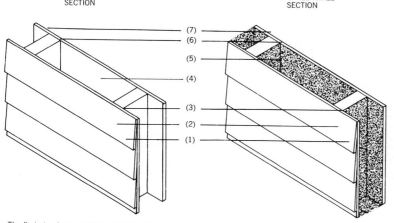

The first step is to establish the R-values of each component of the section and add up the total R_t -value:

Wall Construction	Uninsulated	Insulated
1 Outside surface (film), 15 mph wind	0.17	0.17
2 Wood bevel siding, lapped	0.81	0.81
3 ½″ ins. bd. sheathing, reg. density	1.32	1.32
4 3½″ air space	1.01	
5 R-11 insulation		11.00
6 ½″ gypsumboard	0.45	0.45
7 Inside surface (film)	0.68	0.68
TOTALS (R_t)	4.44	14.43
Taking the inverse, 1/R = U: U-values	0.23	0.069

Figure R·A·3. Comparison of U
values: an insulated wall
compared with an uninsulated
wall.

Once the U-values of all the building sections (windows, walls, roofs, and
floors) have been determined, the calculation of total heat loss can begin. One
approach to this problem is to find the total heat loss of the building at outside
temperatures close to the lowest to be expected; these extreme temperatures are
called "design temperatures." A list of recommended design temperatures for
many cities in the U. S. is given in Appendix B, "Degree Days and Design

Temperatures." The "2½% condition" approach is used to find the number of Btu per hour that the heating system will have to supply to keep a building warm (72°) under all but the most extreme conditions. The design temperature is subtracted from the normal inside temperature to find the temperature difference. Next, the total surface area in square feet of each type of building section is determined. This is multiplied by the temperature difference and the U-value to get the total heat loss of each building section for the hour. The totals for each section are added to arrive at the total hourly heat loss:

> Hourly Heat Loss for a section = (section area) ×
> (temp. diff.) ×
> (section U-value)

To compute the total heat loss for the heating season, multiply the total heating degree days by 24 hours per day to arrive at total heating degree hours (the concept of degree days is explained in Part II). Appendix B on degree days gives typical heating degree days for many cities throughout the country. Once the degree hours are determined, the calculation proceeds as with the worst condition calculation: the area of each section is multiplied by the degree hours and by the U-value for each section. The totals for each section are summed to arrive at the total yearly building heat loss:

> Yearly Heat Loss for a section = (section area) ×
> (24 hrs/day) ×
> (degree days) ×
> (section U-value)

The total cost of heating the building, assuming no "free" heat from the sun, from people, or from other sources such as lights and machines, is the cost of providing the total number of Btu lost by the building over the course of the heating season. The most common method of expressing the cost of heat is in dollars per million Btu (10^6). The actual cost of delivering heat takes into account the price of fuel, the efficiency of delivery, and the number of Btu provided by the particular fuel. Figure R·A·4 aids in determining the cost per million Btu of energy for varying prices of electricity, oil, and gas, and for various efficiencies of heat delivery.

The process of determining total seasonal heat loss is summarized in the "Heat Conduction Cost Chart" (see Figure R·A·5). An example of the use of the chart is included:

1. for a building section with a U-value of 0.58, start at point (1) on the chart;
2. follow the oblique line to the horizontal line representing the total heating degree days for the location, in this case, 7,000 DD;
3. move vertically from this point to find a value of 95,000 Btu/ft² per season for the section;
4. continue vertically to the oblique line representing the total square footage of the section, 90 sq ft;
5. moving horizontally from this point, the total heat loss through the entire surface for the season is 9,600,000 Btu;
6. continue horizontally to the oblique line representing the cost per million

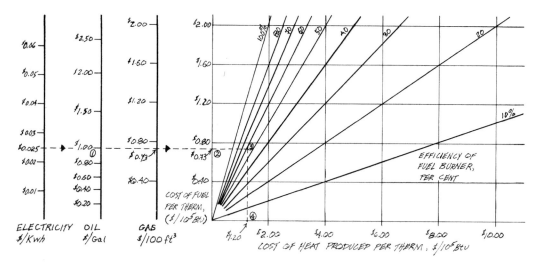

COST PER THERM OF ENERGY

① *FIND POINT ON VERTICAL COLUMN CORRESPONDING TO KNOWN RETAIL PRICE OF FUEL (eg, $1.00/gal for OIL = $0.025/kwh for ELECTRICITY = $0.73/100ft³ for GAS)*
② *MOVE RIGHT TO FIND RETAIL COST OF FUEL PER THERM (eg $0.73/10⁵ Btu)*
③ *STOP AT EFFICIENCY OF FUEL BURNER (eg, 60%)*
④ *MOVE DOWN TO FIND REAL COST OF HEAT PRODUCED PER THERM (eg, $1.20/10⁵ Btu)*
NOTE: THERE ARE 3412 Btu/kwh ELECTRICITY; 135,000 Btu/gal OIL; 1000 Btu/ft³ GAS.
PRICE PER THERM IS OFTEN AVAILABLE FROM UTILITY COMPANIES.
'COST OF HEAT PRODUCED PER THERM' CAN BE USED WITH OTHER CHARTS IN THESIS

HEATING EFFICIENCIES OF SOME FUEL BURNERS, PER CENT

ANTHRACITE COAL, HAND-FIRED	60-75
BITUMINOUS COAL, HAND-FIRED	40-65
BITUMINOUS COAL, STOKER-FIRED	50-70
OIL & GAS FIRED	65-80
DIRECT ELECTRIC HEATING	100

NOTE: POOR FURNACE ADJUSTMENT CAN REDUCE THE ABOVE FIGURES BY 5-10.
ELECTRICITY GENERATION & TRANSMISSION LOSES 2 Btu PER Btu DELIVERED.

Figure R·A·4.   Cost per million Btu of energy.

Btu (10^6 Btu) of heat energy, in this case $9/$10^6$ Btu;

7. moving vertically down from this point, find the total cost for the season of the heat through that section—$86.

The last graph, bottom right, converts this cost to "real cost of energy" through the use of oblique lines called "multiplication factors." This factor can be one consideration or a combination of several.

1. Estimated future cost of energy: design decisions based on present energy costs make little sense as costs soar.

2. Real environmental cost of using fossil fuels: this includes pollution and the depletion of natural resources, both directly as fuels burn and indirectly as they are brought to the consumer from the source.

3. Initial investment cost: use of the proper multiplication factor would give the quantity of increased investment made possible by resultant yearly fuel savings.

In this example, future heat costs may increase by a factor of 10. Continuing down from the last point, intersect the oblique line representing the multiplication factor 10. From the intersection, move left horizontally to arrive at an adjusted seasonal heating cost through the building section of $862. Note that the numerical values of the chart can be changed by a factor of ten. For example,

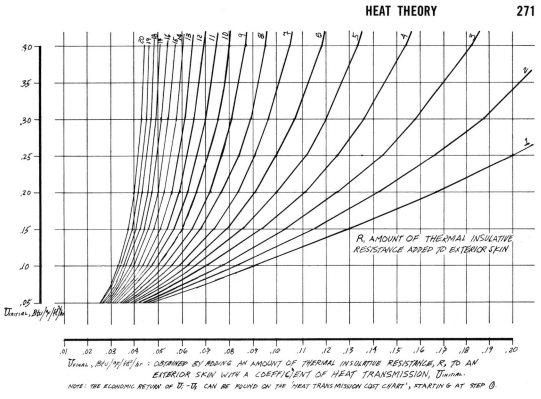

Figure R·A·6. The impact on heat transmission values as a result of a change in thermal resistance.

to determine the heat transfer through a really good exterior wall, U = 0.05, use U = 0.5 on the chart and divide the final answer by ten. Also, each of the graphs can be used independently of one another. For example, knowing a quantity of energy and its price, the top right graph gives the total cost of that energy.

With a bit more information, this chart can be of even greater usefulness. Figure R·A·6 is designed to make readily apparent the change in overall heat transmission for a given building section if insulation is added or removed. For instance, if 3½ inches of fiberglass insulation (R = 11) is added to a wall having a U-value of 0.23, what will be the change in the U-value? (This is the example in Figure R·A·3.) By adding the insulation, the insulating value of the air space, which is 1.01, is removed, so the net increase in resistance is R = 10. To use the chart, then, we begin at 0.23 on the vertical left scale. Moving horizontally, intersect the curve numbered R = 10. Dropping down from this point, find the new U-value 0.069, which agrees with the calculations. With this information, return to the "Heat Conduction Cost Chart" to find the savings resulting from the added insulation. The difference between the original and final U-values is 0.23 − 0.069 = 0.16. This difference can then be run through the cost chart in the same way as the first example. Starting at U = 0.16, assume 4,000 DD, 100 sq ft, and $9 per million Btu; the savings in heating cost for one year is about $17. Since this is more than the cost of the insulation, it is well worth the investment in this case.

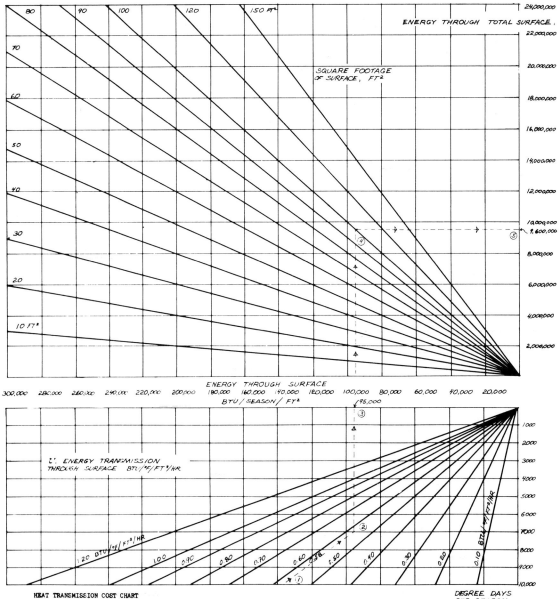

HEAT TRANSMISSION COST CHART

1 Find line corresponding to U (e.g., U=0.58 Btu/$^\circ$F-ft^2-hr)

2 Follow line to degree days per season (e.g., 7,000)

3 Move up vertically to find energy through surface (e.g., 95,000 Btu/season)

4 Stop at square footage of surface (e.g., 90ft^2)

5 Move right to find energy through total surface (e.g., 9,600,000 Btu/season)

6 Stop at cost per million Btu of energy (e.g.,\$9.00/10^6 Btu)

7 Move down vertically to find total cost of energy (e.g.,\$86.00/season)

8 Stop at multiplication factor (e.g., future increased cost of fuel, 10x)

9 Move left to find real cost of energy (e.g., \$862.00/season)

Figure R·A·5. Heat conduction cost chart.

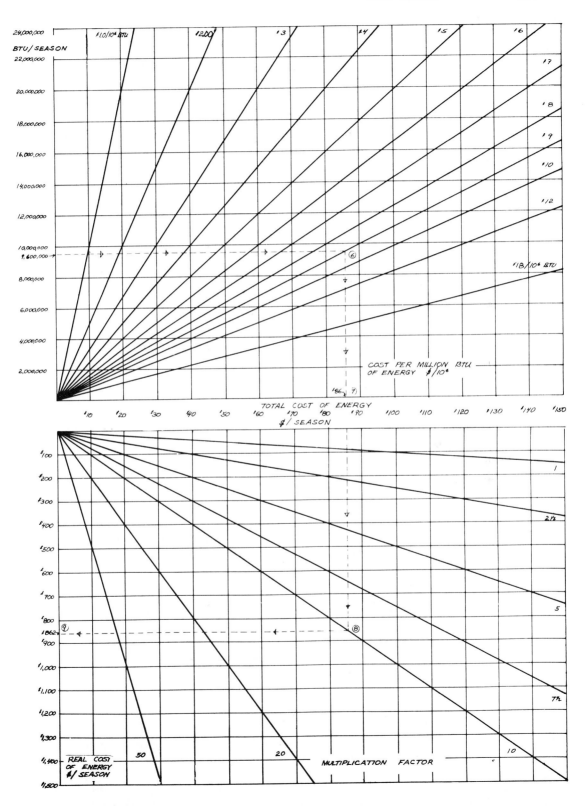

Similar analyses can be made for any situation. It should be remembered, however, that this analysis deals only with conduction heat loss. Radiation and convection heat losses are also quite significant and must be considered. A complete economic analysis, which is beyond this discussion, would include, in addition to the cost of the insulation and the savings in fuel, long-term costing considerations, predictions of future fuel supplies, and moral and social decisions about the use of non-renewable resources.

Note: It can be seen from Figure R·A·6 that it takes much greater amounts of thermal resistance, R, to reduce significantly a small U-value than to reduce a larger one. For example, adding two inches of polyurethane insulation (R = 12) to eight inches of solid concrete reduces U from 0.66 to 0.07. Adding an additional two inches of the same insulation reduces U from 0.07 to 0.038. This is shown mathematically:

The thermal resistance, R, for an initial U-Value, U_i, is:

$$R_i = \frac{1}{U_i}$$

If we add thermal resistance (for example, insulation), R_{in}, to R_i, we have a new resistance, R_f:

$$R_f = R_i + R_{in} = \frac{1}{U_i} + R_{in} = \frac{1}{U_f} \quad \text{and} \quad U_f = \frac{1}{R_f} = \frac{1}{\dfrac{1}{U_i} + R_{in}}$$

$$U_f = \frac{U_i}{1 + U_i R_{in}}$$

CONVECTION

Heat loss by convection is not as readily quantified as loss by conduction. There are three places in the building where convection losses are significant.

The first type of convection loss occurs within the walls and between the layers of glass in the skin of the building. Wherever there is an air space, and whenever there is a temperature difference between the opposing surfaces of the space, there will be a certain amount of natural convection air movement, resulting in the transference of heat across the space. Since this process is not very efficient, air spaces are considered to have insulating value, although it is not very large compared to real insulating materials. The width of the air space must be greater than ¾ inch for the insulating value to be significant, but a quick study of the insulating values of air spaces (see Appendix C, "Insulating Values of Building Materials") shows that increases in the width of the air space do not produce significant increases in insulating value. The freer circulation of air in a wider space offsets the potentially greater insulating value of the thicker air blanket (see Figure R·A·7). The same tables show that the reflective qualities of the surfaces of the air space have an influence on the final insulating value of the space. This radiation effect will be discussed in more detail below.

The movement of air within spaces is significant. The actual insulating capacity of still air is considerable; most of the common forms of insulation simply trap air in tiny spaces and prevent air circulation in the space they occupy. Fiberglass blanket insulation is quite common, but in terms of insulating value it is little better than animal hair, cotton, feathers, or even popcorn. Although these materials have certain characteristics which make them inappropriate for building use, all of them create tiny air pockets to slow down the flow of heat.

A second type of loss by convection works in close conjunction with the conduction loss through the building skin. This is the air movement within the living spaces. Since the interior surfaces of perimeter walls tend to be cooler than other inside surfaces, heat from the room air is drawn to that surface, cooling the

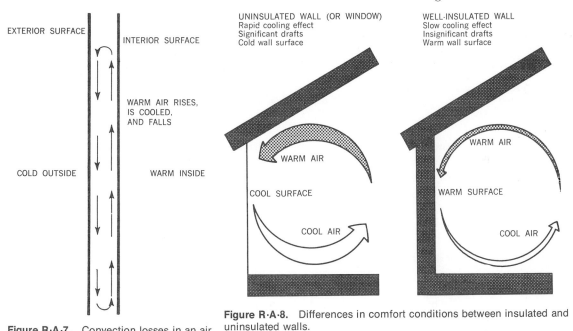

Figure R·A·7. Convection losses in an air space.

Figure R·A·8. Differences in comfort conditions between insulated and uninsulated walls.

air in the space. The cooled air tends to sink down and run across the floor, while warmer air at the top of the room flows in to take its place. In this way, the cooling effect from the entire room to the outside wall is accelerated. A well-insulated perimeter wall will not be a great deal cooler than other inside walls, but the air movement and cooling effects of window surfaces are significant (see Figure R·A·8). Heating units or warm air registers are commonly placed beneath windows in an effort to eliminate the cold draft down from the glass surfaces. This practice greatly improves the comfort of the living space, but does nothing to reduce the heat loss to the outside; in fact, it tends to accelerate the loss through the window, since the effective temperature difference between the warm inside air and the cold outside air is increased.

The insulating value of the air films on either side of a building section, while not extremely insulating, are available at no extra cost and do make a contribution to the overall U-value. Still air provides maximum value, and because of this, the location of the surface makes a difference. In the tables in the Appendix on "Insulating Values of Materials," the air film on a horizontal surface is

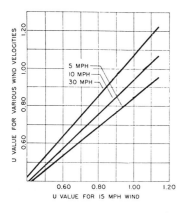

Figure R·A·9. Variation in U value versus wind velocity. Source: (WAG.)

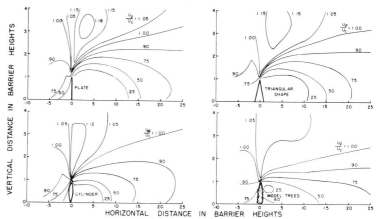

Figure R·A·10. Air flow around four barriers of varying shape. Source: (OLG.)

shown to be more insulating than those on vertical surfaces. This is because the convection air flow, which reduces the effective thickness of the still air insulating film, is greater down a vertical wall than across a horizontal surface. Similarly, the air film on the outside surface is significantly reduced by wind blowing across the surface. Thus heat that leaks through the wall is quickly transmitted to the moving air and carried away. The tables indicate the effect of wind on heat loss. See also Figure R·A·9 for the variation of the U-values for different wind speeds. Heat losses can be reduced by wind screens or plantings which prevent fast moving air from hitting the building skin; Figure R·A·10 shows the effect of various barriers on the air flow.

Air infiltration occurs through openings in buildings (e.g., windows) and through "cracks" (air spaces) around doors and windows. They are the primary convection losses that are not accounted for in conduction loss calculations. These losses are not easy to quantify accurately because they depend greatly on the tightness of construction and the weatherstrip detailing of the windows, doors, and other openings in the building's skin. For calculation purposes, it is assumed that the general construction will be air-tight and that the infiltration through windows and doors is all that need be considered. Great care must be taken in construction to insure that this assumption is valid. Small openings such as holes around outside electrical outlets or outside hose faucets can cause significant air infiltration into the heated spaces, and this cold air must be heated. Air infiltration around windows and doors has been studied enough so that it is somewhat predictable.

In order to determine infiltration, first measure the length of cracks between each window and door and its corresponding frame (jamb), usually the perimeter of the opening. If the seal between the window frame and the wall is not air-tight, the length of this crack must also be considered. Using Figure R·A·11, the volume of air leakage per hour for each type of crack can be approximated. Next, the temperature difference between inside and outside must be determined. Using the density and specific heat of the air leaking in, the amount of

Expressed in cubic feet per ft of crack per hour†

Type of Window	Remarks	Wind Velocity, MPH					
		5	10	15	20	25	30
Double-hung wood sash windows (unlocked)	Around frame in masonry wall—not calked‡	3	8	14	20	27	35
	Around frame in masonry wall—calked‡	1	2	3	4	5	6
	Around frame in wood frame construction‡	2	6	11	17	23	30
	Total for average window, non-weather-stripped, $\frac{1}{16}$-in. crack and $\frac{3}{64}$-in. clearance.§ Includes wood frame leakage‖	7	21	39	59	80	104
	Ditto, weatherstripped‖	4	13	24	36	49	63
	Total for poorly fitted window, non-weather-stripped, $\frac{3}{32}$-in. crack and $\frac{3}{32}$-in. clearance.¶ Includes wood frame leakage‖	27	69	111	154	199	249
	Ditto, weatherstripped‖	6	19	34	51	71	92
Double-hung metal windows**	Non-weather stripped, locked	20	45	70	96	125	154
	Non-weather stripped, unlocked	20	47	74	104	137	170
	Weather stripped, unlocked	6	19	32	46	60	76
Rolled section steel sash windows¶¶	Industrial pivoted, $\frac{1}{16}$-in. crack††	52	108	176	244	304	372
	Architectural projected, $\frac{1}{32}$-in. crack‡‡	15	36	62	86	112	139
	Architectural projected, $\frac{3}{64}$-in. crack‡‡	20	52	88	116	152	182
	Residential casement, $\frac{1}{64}$-in. crack§§	6	18	33	47	60	74
	Residential casement, $\frac{1}{32}$-in. crack§§	14	32	52	76	100	128
	Heavy casement section, projected, $\frac{1}{64}$-in. crack‖‖	3	10	18	26	36	48
	Heavy casement section projected $\frac{1}{32}$-in. crack‖‖	8	24	38	54	72	92
Hollow metal, vertically pivoted window**		30	88	145	186	221	242

From "Heating Ventilating Air-Conditioning Guide 1957." Used by permission.

† The values given in this table, with the exception of those for double-hung and hollow metal windows, are 20 per cent less than test values to allow for building up of pressure in rooms and are based on test data reported in the papers listed in chapter references.

‡ The values given for frame leakage are per foot of sash perimeter as determined for double-hung wood windows. Some of the frame leakage in masonry walls originates in the brick wall itself and cannot be prevented by calking. For the additional reason that calking is not done perfectly and deteriorates with time, it is considered advisable to choose the masonry frame leakage values for calked frames as the average determined by the calked and non-calked tests.

§ The fit of the average double-hung wood window was determined as $\frac{1}{16}$-in. crack and $\frac{3}{64}$-in. clearance by measurements on approximately 600 windows under heating season conditions.

‖ The values given are the totals for the window opening per foot of sash perimeter and include frame leakage and so-called *elsewhere* leakage. The frame leakage values included are for wood frame construction but apply as well to masonry construction assuming a 50 per cent efficiency of frame calking.

¶ A $\frac{3}{32}$-in. crack and clearance represent a poorly fitted window, much poorer than average.

** Windows tested in place in building.

†† Industrial pivoted window generally used in industrial buildings. Ventilators horizontally pivoted at center or slightly above, lower part swinging out.

‡‡ Architecturally projected made of same sections as industrial pivoted except that outside framing member is heavier, and it has refinements in weathering and hardware. Used in semi-monumental buildings such as schools. Ventilators swing in or out and are balanced on side arms. $\frac{1}{32}$-in. crack is obtainable in the best practice of manufacture and installation, $\frac{3}{64}$-in. crack considered to represent average practice.

§§ Of same design and section shapes as so-called *heavy section casement* but of lighter weight. $\frac{1}{64}$-in. crack is obtainable in the best practice of manufacture and installation, $\frac{1}{32}$-in. crack considered to represent average practice.

‖‖ Made of heavy sections. Ventilators swing in or out and stay set at any degree of opening. $\frac{1}{64}$-in. crack is obtainable in the best practice of manufacture and installation, $\frac{1}{32}$-in. crack considered to represent average practice.

¶¶ With reasonable care in installation, leakage at contacts where windows are attached to steel framework and at mullions is negligible. With $\frac{1}{32}$-in. crack, representing poor installation, leakage at contact with steel framework is about one-third and at mullions about one-sixth of that given for industrial pivoted windows in the table.

Figure R·A·11. Air infiltration through windows. Source: (SEV.)

heat required to warm it to room temperature can then be calculated by using this formula:

$$H_{infiltration} = (q)\ (c)\ (d)\ (L)\ (t_i - t_o),$$

where

q = volume of air infiltration per hour per linear ft of crack (ft³/hr/ft); see Figure R·A·11

c = specific heat of air, 0.24 Btu/lb/°F

d = density of air, 0.075 lb/ft³

L = length of crack in ft

$(t_i - t_o)$ = indoor-outdoor temperature difference, °F

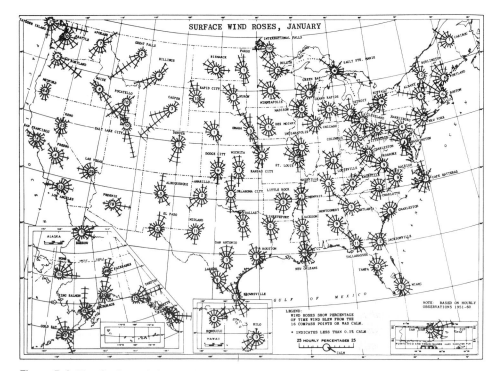

Figure R·A·12.   Surface wind roses for the United States, January. Source: (NOA.)

This formula must be applied to the total crack length for each different type of crack leakage. The total crack length varies with room conditions: for rooms with one exposure, use the entire measured crack length; for rooms with two exposures, use the crack length in the outside wall having the greater amount; for rooms with three or four exposures, use the crack length in the wall having the greatest amount. In no case use less than one-half the total crack length. The temperature difference term may be used in the same way as it was in the conductive heat loss calculation, to figure the worst (*design*) condition loss or the total seasonal heat loss. For the worst condition, the outdoor design temperature and the average wind speed are used. Fortunately, the design temperature is not usually accompanied by the maximum likely wind speed. The total seasonal heat loss due to infiltration is calculated by replacing the temperature difference with the total number of degree hours. The heating degree days are multiplied by 24 to arrive at total degree hours:

$$H_{seasonal} = (q)\ (c)\ (d)\ (L)\ 24\ (Degree\ Days)$$

The surface wind roses (Figure R·A·12 to R·A·14) show the percentage of time wind blows in various directions. The maps of the United States show July, January, and annual conditions.

The "Air Infiltration Cost Chart" (Figure R·A·15) is similar to the chart for heat transmission. Start at point (1) for an air infiltration rate of 45 ft³/hr/ft; (2) follow up the oblique line to the horizontal line representing the total heating

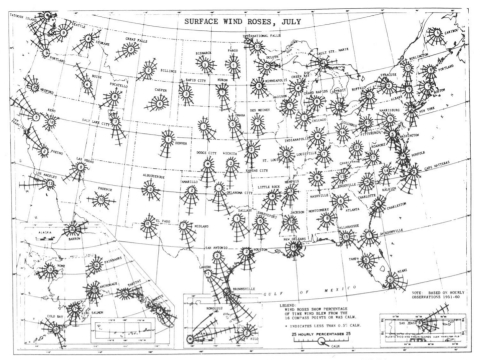

Figure R·A·13.   Surface wind roses for the United States, July. Source: (NOA.)

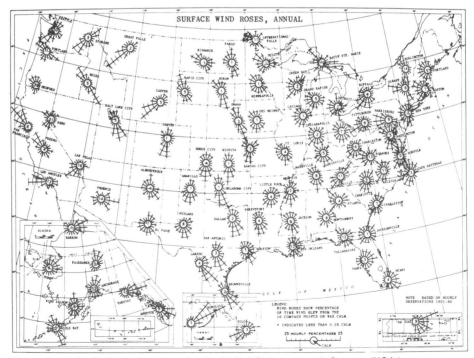

Figure R·A·14.   Surface wind roses for the United States, annual. Source: (NOA.)

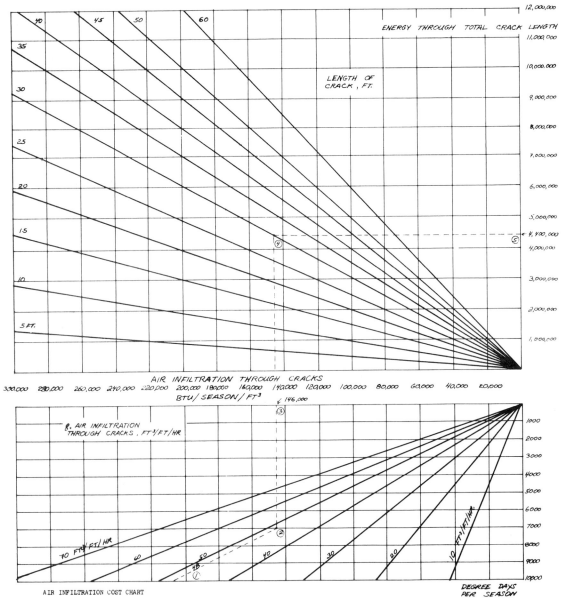

AIR INFILTRATION COST CHART

1 Find line corresponding to known q (e.g., q=48 ft³/ft-hr)

2 Follow line to degree days per season(e.g., 7,000)

3 Move up vertically to find air infiltration through cracks (e.g. 146,000 Btu/season-ft)

4 Stop at known length of crack (e.g., 30ft)

5 Move right to find energy through total crack length (e.g.,4,400,000 Btu/season)

6 Stop at cost per million Btu of energy ($6.9×10⁶ Btu)

7 Move down vertically to find total cost of energy (e.g., $26.75/season)

8 Stop at multiplication factor (e.g., future increased cost of fuel, 10x)

9 Move left to find real cost of energy (e.g.,$270.00/season)

Figure R·A·15. Air infiltration cost chart.

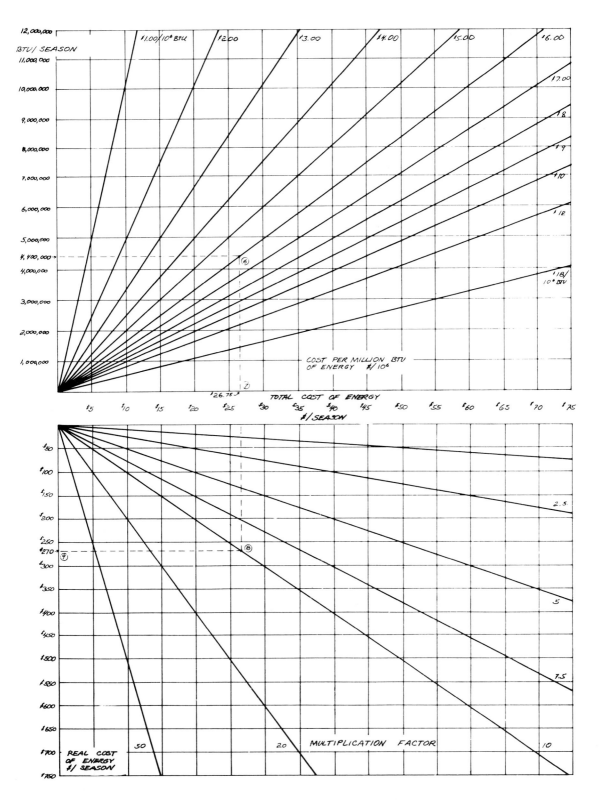

degree days for the location, in this case, 7,000 DD; (3) move vertically from this point to find that 146,000 Btu of energy are consumed per season for each foot of crack length; (4) continue vertically to the oblique line of the known crack length, for example, 30 feet; (5) move horizontally from this point to the total seasonal heat loss through the window crack, 4,400,000 Btu; (6) continue horizontally to the oblique line representing the cost per million Btu (10^6) of heat energy, in this case, $6.00/$10^6$ Btu (see Figure R·A·4 for assistance in obtaining this figure); (7) move vertically down from this point to the total cost for heat for the season through the crack, $26.75.

As with the "Heat Conduction Cost Chart" (see Figure R·A·5), the bottom right graph converts the cost to "real cost of energy" through the use of "multiplication factors." In this example, a factor of 10 is used. Continuing down from the last point, intersect the oblique line representing the multiplication factor 10. From the intersection, move horizontally left to arrive at an adjusted seasonal heating cost of $270 per season.

This chart can be used to make quick evaluations of the savings resulting from changes in air infiltration. For example, if a wooden double-hung window is weatherstripped, the air infiltration rate will change from 39 to 24 ft³/hr/ft. By moving through the chart from a starting point of 15 (39 − 24 = 15 ft³/hr/ft), we can arrive at the savings resulting from the weatherstripping. Assuming 5,000 degree days, 15 ft of crack length, and $6.00/$10^6$ Btu costs, we arrive at a savings of about $4.00 per season, without making any multiplier decisions. Since weatherstripping is inexpensive, it will quickly pay for itself in this case.

RADIATION

Compared with conduction and convection heat losses, seasonal radiational losses are almost impossible to compute, but they are, fortunately, a relatively small portion of the total heat loss. Appendix E, "Emittances and Absorptances of Materials," briefly covers the important aspects of radiation. Of special relevance to solar cooling is the concept of nocturnal radiational cooling. Under proper conditions, as much as 30 Btu per hour per square foot of surface can be radiated to a cool, clear sky when the dewpoint temperature is low. Figure R·A·16 gives the radiational effects for horizontal black surfaces as outdoor temperature near the ground and specific humidity vary.

Harold Hay's house in Atascadero, California, is the most significant solar project to make use of this phenomenon. Yanagimachi, Bliss, and Thomason also used nocturnal radiation systems in solar projects.

Temp		Specific humidity, mm Hg										
		1	2	3	4	5	6	7	8	10	12	15
°C	°F	Radiation, cal per cm² per min.										
−20	− 4	0.11	—	—	—	—	—	—	—	—	—	—
−15	+ 5	0.12	—	—	—	—	—	—	—	—	—	—
−10	14	0.13	0.12	—	—	—	—	—	—	—	—	—
− 5	23	0.14	0.13	0.12	—	—	—	—	—	—	—	—
0	32	0.15	0.14	0.13	0.12	—	—	—	—	—	—	—
+ 5	41	0.16	0.15	0.14	0.13	0.13	0.12	—	—	—	—	—
10	50	0.17	0.16	0.15	0.14	0.14	0.13	0.12	0.11	—	—	—
15	59	—	0.17	0.16	0.15	0.15	0.14	0.13	0.12	0.11	0.10	—
20	68	—	—	0.17	0.16	0.16	0.15	0.14	0.13	0.12	0.11	—
25	77	—	—	—	0.17	0.17	0.16	0.15	0.14	0.13	0.12	0.10
30	86	—	—	—	—	0.18	0.17	0.16	0.15	0.14	0.13	0.11

Figure R·A·16. Effective outgoing radiation with a cloudless sky. Source: (BUD.)

RESOURCE SECTION R·B

Insolation:
The Phenomena Of Sunshine

by Douglas Mahone,
Associate,
Total Environmental Action, Inc.

Arriving at a quantitative description of solar weather and the amount of radiation available for use in heating is a difficult task. There is a relative dearth of good data, but most of the difficulty revolves around the large number of highly variable factors that affect radiation availability at a collector location. Some of these factors are directly quantifiable, but most must be treated as statistical factors based on long-term averages of recorded data.

The least modified and most directly useable statistics on solar radiation are available from the Weather Bureau in Asheville, North Carolina. Measurements are recorded in units of langleys striking a horizontal surface over a given time period. A langley is 1 calorie per square centimeter, and 3.69 langleys are equivalent to 1 Btu per square foot. These radiation figures are a measure of total radiation, direct sun radiation and diffuse sky radiation. Because the surface which records solar radiation is horizontal, it is sensitive to variations in the altitude of the sun. Thus the same intensity of radiation would be recorded as less in winter than in summer because, with the winter sun lower in the sky, the angle of incidence to the horizontal surface is lower. This increases reflection and lowers flux density. Trigonometric conversions of this data must be performed to apply it to tilted surfaces.

Less directly indicative of solar energy, but nevertheless useful, are Weather Bureau records of sunshine and cloudiness. Sunshine is recorded as "hours of sunshine" and "percentage of possible sunshine." A chart records the number of hours when there is enough sunshine to "cast a shadow." This then is compared to the total hours from sunrise to sunset to get the percentage of possible sunshine. Cloudiness is an expression of the amount of sky covered by clouds. It is given as tenths of the sky obscured, from 0.0 to 1.0. This judgment is based on human observations of sky conditions, but accuracy is probably within acceptable limits.

The quality, or intensity, of radiation falling on a surface is a function of many factors. As the earth travels in its orbit, its distance from the sun changes slightly:

it is closest in January and farthest in July. The actual radiation striking the outside of the earth's atmosphere, then, is most intense in January. The declination of the sun, however, is changing at the same time. The sun moves north in the sky between March and September, while it is farther south from September through March. In the northern hemisphere, this means the sun is lower in the sky in the winter. As a result, the sun's rays have a greater distance to travel through the atmosphere and they are less dense per square foot on the horizontal (see Figure R·B·1). This tends to balance out the radiational intensity through the year.

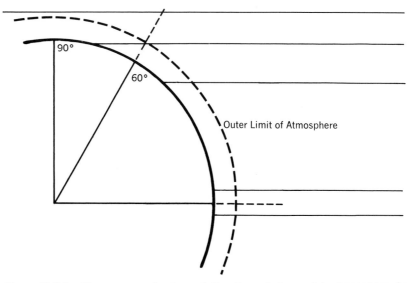

Figure R·B·1. The passage of solar radiation through the earth's atmosphere. As latitude increases, the angle of the sun's rays relative to the earth's surface decreases, and it therefore has to travel a greater distance through the atmosphere and is spread over a greater area when it reaches the surface. Source: (MOO.)

Other factors tend to vary both with season and from day to day. Atmospheric clearness, in general, is greater in winter because there is less dust, pollen, and haze in the air in winter than in summer. The reflection of radiation from surroundings also varies: snow in the winter is considerably more reflective than grass in the summer. Location can influence both clearness and reflection: a high altitude has clearer air than an industrial site; and a beach house gets more reflection than a forest dwelling. The influence of these factors is quite difficult to predict in any but the most general way, but they cannot be ignored.

Factors affecting the quantity of radiation are likewise quite variable. The obvious one is cloudiness, which not only changes from day to day, but from hour to hour. Weather Bureau sunshine and cloudiness data are helpful in predicting the influence of clouds on long-term radiation, but most of the published cloudiness data are daily, or even monthly, averages which say nothing about what time of day the clouds appeared. Since radiation alone varies considerably during the course of a day, the lack of data on daily cloudiness variation can introduce considerable error into radiation predictions.

A related variable affecting radiation intensity is the ratio of diffuse sky radiation to direct radiation. The direct component consists of the straight rays of the sun, which cast shadows. The diffuse component is a result of the scattering of light by air molecules, dust, clouds, ozone, water vapor, etc. This scattering makes the sky blue on clear days and grey on hazy days, and it is fairly uniformly distributed across the sky. It is difficult to measure diffuse radiation, and little is known about its strength and variability, although it can amount to 10% – 100% of incoming radiation.

The two common methods of dealing with such widely changeable variables as cloudiness and atmospheric clearness are either to apply statistically computed multiplication factors or to use records of actual radiation on a surface over a period of years. The actual data would reflect all the variables except diffuse radiation, since Weather Bureau measurements do not distinguish between direct and diffuse. The distinction is important because, on a surface of given orientation, the angle of incidence of the direct component determines how much of it can be collected. (Radiation perpendicular to the surface is almost completely absorbed, while radiation at low angles is substantially reflected.) Diffuse radiation, on the other hand, is treated as if it is uniformly distributed across the sky; it varies only in intensity, depending on atmospheric conditions. At present, a statistical approach is the only method available for separating out the diffuse component of total insolation.

In actual practice, statistical methods are widely used to deal with all the other variables too, because Weather Bureau records of radiation data represent conditions at only the 80 weather stations across the country. Unless a particular project is located quite close to one of these stations, the recorded data would not apply. (Great variations in intensity can occur over very short distances, so interpolation between recording stations is not considered strictly valid.) Statistical methods, then, provide the best radiation predictions in most cases, but the predicted values will be approximate at best.

The actual recorded radiation data is often characterized by inaccuracy and gaps of record. This is primarily due to the fact that measuring devices are difficult to calibrate and lose their sensitivity over time. The glass in older instruments was not uniformly transparent to all wavelengths, and the devices were also subject to ambient air temperature errors. Data was recorded on a chart recorder, which was then analyzed by hand, thus making human error a factor. Most of these problems are being solved by better instruments and automatic recording devices, but changeover at weather stations has been slow. Errors and inaccuracies in recorded data can run as high as 20%. Because of this, the Weather Service stopped publishing radiation summaries in 1972. Data is still being recorded at stations with operable instruments, but the entire radiation measurement system is changing. Improvements will result in a more widespread and accurate network to collect the basic information for solar energy work.

In general the data is summarized in either table form, graphs, or national maps. Reprinted in Figures R·B·2 to R·B·6 are several maps summarizing some relevant types of data, based on long-term means. Maps are useful for

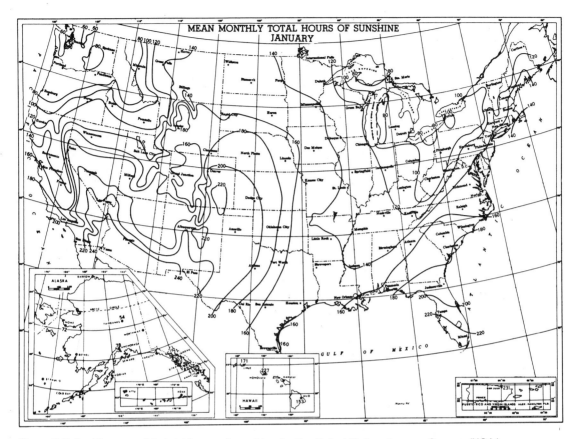

Figure R·B·2. Mean monthly total hours of sunshine for the United States, January. Source: (NOA.)

presenting an overview of the relative effects of locale, and for many locations they may be the only way of finding a particular value. As a rule, however, they should not be relied upon unless other data is unavailable. Many local factors can have significant influence, so great care and judgment are called for in the use of interpolated data from national maps.

Insolation data can be used to give very rough, preliminary information, as in this example. A 50 ft, 1-inch diameter black garden hose is sitting on a flat roof in New York City on an average May 21. Assuming 30% of the radiation striking the hose is transferred to the water, how many gallons of 125°F water can be obtained through the hose during the day?

Assume the hose draws heat from an area of 1-inch on either side of it. This makes an absorbing area of $0.25' \times 50' = 12.5$ sq ft.

1. In Weather Bureau Technical Paper No. 11, "Weekly Mean Values of Daily Solar and Sky Radiation," the long-term average for New York City on May 21 is 450 langleys/day
2. 450 langleys/day × 3.69 Btu/ft²/langley = 1660 Btu/ft²/day
3. 1660 Btu/ft²/day × 12.5 ft² = 20,750 Btu/day
4. 20,750 Btu/day × 30% = 6225 Btu/day collected

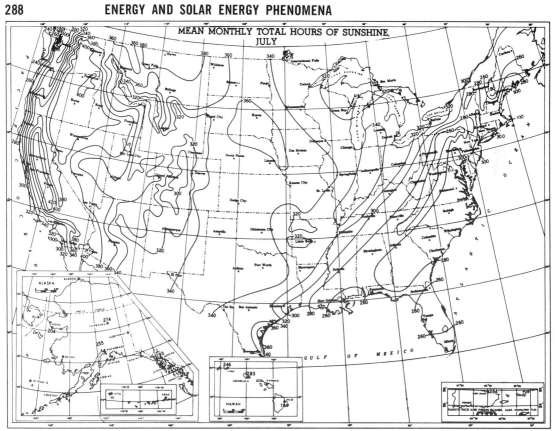

Figure R·B 3. Mean monthly total hours of sunshine for the United States, July. Source: (NOA.)

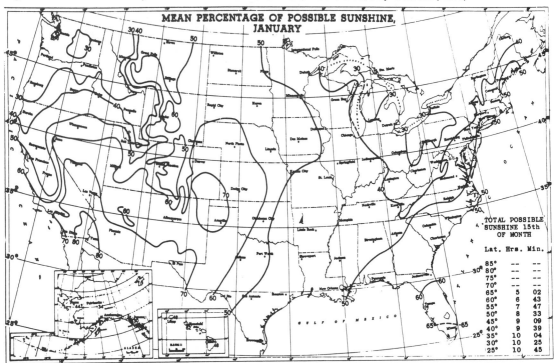

Figure R· ·4. Mean percentage of possible sunshine for the United States. Source: (NOA.)

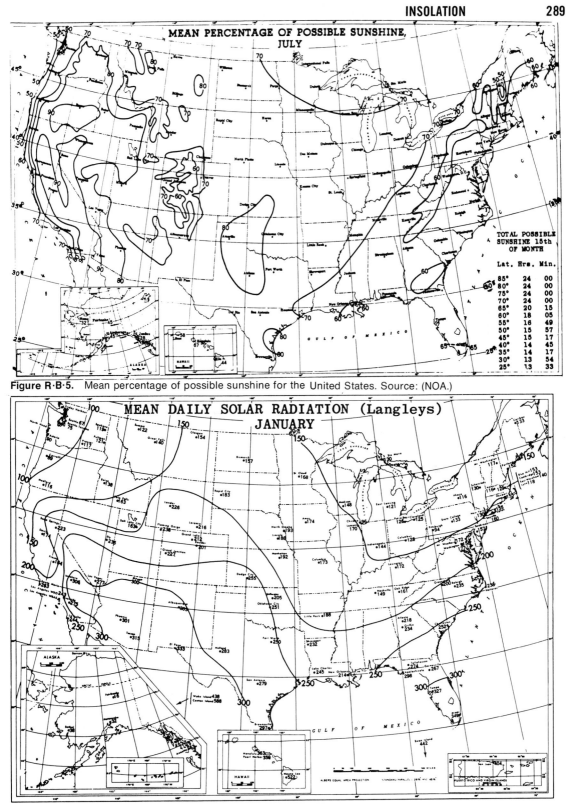

Figure R·B·5. Mean percentage of possible sunshine for the United States. Source: (NOA.)

Figure R·B·6. Mean daily solar radiation (Langleys) for the United States. Source: (NOA.)

5. If incoming water is 50°F, the temperature change will be 125°F − 50°F = 75°F temperature change

6. 6225 lb − °F/day ÷ 75°F = 83 lb/day water heated

7. 83 lb/day ÷ 8.35 lb/gal = 10 gal/day hot water

This type of analysis ignores many factors and should be used only to arrive at preliminary figures. More precise calculation procedures are described below.

In addition to the data published by the Weather Bureau, there are two types published by ASHRAE (American Society of Heating, Refrigerating, and Air Conditioning Engineers): Solar Heat Gain Factors (SHGF), and the newer Clear Day Insolation Data.

The SHGF data has certain serious limitations. It was developed to determine maximum heat gain through windows as an aid in the sizing of air conditioning systems. Because of this, some assumptions are made:

1. radiation has passed through a single sheet of double strength glass with a specified transmittance, reflectance, and absorptance; and

2. values are for a typical, cloudless condition on the 21st day of the given month.

The calculations which produce the SHGF take into account solar altitude and azimuth, direct and diffuse sky radiation, and an average ground reflectance. They are given for vertical wall surfaces at various orientations and for horizontal surfaces. For sizing air conditioning loads, SHGF figures are invaluable, but they have little applicability to tilted surfaces and almost none whatsoever to long-term predictions of heat gain. In addition, the considerations for double strength glass are inappropriate for most collectors. Their applicability is discussed in the Resource Section on "Solar Heat Gain Through Windows." In recognition of these limitations, ASHRAE has developed insolation figures for tilted, south-facing surfaces, such as those used for collecting solar energy. The format of this data is similar (by the hour on the 21st of each month) to that of the SHGF, but the surface orientations are different. For latitudes 24°, 32°, 40°, 48°, 56° and 64°, values are given for tilts equal to the latitude, latitude ± 10°, latitude + 20°, and vertical. These tables are reproduced in Figures R·B·7 to R·B·12 along with examples of their application. Example: Determine the most favorable tilt angle for a flat plate solar collector located in Atlanta, Georgia (32°N latitude). The tilt angle is to be selected so as to maximize the surface insolation for the following time periods: annual, heating season, and cooling season. Proceed as follows:

1. The heating season in Atlanta is from October through April; the cooling season is from May through September.

2. Referring to the 32°N table, sum the surface daily totals for the 22° tilt for the months October through April. This totals 14,469 Btu/hr. Perform the same operation for the 32°, 42°, 52°, and 90° tilts. These totals, respectively, are 15,142 Btu/hr, 15,382 Btu/hr, 15,172 Btu/hr and 10,588 Btu/hr.

3. Comparing the totals shows that the 42° tilt, or latitude + 10°, provides the best surface for heating collection.

4. For the cooling season, a similar set of totals is generated, but for the months May through September. These are 11,987 Btu/hr for 22°, 11,372 for 32°, 10,492 for 42°, 9,320 for 52°, and 3,260 for 90° tilt.

Figure R-B-7. Clear day solar position and insolation values for 24 degrees N latitude. (Reprinted by permission of ASHRAE.)

BTUH/SQ. FT. TOTAL INSOLATION ON SURFACES

DATE	AM	PM	SOLAR POSITION ALT	SOLAR POSITION AZM	NORMAL	HORIZ	South Facing Surface Angle with Horiz. 14	24	34	54	90
JAN 21	7	5	4.8	65.6	71	10	17	21	25	28	31
	8	4	16.9	58.3	239	83	110	126	137	145	127
	9	3	27.9	48.8	288	151	188	207	221	228	176
	10	2	37.2	36.1	308	204	246	268	282	287	207
	11	1	43.6	19.6	317	237	283	306	319	324	226
	12		46.0	0.0	320	249	296	319	332	336	232
			SURFACE DAILY TOTALS		2766	1622	1984	2174	2300	2360	1766
FEB 21	7	5	9.3	74.6	158	35	44	49	53	56	46
	8	4	22.3	67.2	263	116	135	145	150	151	102
	9	3	34.4	57.6	298	187	213	225	230	228	141
	10	2	45.1	44.2	314	241	273	286	291	287	168
	11	1	53.0	25.0	324	276	310	324	328	323	185
	12		56.0	0.0	324	288	323	337	341	335	191
			SURFACE DAILY TOTALS		3036	1998	2276	2396	2446	2424	1476
MAR 21	7	5	13.7	83.8	194	60	63	64	62	59	27
	8	4	27.2	67.9	267	141	150	152	149	142	64
	9	3	40.2	54.8	295	212	226	229	225	214	95
	10	2	52.3	41.9	309	266	285	288	283	270	120
	11	1	61.9	21.9	315	300	322	326	320	305	135
	12		66.0	0.0	317	312	334	339	333	317	140
			SURFACE DAILY TOTALS		3078	2270	2428	2456	2412	2298	1022
APR 21	6	6	4.7	100.6	40	7	7	4	3	3	2
	7	5	18.3	94.9	203	83	77	70	62	51	10
	8	4	32.0	89.0	256	160	157	149	137	122	16
	9	3	45.6	81.9	280	228	227	220	206	186	41
	10	2	59.0	71.8	292	278	282	275	259	237	61
	11	1	71.1	51.6	298	310	316	309	293	269	74
	12		77.6	0.0	299	322	328	321	305	280	79
			SURFACE DAILY TOTALS		3036	2454	2458	2374	2228	2016	488
MAY 21	6	6	8.0	108.4	86	22	15	10	9	9	5
	7	5	21.2	103.2	203	98	85	73	59	44	12
	8	4	34.6	98.5	248	171	159	145	127	106	15
	9	3	48.3	93.6	269	233	224	210	190	165	16
	10	2	62.0	87.7	280	281	275	261	239	211	22
	11	1	75.5	76.9	286	307	307	293	270	240	34
	12		86.0	0.0	288	317	317	304	281	250	37
			SURFACE DAILY TOTALS		3032	2556	2447	2286	2072	1800	246
JUN 21	6	6	9.3	111.6	97	29	20	12	11	11	7
	7	5	22.3	106.8	201	103	87	73	58	41	13
	8	4	35.5	102.6	242	173	158	142	122	99	16
	9	3	49.0	98.7	263	234	221	204	182	155	18
	10	2	62.6	95.0	274	280	269	253	229	199	18
	11	1	76.3	90.8	279	309	300	283	259	227	19
	12		89.4	0.0	281	319	310	294	269	236	22
			SURFACE DAILY TOTALS		2994	2574	2422	2230	1992	1700	204

DATE	AM	PM	SOLAR POSITION ALT	SOLAR POSITION AZM	NORMAL	HORIZ	South Facing Surface Angle with Horiz. 14	24	34	54	90
JUL 21	6	6	8.2	109.0	81	23	16	11	10	9	6
	7	5	21.4	103.8	195	98	85	73	59	44	13
	8	4	34.8	99.2	239	169	157	143	125	104	16
	9	3	48.4	94.5	261	231	221	207	187	161	18
	10	2	62.1	89.0	272	278	270	256	235	206	21
	11	1	75.7	79.2	278	307	302	287	265	235	32
	12		86.6	0.0	280	317	312	298	275	245	36
			SURFACE DAILY TOTALS		2932	2526	2412	2250	2036	1766	246
AUG 21	6	6	5.0	101.3	35	7	5	4	4	4	2
	7	5	18.5	95.6	186	82	76	69	60	50	11
	8	4	32.2	89.7	241	158	154	146	134	118	16
	9	3	45.9	82.9	265	223	222	214	200	181	39
	10	2	59.3	73.0	278	273	275	268	252	230	58
	11	1	71.6	53.2	284	304	309	303	287	261	71
	12		78.3	0.0	286	315	320	313	296	272	75
			SURFACE DAILY TOTALS		2864	2408	2402	2316	2168	1958	470
SEP 21	7	5	13.7	83.8	173	57	60	60	59	56	26
	8	4	27.2	67.9	248	136	144	146	143	136	62
	9	3	40.2	54.8	278	205	218	221	217	206	93
	10	2	52.3	41.9	292	258	275	278	273	261	116
	11	1	61.9	21.9	299	291	315	315	309	295	131
	12		66.0	0.0	301	302	320	323	321	306	136
			SURFACE DAILY TOTALS		2878	2194	2366	2392	2322	2212	992
OCT 21	7	5	9.1	74.1	138	32	40	45	48	50	42
	8	4	22.0	66.7	247	111	129	139	144	145	99
	9	3	34.1	57.1	284	180	206	217	223	221	137
	10	2	44.7	43.8	301	234	265	277	282	279	164
	11	1	52.5	24.7	309	268	302	315	319	314	181
	12		55.5	0.0	311	279	313	327	332	327	187
			SURFACE DAILY TOTALS		2868	1928	2198	2314	2364	2346	1442
NOV 21	7	5	4.9	65.8	67	10	16	20	24	27	29
	8	4	17.0	58.4	232	82	108	123	135	142	124
	9	3	28.0	48.9	282	150	186	205	217	224	172
	10	2	37.3	36.3	303	203	244	265	278	283	204
	11	1	43.8	19.7	312	236	280	302	315	320	222
	12		46.2	0.0	315	247	293	315	328	332	228
			SURFACE DAILY TOTALS		2706	1610	1962	2146	2268	2324	1730
DEC 21	7	5	3.2	62.6	30	3	7	11	12	14	14
	8	4	14.9	55.3	225	71	99	116	129	139	130
	9	3	25.5	46.0	281	137	176	198	214	223	184
	10	2	34.3	33.7	304	189	234	258	275	283	217
	11	1	40.4	18.2	314	221	270	295	312	320	236
	12		42.6	0.0	317	232	282	308	325	332	243
			SURFACE DAILY TOTALS		2624	1474	1852	2058	2204	2286	1808

NOTE:
1) BASED ON DATA IN TABLE 1, pp 387 in 1972 ASHRAE HANDBOOK OF FUNDAMENTALS; 0% GROUND REFLECTANCE; 1.0 CLEARNESS FACTOR.
2) SEE FIG. 4, pp 394 in 1972 ASHRAE HANDBOOK OF FUNDAMENTALS FOR TYPICAL REGIONAL CLEARNESS FACTORS.
3) GROUND REFLECTION NOT INCLUDED ON NORMAL OR HORIZONTAL SURFACES.

Figure R-B-8. Clear day solar position and insolation values for 32 degrees N latitude.

DATE	SOLAR TIME AM	PM	SOLAR POSITION ALT	AZM	BTUH/SQ. FT. TOTAL INSOLATION ON SURFACES — NORMAL	HORIZ.	SOUTH FACING SURFACE ANGLE WITH HORIZ. 22	32	42	52	90
JAN 21	7	5	1.4	65.2	1						
	8	4	12.5	56.5	203	56	93	106	116	123	115
	9	3	22.5	46.0	269	118	175	193	206	212	181
	10	2	30.6	33.1	295	167	235	256	269	274	221
	11	1	36.1	17.5	306	198	273	295	308	312	245
	12		38.0	0.0	310	209	285	308	321	324	253
	SURFACE DAILY TOTALS				2458	1288	1839	2008	2118	2166	1779
FEB 21	7	5	7.1	73.5	121	22	34	37	40	42	38
	8	4	19.0	64.4	247	95	127	136	140	141	108
	9	3	29.9	53.4	288	161	206	217	222	220	158
	10	2	39.1	39.4	306	212	266	278	283	279	193
	11	1	45.6	21.4	315	244	304	317	321	315	214
	12		48.0	0.0	317	255	316	330	334	328	222
	SURFACE DAILY TOTALS				2872	1724	2188	2300	2345	2322	1644
MAR 21	7	5	12.7	81.9	185	54	60	60	59	56	32
	8	4	25.1	73.0	260	129	146	147	144	137	78
	9	3	36.8	62.1	290	194	222	224	220	209	119
	10	2	47.3	47.5	304	245	280	283	278	265	150
	11	1	55.0	26.8	311	277	317	321	315	300	170
	12		58.0	0.0	313	287	329	333	327	312	177
	SURFACE DAILY TOTALS				3012	2084	2378	2403	2358	2246	1276
APR 21	6	6	6.1	99.9	66	14	9	6	6	6	3
	7	5	18.8	92.2	206	86	78	71	62	51	10
	8	4	31.5	84.0	255	158	156	148	136	120	35
	9	3	43.9	74.2	278	220	225	217	203	183	68
	10	2	55.7	60.3	290	267	279	272	256	234	95
	11	1	65.4	37.5	295	297	313	306	290	265	112
	12		69.6	0.0	297	307	325	318	301	276	118
	SURFACE DAILY TOTALS				3076	2390	2444	2356	2206	1994	764
MAY 21	6	6	10.4	107.2	119	36	13	8	7	7	9
	7	5	22.8	100.1	211	107	104	95	82	67	14
	8	4	35.4	92.9	250	175	175	159	143	120	16
	9	3	48.1	84.7	269	233	236	224	203	174	33
	10	2	60.6	73.3	280	277	290	277	252	219	56
	11	1	72.0	51.9	285	305	321	306	279	237	72
	12		78.0	0.0	286	315	331	318	290	247	77
	SURFACE DAILY TOTALS				3112	2582	2454	2284	2064	1788	469
JUN 21	6	6	12.2	110.2	131	45	15	14	14	14	9
	7	5	24.3	103.4	210	115	91	76	59	41	14
	8	4	36.9	96.8	245	180	159	143	122	99	16
	9	3	49.6	89.4	264	236	221	204	181	153	19
	10	2	62.2	79.7	274	279	268	251	227	197	41
	11	1	74.2	60.9	279	306	299	282	257	224	56
	12		81.5	0.0	280	315	309	292	267	234	60
	SURFACE DAILY TOTALS				3084	2634	2436	2234	1990	1690	370

DATE	SOLAR TIME AM	PM	SOLAR POSITION ALT	AZM	BTUH/SQ. FT. TOTAL INSOLATION ON SURFACES — NORMAL	HORIZ.	SOUTH FACING SURFACE ANGLE WITH HORIZ. 22	32	42	52	90
JUL 21	6	6	10.7	107.7	113	37	22	14	13	12	8
	7	5	23.1	100.6	203	107	87	75	60	44	14
	8	4	35.7	93.6	241	174	158	143	125	104	16
	9	3	48.4	85.5	261	231	220	205	185	159	31
	10	2	60.9	74.3	271	274	269	254	232	204	54
	11	1	72.4	53.3	277	302	300	285	262	232	69
	12		78.6	0.0	279	311	310	296	273	242	74
	SURFACE DAILY TOTALS				3012	2558	2422	2250	2030	1754	458
AUG 21	6	6	6.5	100.5	59	10	10	6	6	6	4
	7	5	19.1	92.8	190	85	77	69	60	50	12
	8	4	31.8	84.7	240	152	144	132	116	99	33
	9	3	44.3	75.0	263	216	220	212	197	178	65
	10	2	56.1	61.3	282	246	272	264	249	226	91
	11	1	66.0	38.4	284	292	305	298	281	257	107
	12		70.3	0.0	284	302	317	309	292	268	113
	SURFACE DAILY TOTALS				2902	2352	2388	2296	2144	1934	736
SEP 21	7	5	12.7	81.9	163	51	56	56	55	52	30
	8	4	25.1	73.0	240	124	140	141	138	131	75
	9	3	36.8	62.1	272	188	213	215	211	201	114
	10	2	47.3	47.5	287	237	273	273	268	255	145
	11	1	55.0	26.8	294	268	306	309	303	289	164
	12		58.0	0.0	296	278	318	321	315	300	171
	SURFACE DAILY TOTALS				2708	2014	2308	2264	2154	2154	1226
OCT 21	7	5	6.8	73.1	99	19	29	32	34	36	32
	8	4	18.7	64.0	229	90	120	128	133	134	104
	9	3	29.5	53.0	273	155	198	208	212	212	153
	10	2	38.7	39.1	293	204	257	269	273	270	188
	11	1	45.1	21.1	302	236	294	307	311	306	209
	12		47.5	0.0	304	247	306	320	324	318	217
	SURFACE DAILY TOTALS				2696	1654	2100	2208	2252	2232	1588
NOV 21	7	5	1.5	65.4	2	0	1	1	1	1	1
	8	4	12.7	56.6	196	55	91	104	113	119	111
	9	3	22.6	46.1	263	118	173	190	202	208	176
	10	2	30.8	33.2	289	166	233	252	265	270	217
	11	1	36.2	17.6	301	197	270	291	303	307	241
	12		38.2	0.0	304	207	282	304	316	320	249
	SURFACE DAILY TOTALS				2406	1280	1816	1980	2084	2130	1742
DEC 21	8	4	10.3	53.8	176	41	77	90	101	108	107
	9	3	19.8	43.6	257	102	161	180	195	204	183
	10	2	27.6	31.2	288	150	221	244	259	267	226
	11	1	32.7	16.4	301	180	258	282	298	305	251
	12		34.6	0.0	304	190	271	295	311	318	259
	SURFACE DAILY TOTALS				2348	1136	1704	1888	2016	2086	1794

NOTE: 1) BASED ON DATA IN TABLE 1, pp 387 in 1972 ASHRAE HANDBOOK OF FUNDAMENTALS, 0% GROUND REFLECTANCE, 1.0 CLEARNESS FACTOR.
2) SEE FIG. 4, pp 394 in 1972 ASHRAE HANDBOOK OF FUNDAMENTALS FOR TYPICAL REGIONAL CLEARNESS FACTORS.
3) GROUND REFLECTION NOT INCLUDED ON NORMAL OR HORIZONTAL SURFACES.

Figure R-B-8. Clear day solar position and insolation values for 32 degrees N latitude. (Reprinted by permission of ASHRAE.)

BTUH/SQ. FT. TOTAL INSOLATION ON SURFACES — Solar position and insolation values, 40° N latitude.

JAN 21 – JUN 21

DATE	AM	PM	ALT	AZM	NORMAL	HORIZ.	30	40	50	60	90
JAN 21	8	4	8.1	55.3	142	28	65	74	81	85	84
	9	3	16.8	44.0	239	83	155	171	182	187	171
	10	2	23.8	30.9	274	127	218	237	249	254	223
	11	1	28.4	16.0	289	154	257	277	290	293	253
	12		30.0	0.0	294	164	270	291	303	306	263
	SURFACE DAILY TOTALS				2182	948	1660	1810	1906	1944	1726
FEB 21	7	5	4.8	72.7	69	10	19	21	23	24	22
	8	4	15.4	62.2	224	73	114	122	126	127	107
	9	3	25.0	50.2	274	132	195	209	209	208	167
	10	2	32.8	35.9	295	178	256	267	271	267	208
	11	1	38.1	18.9	305	206	293	306	310	304	236
	12		40.0	0.0	308	216	306	319	323	317	245
	SURFACE DAILY TOTALS				2640	1414	2060	2162	2202	2176	1730
MAR 21	7	5	11.4	80.2	171	46	55	54	51	47	35
	8	4	22.5	69.6	250	114	140	141	138	131	89
	9	3	32.8	57.3	282	173	215	217	213	203	138
	10	2	41.6	41.9	297	218	273	276	271	258	176
	11	1	47.7	22.6	305	247	310	313	307	293	200
	12		50.0	0.0	307	257	322	326	320	305	208
	SURFACE DAILY TOTALS				2916	1852	2308	2330	2284	2174	1484
APR 21	6	6	7.4	98.9	89	20	11	8	7	7	4
	7	5	18.9	89.5	206	87	77	70	61	50	12
	8	4	30.3	79.3	252	152	153	145	133	117	53
	9	3	41.3	67.2	274	207	221	213	199	179	93
	10	2	51.2	51.4	286	250	275	267	252	229	126
	11	1	58.7	29.2	292	277	308	301	285	260	147
	12		61.6	0.0	293	287	320	313	296	271	154
	SURFACE DAILY TOTALS				3092	2274	2360	2258	2104	1894	994
MAY 21	5	7	1.9	114.7	1	0	0	0	0	0	0
	6	6	12.7	105.6	144	49	25	14	13	13	9
	7	5	24.0	96.6	216	114	89	76	60	44	12
	8	4	35.4	87.2	250	175	158	144	125	104	25
	9	3	46.8	76.0	267	227	221	206	186	160	60
	10	2	57.5	60.9	277	267	270	255	233	205	89
	11	1	66.2	37.1	283	293	296	281	264	234	108
	12		70.0	0.0	284	301	312	297	274	243	114
	SURFACE DAILY TOTALS				3160	2552	2442	2264	2040	1760	724
JUN 21	5	7	4.2	117.3	22	4	2	2	2	2	1
	6	6	14.8	108.4	155	60	30	18	16	16	10
	7	5	26.0	99.7	216	123	92	77	59	41	14
	8	4	37.4	90.7	246	182	159	142	121	97	16
	9	3	48.8	80.2	263	233	219	202	179	151	47
	10	2	59.8	65.8	272	272	266	248	224	194	74
	11	1	69.2	41.9	277	296	293	274	247	221	82
	12		73.5	0.0	279	304	306	287	260	230	90
	SURFACE DAILY TOTALS				3180	2648	2434	2224	1974	1670	610

JUL 21 – DEC 21

DATE	AM	PM	ALT	AZM	NORMAL	HORIZ.	30	40	50	60	90
JUL 21	5	7	2.3	115.2	2	0	0	0	0	0	0
	6	6	13.1	106.1	138	50	26	17	15	14	9
	7	5	24.3	97.2	208	114	89	75	60	44	14
	8	4	35.8	87.8	241	174	157	142	124	102	24
	9	3	47.2	76.7	259	225	218	203	182	157	58
	10	2	57.9	61.7	269	265	266	251	229	200	86
	11	1	66.7	37.9	275	290	296	281	258	228	104
	12		70.6	0.0	276	298	307	292	269	238	111
	SURFACE DAILY TOTALS				3062	2534	2409	2230	2006	1728	702
AUG 21	6	6	7.9	99.5	81	21	12	9	8	7	5
	7	5	19.3	90.0	191	87	76	69	60	49	12
	8	4	30.7	79.9	237	150	150	141	129	113	50
	9	3	41.8	67.9	260	205	216	207	193	173	89
	10	2	51.7	52.1	272	246	267	259	244	221	120
	11	1	59.3	29.7	278	273	300	292	276	252	140
	12		62.3	0.0	280	282	311	303	287	262	147
	SURFACE DAILY TOTALS				2916	2244	2354	2258	2104	1894	978
SEP 21	7	5	11.4	80.2	149	43	51	51	49	47	32
	8	4	22.5	69.6	230	109	133	134	131	124	84
	9	3	32.8	57.3	263	167	206	208	203	193	132
	10	2	41.6	41.9	280	211	262	265	260	247	168
	11	1	47.7	22.6	287	239	298	301	295	281	192
	12		50.0	0.0	290	249	310	313	307	292	200
	SURFACE DAILY TOTALS				2708	1788	2210	2228	2182	2074	1416
OCT 21	7	5	4.5	72.3	48	7	14	15	17	17	16
	8	4	15.0	61.9	204	68	106	113	117	118	100
	9	3	24.5	49.8	257	126	185	195	200	198	160
	10	2	32.4	35.6	280	170	245	257	261	257	203
	11	1	37.6	18.7	291	199	283	295	299	294	229
	12		39.5	0.0	294	208	295	308	312	306	238
	SURFACE DAILY TOTALS				2454	1348	1962	2060	2098	2074	1654
NOV 21	8	4	8.2	55.4	136	28	63	72	78	82	81
	9	3	17.0	44.1	232	82	152	167	178	183	167
	10	2	24.0	31.0	268	126	215	233	245	249	219
	11	1	28.6	16.1	283	153	254	273	285	288	248
	12		30.2	0.0	288	163	267	287	298	301	258
	SURFACE DAILY TOTALS				2128	942	1636	1778	1870	1908	1686
DEC 21	8	4	5.5	53.0	89	14	39	45	50	54	56
	9	3	14.0	41.9	217	65	135	152	164	171	163
	10	2	20.7	29.4	261	107	200	221	235	242	221
	11	1	25.0	15.2	280	134	239	262	276	283	252
	12		26.6	0.0	285	143	253	275	290	296	263
	SURFACE DAILY TOTALS				1978	782	1480	1634	1740	1796	1646

NOTE: 1) BASED ON DATA IN TABLE 1, pp 387 in 1972 ASHRAE HANDBOOK OF FUNDAMENTALS; 0% GROUND REFLECTANCE; 1.0 CLEARNESS FACTOR.
2) SEE FIG. 4, pp 394 in 1972 ASHRAE HANDBOOK OF FUNDAMENTALS FOR TYPICAL REGIONAL CLEARNESS FACTORS.
3) GROUND REFLECTION NOT INCLUDED ON NORMAL OR HORIZONTAL SURFACES.

Figure R·B·9. Clear day solar position and insolation values for 40 degrees N latitude. (Reprinted by permission of ASHRAE.)

Figure R·B·10 — Clear day solar position and insolation values for 48 degrees N latitude.

BTUH/SQ. FT. TOTAL INSOLATION ON SURFACES — SOUTH FACING SURFACE ANGLE WITH HORIZ.

DATE	AM	PM	ALT	AZM	NORMAL	HORIZ.	38	48	58	68	90
JAN 21	8	4	3.5	54.6	37	4	17	19	21	22	22
	9	3	11.0	42.6	185	46	120	132	140	145	139
	10	2	16.9	29.4	239	83	190	206	216	220	206
	11	1	20.7	15.1	261	107	231	249	260	263	243
		12	22.0	0.0	267	115	245	264	275	278	255
	SURFACE DAILY TOTALS				1710	596	1360	1478	1550	1578	1478
FEB 21	7	5	2.4	72.2	12	3	4	4	4	4	4
	8	4	11.6	60.5	188	49	95	102	106	106	96
	9	3	19.7	47.7	251	100	178	187	193	190	167
	10	2	26.2	33.3	278	139	240	251	257	251	217
	11	1	30.5	17.2	290	165	278	290	297	288	247
		12	32.0	0.0	293	173	291	304	310	301	258
	SURFACE DAILY TOTALS				2330	1080	1880	1972	2024	1978	1720
MAR 21	7	5	10.0	78.7	153	37	45	47	49	45	35
	8	4	19.5	66.8	236	96	131	132	129	122	96
	9	3	28.2	53.4	270	147	205	208	203	193	152
	10	2	35.4	37.8	287	187	266	267	261	248	195
	11	1	40.3	19.8	295	212	300	303	297	283	223
		12	42.0	0.0	298	220	312	314	309	294	232
	SURFACE DAILY TOTALS				2780	1578	2208	2228	2182	2074	1632
APR 21	6	6	8.6	97.8	108	27	13	9	8	7	2
	7	5	18.6	86.7	205	85	76	69	59	48	21
	8	4	28.5	74.9	247	142	149	141	129	113	69
	9	3	37.8	61.2	268	191	216	208	194	174	115
	10	2	45.8	44.6	280	228	268	260	245	223	152
	11	1	51.5	24.0	286	252	301	294	278	254	177
		12	53.6	0.0	288	260	313	305	289	264	185
	SURFACE DAILY TOTALS				3076	2106	2358	2266	2114	1902	1262
MAY 21	5	7	5.2	114.3	41	9	4	4	3	3	2
	6	6	14.7	103.7	162	61	27	16	15	13	9
	7	5	24.6	93.0	219	118	89	75	60	43	13
	8	4	34.7	81.6	248	171	156	142	123	101	45
	9	3	44.3	68.3	264	217	217	202	182	156	86
	10	2	53.0	51.3	274	252	265	251	229	200	120
	11	1	59.5	28.6	279	274	296	281	258	228	141
		12	62.0	0.0	280	281	306	292	269	238	149
	SURFACE DAILY TOTALS				3254	2482	2418	2234	2010	1728	982
JUN 21	5	7	7.9	116.5	77	21	15	13	11	10	5
	6	6	17.2	106.2	172	74	38	20	16	14	12
	7	5	27.0	95.8	220	129	95	78	61	40	15
	8	4	37.1	84.6	246	181	154	138	120	96	35
	9	3	46.9	71.6	261	225	212	196	175	145	74
	10	2	55.8	54.8	269	259	259	244	221	186	105
	11	1	62.7	31.2	274	280	289	272	241	213	125
		12	65.5	0.0	275	287	298	283	260	236	133
	SURFACE DAILY TOTALS				3312	2626	2420	2204	1950	1644	874

DATE	AM	PM	ALT	AZM	NORMAL	HORIZ.	38	48	58	68	90
JUL 21	5	7	5.7	114.7	43	10	5	5	4	4	3
	6	6	15.2	104.1	156	62	28	18	16	15	11
	7	5	25.1	93.5	211	118	89	75	59	42	14
	8	4	35.1	82.1	240	171	154	140	121	99	43
	9	3	44.8	68.8	256	215	214	199	178	153	83
	10	2	53.5	51.9	266	250	261	246	224	195	116
	11	1	60.1	29.0	271	272	291	276	253	223	137
		12	62.6	0.0	272	279	301	286	263	232	144
	SURFACE DAILY TOTALS				3158	2474	2386	2200	1974	1694	956
AUG 21	6	6	9.1	98.3	99	28	14	10	9	8	6
	7	5	19.1	87.2	190	85	75	67	58	47	20
	8	4	29.0	75.4	232	141	145	137	125	109	65
	9	3	38.4	61.8	254	189	210	201	187	168	110
	10	2	46.4	45.1	266	225	260	252	237	214	146
	11	1	52.2	24.3	272	248	293	285	268	244	169
		12	54.3	0.0	274	256	304	296	279	255	177
	SURFACE DAILY TOTALS				2898	2086	2300	2200	2046	1836	1208
SEP 21	7	5	10.0	78.7	131	35	43	44	44	43	31
	8	4	19.5	66.8	215	92	124	123	115	110	90
	9	3	28.2	53.4	251	142	196	198	193	184	143
	10	2	35.4	37.8	269	181	251	254	248	236	185
	11	1	40.3	19.8	278	205	287	290	284	269	212
		12	42.0	0.0	280	213	299	301	296	281	221
	SURFACE DAILY TOTALS				2568	1522	2102	2118	2070	1966	1546
OCT 21	7	5	2.0	71.9	17	3	7	7	8	8	7
	8	4	11.2	60.2	165	44	86	91	95	95	87
	9	3	19.3	47.4	233	94	167	176	180	178	157
	10	2	25.7	33.1	262	133	228	239	242	239	207
	11	1	30.0	17.1	274	157	266	277	281	276	237
		12	31.5	0.0	278	166	279	291	294	288	247
	SURFACE DAILY TOTALS				2154	1022	1774	1860	1890	1866	1626
NOV 21	8	4	3.6	54.7	36	5	17	19	21	22	22
	9	3	11.2	42.7	179	46	117	129	137	141	135
	10	2	17.1	29.5	233	83	186	202	212	215	201
	11	1	20.9	15.1	255	107	227	245	255	258	238
		12	22.2	0.0	261	115	241	259	270	272	250
	SURFACE DAILY TOTALS				1668	596	1336	1448	1518	1544	1442
DEC 21	9	3	8.0	40.9	140	27	87	98	105	110	109
	10	2	13.6	28.2	214	63	164	180	192	197	190
	11	1	17.3	14.4	242	86	207	226	239	244	231
		12	18.6	0.0	250	94	222	241	254	260	244
	SURFACE DAILY TOTALS				1444	446	1136	1250	1326	1364	1304

NOTE: 1) BASED ON DATA IN TABLE 1, pp 387, in 1972 ASHRAE HANDBOOK OF FUNDAMENTALS.
GROUND REFLECTANCE = 0.2   CLEARNESS FACTOR
2) SEE FIG. 4, pp 394, in 1972 ASHRAE HANDBOOK OF FUNDAMENTALS FOR TYPICAL REGIONAL
CLEARNESS FACTORS.
3) GROUND REFLECTION NOT INCLUDED ON NORMAL OR HORIZONTAL SURFACES.

Figure R·B·10. Clear day solar position and insolation values for 48 degrees N latitude. (Reprinted by permission of ASHRAE.)

BTUH/SQ. FT. TOTAL INSOLATION ON SURFACES — 56°N Latitude (JAN–JUN)

DATE	AM	PM	ALT	AZM	NORMAL	HORIZ.	46	56	66	76	90
JAN 21	9	3	5.0	41.8	78	11	50	55	59	60	60
	10	2	9.9	28.5	170	39	135	146	154	156	153
	11	1	12.9	14.5	207	58	183	197	206	208	201
	12		14.0	0.0	217	65	198	214	225	225	217
	SURFACE DAILY TOTALS				1126	282	934	1010	1058	1074	1044
FEB 21	8	4	7.6	59.4	129	25	65	69	72	72	69
	9	3	14.2	45.9	214	65	151	159	162	161	151
	10	2	19.4	31.5	250	98	215	225	228	224	208
	11	1	22.8	16.1	266	119	254	265	268	263	243
	12		24.0	0.0	270	126	268	279	282	276	255
	SURFACE DAILY TOTALS				1986	740	1640	1716	1742	1716	1598
MAR 21	7	5	8.3	77.5	128	28	40	40	39	37	32
	8	4	16.2	64.4	215	75	119	120	117	111	97
	9	3	23.3	50.3	253	118	192	193	189	180	154
	10	2	29.0	34.9	272	151	249	251	246	234	205
	11	1	32.7	17.9	282	172	285	288	282	268	236
	12		34.0	0.0	284	179	297	300	294	280	246
	SURFACE DAILY TOTALS				2586	1268	2066	2084	2040	1938	1700
APR 21	5	7	1.4	108.8	0	0	0	0	0	0	0
	6	6	9.6	96.5	122	32	14	9	8	7	6
	7	5	18.0	84.1	201	81	74	66	57	46	29
	8	4	26.1	70.9	239	129	143	135	123	108	82
	9	3	33.1	56.3	260	169	208	200	186	167	133
	10	2	39.9	39.7	272	201	259	251	236	214	174
	11	1	44.1	20.7	278	220	292	284	268	245	200
	12		45.6	0.0	280	227	303	295	279	255	209
	SURFACE DAILY TOTALS				3024	1892	2282	2186	2038	1830	1458
MAY 21	4	8	1.2	125.5	0	0	0	0	0	0	0
	5	7	8.5	113.4	93	25	10	9	8	7	6
	6	6	16.5	101.5	175	71	28	17	15	13	11
	7	5	24.8	89.3	219	119	88	74	58	41	16
	8	4	33.1	76.3	244	163	153	138	119	98	57
	9	3	40.9	61.6	259	201	212	197	176	147	105
	10	2	47.6	44.2	268	231	259	244	222	194	148
	11	1	52.3	23.4	273	249	288	274	251	225	174
	12		54.0	0.0	275	255	299	284	261	232	182
	SURFACE DAILY TOTALS				3340	2374	2374	2188	1962	1682	1218
JUN 21	4	8	4.2	127.2	21	4	2	2	2	2	1
	5	7	11.4	115.3	122	40	14	13	11	10	8
	6	6	19.3	103.6	185	86	34	19	17	15	12
	7	5	27.6	91.7	222	132	92	75	57	38	15
	8	4	35.9	78.8	243	175	154	137	116	92	55
	9	3	43.8	64.1	257	212	211	193	170	143	98
	10	2	50.7	46.4	265	240	255	238	214	184	133
	11	1	55.6	24.9	269	258	284	267	242	210	156
	12		57.5	0.0	271	264	294	276	251	219	164
	SURFACE DAILY TOTALS				3438	2562	2388	2166	1910	1606	1120

BTUH/SQ. FT. TOTAL INSOLATION ON SURFACES — 56°N Latitude (JUL–DEC)

DATE	AM	PM	ALT	AZM	NORMAL	HORIZ.	46	56	66	76	90
JUL 21	4	8	1.7	125.8	0	0	0	0	0	0	0
	5	7	9.0	113.7	91	27	11	10	9	8	6
	6	6	17.0	101.9	169	72	30	18	16	14	12
	7	5	25.3	89.7	212	119	88	74	58	41	15
	8	4	33.6	76.7	237	163	151	136	117	96	61
	9	3	41.4	62.0	252	201	208	193	173	147	106
	10	2	48.2	44.6	261	230	254	239	217	189	142
	11	1	52.9	23.7	265	248	283	268	245	216	165
	12		54.6	0.0	267	254	293	278	255	225	173
	SURFACE DAILY TOTALS				3240	2372	2342	2152	1926	1646	1186
AUG 21	5	7	2.0	109.2	0	0	0	0	0	0	0
	6	6	10.2	97.0	112	34	16	11	10	9	7
	7	5	18.5	84.5	187	82	73	65	56	45	28
	8	4	26.7	71.3	225	128	140	131	119	104	78
	9	3	34.3	56.7	246	168	202	193	179	160	126
	10	2	40.5	40.0	258	199	251	242	227	206	166
	11	1	44.8	20.9	264	218	282	274	258	235	191
	12		46.3	0.0	266	225	293	285	269	245	200
	SURFACE DAILY TOTALS				2850	1884	2218	2118	1966	1760	1392
SEP 21	7	5	8.3	77.5	107	25	36	36	34	32	28
	8	4	16.2	64.4	194	72	111	112	108	102	89
	9	3	23.3	50.3	233	114	181	182	178	168	147
	10	2	29.0	34.9	253	146	236	237	232	221	193
	11	1	32.7	17.9	263	166	271	273	267	254	223
	12		34.0	0.0	266	173	283	285	279	265	233
	SURFACE DAILY TOTALS				2368	1220	1950	1962	1918	1820	1594
OCT 21	8	4	7.1	59.1	104	20	53	57	59	59	57
	9	3	13.8	45.7	193	60	138	145	148	147	138
	10	2	19.0	31.3	231	92	201	210	213	210	195
	11	1	22.3	16.0	248	112	240	250	253	248	230
	12		23.5	0.0	253	119	253	263	266	261	241
	SURFACE DAILY TOTALS				1804	688	1516	1586	1612	1588	1480
NOV 21	9	3	5.0	41.9	76	12	49	54	57	59	58
	10	2	10.0	28.5	165	39	132	143	149	152	148
	11	1	13.1	14.2	201	58	179	193	201	203	196
	12		14.2	0.0	211	65	194	209	217	219	211
	SURFACE DAILY TOTALS				1094	284	914	986	1032	1046	1016
DEC 21	9	3	1.9	40.5	5	0	3	4	4	4	4
	10	2	6.6	27.5	113	19	86	95	101	104	103
	11	1	9.5	13.9	166	37	141	154	163	167	164
	12		10.6	0.0	180	43	159	173	182	186	182
	SURFACE DAILY TOTALS				748	156	620	678	716	734	722

NOTE: 1) BASED ON DATA IN TABLE 1, pp. 387 in 1972 ASHRAE HANDBOOK OF FUNDAMENTALS; 0.0 GROUND REFLECTANCE; 1.0 CLEARNESS FACTOR.
2) SEE FIG. 4 in 1972 ASHRAE HANDBOOK OF FUNDAMENTALS FOR TYPICAL REGIONAL CLEARNESS FACTORS.
3) GROUND REFLECTION NOT INCLUDED ON NORMAL OR HORIZONTAL SURFACES.

Figure R·B-11. Clear day solar position and insolation values for 56 degrees N latitude. (Reprinted by permission of ASHRAE.)

BTUH/SQ. FT. TOTAL INSOLATION ON SURFACES — 64° N latitude (July–December)

DATE	AM	PM	ALT	AZM	NORMAL	HORIZ.	54	64	74	84	90
JUL 21	4	8	6.4	125.3	53	13	6	3	4	4	4
	5	7	12.1	112.4	128	44	14	13	11	10	9
	6	6	18.4	99.4	179	81	30	17	16	13	12
	7	5	25.0	86.0	211	118	86	72	56	38	28
	8	4	31.4	71.8	231	152	146	131	113	91	77
	9	3	37.3	56.3	245	182	201	186	166	141	124
	10	2	42.2	39.2	253	204	245	230	208	181	162
	11	1	45.4	20.2	257	218	273	258	236	207	187
	12		46.6	0.0	259	223	282	267	245	216	195
			SURFACE DAILY TOTALS		3372	2248	2280	2090	1864	1588	1400
AUG 21	5	7	4.6	108.8	29	6	3	3	2	2	2
	6	6	11.0	95.5	123	39	16	11	10	8	7
	7	5	17.6	81.9	181	77	69	61	52	42	35
	8	4	23.9	67.8	214	113	132	123	112	97	87
	9	3	29.6	52.6	234	144	190	182	169	150	138
	10	2	34.2	36.2	246	168	237	229	215	194	179
	11	1	37.8	18.5	252	183	268	260	244	222	205
	12		38.3	0.0	254	188	278	270	255	232	215
			SURFACE DAILY TOTALS		2808	1646	2108	2008	1838	1662	1522
SEP 21	7	5	6.5	76.5	77	16	25	25	24	24	21
	8	4	12.7	62.6	163	51	92	92	90	85	81
	9	3	18.1	48.1	206	83	159	159	156	147	141
	10	2	22.3	32.7	229	108	212	201	209	198	189
	11	1	25.1	16.6	240	124	246	248	243	230	220
	12		26.0	0.0	244	129	258	260	254	241	230
			SURFACE DAILY TOTALS		2074	892	1726	1736	1696	1608	1532
OCT 21	8	4	3.0	58.5	17	2	18	20	20	21	21
	9	3	8.1	44.6	122	26	86	91	93	92	90
	10	2	12.1	30.2	176	50	152	159	161	159	155
	11	1	14.6	15.2	201	65	193	201	203	200	195
	12		15.5	0.0	208	71	207	215	217	213	208
			SURFACE DAILY TOTALS		1238	358	1088	1136	1152	1134	1106
NOV 21	10	2	3.0	28.1	23	3	18	20	21	21	21
	11	1	5.4	14.2	79	12	70	76	79	80	79
	12		6.2	0.0	97	17	89	96	101	101	100
			SURFACE DAILY TOTALS		302	46	266	286	298	302	300
DEC 21	11	1	1.8	13.7	16	0	14	15	16	17	17
	12		2.6	0.0	24	2	20	22	24	24	24
			SURFACE DAILY TOTALS		24	2	28	30	32	32	32

BTUH/SQ. FT. TOTAL INSOLATION ON SURFACES — 64° N latitude (January–June)

DATE	AM	PM	ALT	AZM	NORMAL	HORIZ.	54	64	74	84	90
JAN 21	10	2	2.8	28.1	81	12	17	19	20	20	20
	11	1	5.2	14.1	100	16	72	77	80	81	81
	12		6.0	0.0	100	16	91	98	102	103	103
			SURFACE DAILY TOTALS		306	45	268	290	302	306	304
FEB 21	8	4	3.4	58.7	35	4	17	19	19	19	19
	9	3	8.6	44.8	147	31	103	108	111	110	107
	10	2	12.6	30.3	199	55	170	178	181	178	173
	11	1	15.1	15.3	222	71	212	220	223	219	213
	12		16.0	0.0	228	77	225	235	237	232	226
			SURFACE DAILY TOTALS		1432	400	1230	1286	1302	1282	1252
MAR 21	7	5	6.5	76.5	95	18	30	29	29	27	25
	8	4	20.7	62.6	185	54	101	102	99	94	89
	9	3	23.3	48.1	227	87	171	172	169	160	153
	10	2	25.1	32.7	249	112	227	229	224	213	203
	11	1	26.0	16.6	260	134	262	265	259	246	235
	12		26.0	0.0	263	134	274	277	271	258	246
			SURFACE DAILY TOTALS		2296	932	1856	1870	1830	1736	1656
APR 21	5	7	4.0	108.5	27	2	2	2	2	1	1
	6	6	10.4	95.1	133	37	15	9	8	7	6
	7	5	17.3	81.6	194	76	70	63	54	43	37
	8	4	23.3	67.5	228	112	136	128	116	102	91
	9	3	29.0	52.3	248	144	197	189	176	158	145
	10	2	33.5	36.0	260	169	246	239	224	203	188
	11	1	36.5	18.4	266	184	278	270	255	233	216
	12		37.6	0.0	268	190	289	281	266	243	225
			SURFACE DAILY TOTALS		2982	1644	2176	2082	1936	1736	1594
MAY 21	4	8	5.8	125.1	51	11	5	4	4	3	3
	5	7	11.6	112.1	132	42	13	11	10	9	8
	6	6	17.9	99.1	185	79	29	29	16	12	11
	7	5	24.5	85.7	218	117	86	72	56	39	28
	8	4	30.9	71.5	239	152	148	133	115	94	80
	9	3	36.8	56.1	252	182	204	190	170	145	128
	10	2	41.6	38.8	261	205	249	235	213	186	167
	11	1	44.9	20.1	265	219	278	264	242	213	193
	12		46.0	0.0	267	235	288	274	251	222	201
			SURFACE DAILY TOTALS		3470	2236	2312	2124	1898	1624	1436
JUN 21	3	9	4.2	139.4	21	4	2	2	2	2	2
	4	8	9.0	126.4	93	27	10	9	8	7	7
	5	7	14.7	113.6	154	60	16	13	13	11	10
	6	6	21.0	100.8	194	96	34	19	17	14	13
	7	5	27.5	87.5	221	132	91	74	55	36	23
	8	4	34.0	73.3	239	166	150	133	112	88	73
	9	3	39.9	57.8	251	195	204	187	164	137	119
	10	2	44.9	40.4	258	217	247	230	206	177	157
	11	1	48.3	20.9	262	231	275	258	233	202	181
	12		49.5	0.0	263	235	284	267	242	211	189
			SURFACE DAILY TOTALS		3650	2488	2342	2118	1862	1558	1356

NOTE: 1) BASED ON DATA IN TABLE 1, p. 387 IN 1972 ASHRAE HANDBOOK OF FUNDAMENTALS; 01
    2) GROUND REFLECTANCE, 1.0; CLEARNESS FACTOR
      CLEARNESS FCTR p. 394 IN 1972 ASHRAE HANDBOOK OF FUNDAMENTALS FOR TYPICAL REGIONAL
      CLEARNESS FACTORS.
    3) GROUND REFLECTION NOT INCLUDED ON NORMAL OR HORIZONTAL SURFACES.

Figure R·B-12. Clear day solar position and insolation values for 64 degrees N latitude. (Reprinted by permission of ASHRAE.)

5. Comparing the totals shows that the 22° tilt, or latitude − 10°, is best for cooling.

6. The same process for year-round collection yields totals of: 26,456 Btu/hr for 22°, 26,514 for 32°, 25,874 for 42°, 24,492 for 52°, and 13,848 for 90° tilt.

7. Comparing the totals reveals that the 32° tilt, or latitude, is best for year-round collection.

These conclusions are useful for the designer as they stand, but closer scrutiny of the figures will provide further guidance. For instance, the 42° tilt was judged best for heating, but the totals for 32° and 52° are within 2% of the 42° total. Other considerations of the design (building layout, structural framing, height restrictions, etc.) can, therefore, come into the decision process without seriously affecting the final efficiency due to tilt.

The Clear Day Insolation Data is an extremely valuable tool for analysis, but its limitations must be kept in mind. For instance, there is no consideration of ground reflectance. In the heating season example above, the 90° tilt total is about 30% below the 42° maximum. In reality, values of insolation on a vertical surface are only about 10–20% lower than optimum during the heating season because of the contribution of ground reflection. This is especially true at higher latitudes. Another limitation is the assumption of an average clear day. Many locations are clearer than this (high altitudes, deserts, etc.), and many are less clear (industrial areas, dust areas, etc.). In addition, the data does not account for cloud conditions, which become quite important for long-term predictions. Thus, careful judgment must be exercised in the use of these extremely valuable tables.

A nomograph providing useful general information about solar radiation is included in Figure R·B·13. First is the method for determining hour of sunset

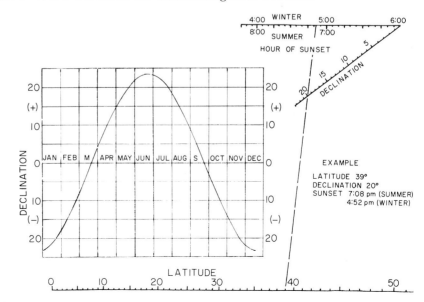

Figure R·B·13. Nomograph for determining length of collection day. Source: (WHI.)

for any location and time of year. The following example illustrates the use of the chart. Determine the time of sunset on May 20 in Lisbon, Portugal (39°N latitude). The declination graph shows that on May 20, declination is 20°. By placing a straightedge across the 39° latitude point and the 20° declination point, we can read the sunset time. May is summer in the northern hemisphere, so the sunset is 7:08 pm. On January 20, when declination is also 20°, sunset time would be 4:52 pm. These times are in local solar time.

Next is a graph (Figure R·B·14) showing the fraction of the day's solar radiation that is received on a horizontal surface between specified hours. To use this graph, one needs to know the mean daily solar radiation and the length of the day (determined from the previous nomograph). For south-facing (north-facing in the southern hemisphere) tilted surfaces, an "apparent length" of day must be determined. This is done by subtracting the angle of tilt from the actual latitude to arrive at an apparent latitude. This, then, is used in the first nomograph as the latitude. This use for tilted surfaces is valid only if the angle of tilt does not greatly exceed the latitude. This nomograph can be applied in the following manner. What fraction of the day's solar radiation on a horizontal surface is received between 9 am and 3 pm on May 20 in Lisbon? From the previous example, we know the sunset is at 7:08 pm; thus, the length of day is 5 am to 7pm. A vertical line drawn from the "9 to 3" point up to the 5 am to 7 pm

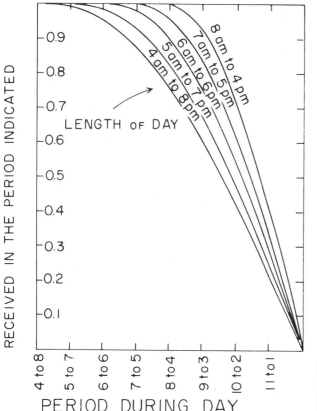

Figure R·B·14.
Fraction of the day's solar radiation that is received between specified hours.
Source: (WHI.)

curve corresponds to 0.67 as the fraction received during that period. It is interesting to note from this graph that 90% of any day's radiation (on a clear day) is received during the middle two-thirds of the day. Many such useful insights can be provided by studying this graph.

In addition to the direct data available, some common methods of operating on the data can produce further information. These are generally based on the statistical considerations of long-term average conditions. The full detail of these calculations, not included here, is contained in the article "Availability of Solar Energy for Flat Plate Solar Heat Collectors" by Y. H. Liu and R. C. Jordan, in *Low Temperature Engineering Application of Solar Energy,* published by ASHRAE. Isolating the diffuse component of total radiation is not an exact operation. The method recommended by Liu and Jordan is based on the ratio of average daily total radiation on a horizontal surface (measured at a given weather station) to the extraterrestrial daily radiation on a horizontal surface. This data (usually called percent of Extraterrestrial Radiation, or % ETR), while not generally published with radiation summaries, is available from the National Weather Records Center. Data for 80 stations is also included in an appendix to the Liu and Jordan article. This ratio, in combination with a simple graph, gives the relationship of diffuse to total radiation for the day (see Figure R·B·15).

For example, the graph can be used to find the portion of daily radiation that is diffuse if the % ETR is 50 and the daily total radiation is 1350 Btu/ft². On the graph, 50% corresponds to a ratio of .38, diffuse to total. $1350 \times 0.38 = 513$ Btu/

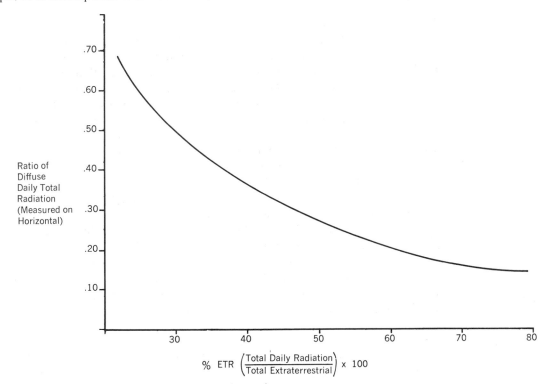

Figure R·B·15. Relationship of diffuse to total radiation. Source: (JOR.)

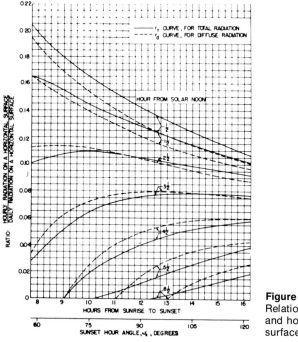

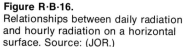

Figure R·B·16.
Relationships between daily radiation and hourly radiation on a horizontal surface. Source: (JOR.)

ft² diffuse for the day. The direct component, then, is $1350 - 513 = 837$ Btu/ft²/ day. Many calculations require knowledge of the relationships between hourly and daily radiation on a horizontal surface. This information is needed for both total radiation and its two components. The chart (Figure R·B·16) taken from Liu and Jordan gives the hourly to daily relationships for the total and the diffuse, based on hours from sunrise to sunset (found from the above nomograph). Using Figure R·B·16 find the amount of diffuse and direct radiation on a horizontal surface at 10:30 am on a 12 hour day, if the daily total radiation is 1350 Btu/ft², and the % of ETR is 50. From the above example, daily diffuse is 513 Btu/ft²/day. 10:30 am is one and one half hours from solar noon. The diffuse curve for this hour intersects the 12 hours (vertical) line at a point corresponding to 0.120 as the ratio between hourly and daily radiation. Thus, $513 \times 0.120 = 62$ Btu/ft²/hr, the diffuse component. Similarly, the total radiation curve indicates a ratio of 0.128. Thus, $1350 \times 0.128 = 173$ Btu/ft²/hr, the total radiation for the hour. The direct component, then, is $173 - 62 = 91$ Btu/ft²/hr.

Two trigonometric conversions, one for hourly radiation on any surface and one for daily radiation on south-facing surfaces, are used to arrive at the ratio of direct energy incident on a surface of given orientation to the direct energy incident on a horizontal surface. A graphic example of the effect of this ratio is given in Figure R·B·17. Ratios were computed by month for the indicated tilts and applied to the corresponding month's average daily radiation. In many cases, the radiation on the tilted surface is considerably greater than that on the horizontal. For instance, in November there is over twice as much direct radiation on a surface tilted to latitude + 20° than there is on a horizontal surface. It

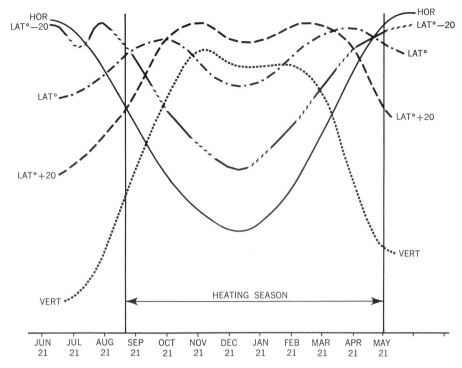

Figure R·B·17. South-facing surfaces: effect of tilt on direct radiation for 40° N latitude. Source: (JOR.)

should be stressed that the diffuse component and any reflected energy from surrounding surfaces are treated differently. Only the direct component can be converted trigonometrically.

The conversion factor, R_D, for daily direct radiation on south-facing surfaces (north-facing in the southern hemisphere) depends on latitude (L), tilt (β), and sunset hour angle (ω) of both the horizontal and the tilted surfaces. This angle is, in turn, a function of latitude (L), tilt (β), and declination (δ). For a horizontal surface, the sunset hour angle is $\omega_s = -\tan L \tan \delta$. For a tilted surface, it is $\omega_s' = -\tan(L-\beta)\tan\delta$. Declination ($\delta$) is found from the nomograph above, and ω_s and ω_s' are computed. The conversion factor is slightly different, depending on the two hour angles:

$$\text{if } \omega_s \leq \omega_s' \text{ then } R_D = \frac{\cos(L-\beta)\,\sin\omega_s - \omega_s\cos\omega_s'}{\cos I\,\sin\omega_s - \omega_s\cos\omega_s}$$

$$\text{if } \omega_s' < \omega_s \text{ then } R_D = \frac{\cos(L-\beta)}{\cos L} \times \frac{\sin\omega_s' - \omega_s'\cos\omega_s'}{\sin\omega_s - \omega_s\cos\omega_s}$$

The conversion factor, $R_{\gamma\beta}$, for hourly direct radiation on a surface of any orientation, depends on surface tilt (β), surface azimuth (γ), sun altitude (α), and sun azimuth (ϕ). The sun angles are found in the Resource Section on "Solar

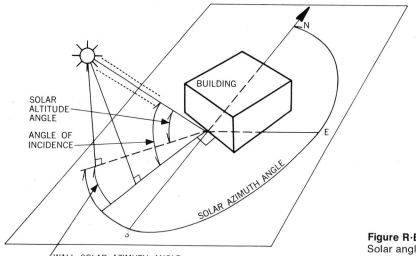

Figure R·B·18.
Solar angles for vertical and horizontal surfaces.

Angles and Shading." The surface tilt angles are described in Figure R·B·18. Once the four factors are known, the conversion can be calculated:

$$R_{\gamma\beta} = \cos\beta + \sin\beta\, \frac{\cos(\phi - \gamma)}{\tan\alpha}$$

The treatment of diffuse and reflected energy is somewhat different. It is generally assumed that diffuse radiation is uniform from all parts of the sky, so it is simply a matter of determining what fraction of the sky a given surface "sees" and reducing the total diffuse accordingly (see Figure R·B·19).

The diffuse radiation on a given surface, then, is

$$\frac{1 + \cos\beta}{2}\, D$$

where β is the tilt from horizontal and D is the diffuse radiation for the time period in question (see above).

The reflected radiation depends on the reflectivity (ρ) of the ground, on the fraction of the ground the surface "sees" (Figure R·B·19), and is a function of total radiation. A common average value is $\rho = 0.2$, although this varies with snow, grass, water, etc. Not considered in this is the effect of reflection from surrounding skyscrapers or other objects, which can be significant.

The reflected energy on a given surface is

$$\rho\left(\frac{1 - \cos\beta}{2}\right) H$$

where ρ is ground reflectance and H is total radiation on a horizontal surface over the time period in question.

Two examples of computations for actual insolation use the following Weather

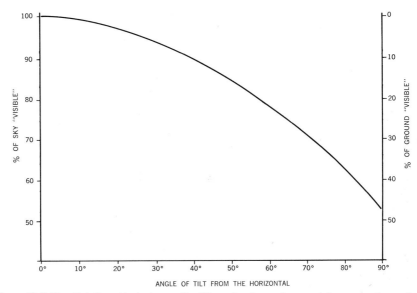

Figure R·B·19.   Relationship between the tilt angle of a surface and the percentage of sky which that surface "sees."

Bureau data for mean daily solar radiation on a horizontal surface for each month in New York City (JOR.).

			MONTHLY AVERAGE DAILY TOTAL RADIATION ON A HORIZONTAL SURFACE, Btu/day/ft² New York, New York Lat. 40°46′N.								
JAN	**FEB**	**MAR**	**APR**	**MAY**	**JUN**	**JUL**	**AUG**	**SEP**	**OCT**	**NOV**	**DEC**
540	791	1180	1426	1738	1994	1939	1606	1350	978	598	476

Data is recorded in Central Park, which is probably reasonably representative for the entire city. Areas of clearer air would get better values; heavily polluted or foggy areas less. There is, however, no other actual data with which to make meaningful adjustments.

The first example shows a method for determining insolation on vertical walls of varying compass azimuths. The first step in computation is to separate out the diffuse component of the total radiation received. Shown above are the curves for the ratio of total and diffuse radiation for times on either side of solar noon (plus and minus ½ hour, 1½ hours, 2½ hours, 3½ hours, 4½ hours, 5½ hours, and 6½ hours). Determine values for the 16th day of each month (15th in February), a more representative day than the 21st day used by ASHRAE. This accounts for the changing length of days and, in a general way, for the sky conditions that create the diffuse component. This is an approximation at best, but it is a reasonable one since it is considered valid for any location and is close to actual conditions. The diffuse component is a small part of the total, so a reasonably close approximation such as this will not seriously affect the overall accuracy.

The diffuse component is subtracted from the total to get the direct component on an hourly basis. By the above trigonometry, which takes into account the

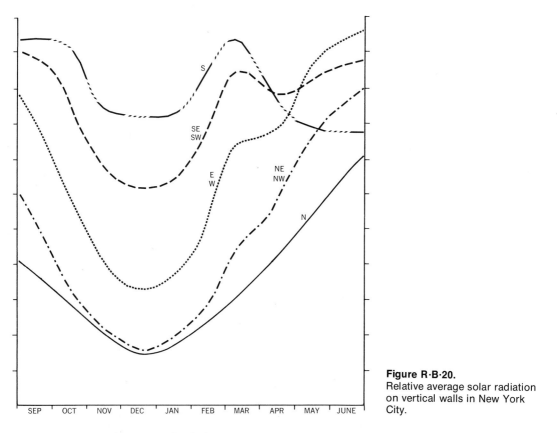

Figure R·B·20.
Relative average solar radiation
on vertical walls in New York
City.

hour, angle, latitude, solar altitude, and solar azimuth, along with the tilt and
orientation of the surface in question, the hourly values are obtained for direct
solar insolation on various surfaces throughout the day. The diffuse component
is treated as if sky radiation is uniform across the sky. For vertical walls, then, one
half of the sky is "visible" to the wall. Likewise, reflected radiation can be treated
as uniform from all directions. If half of the possible is assumed to be "visible"
and the ground has a reflectivity of 0.2 (average value), one tenth of the total
radiation is reflected on the wall. These values for diffuse and reflected energy
are added to the adjusted direct component values to give hourly values for
various months and orientations. A simple summation for the day gives the
expected average gain on vertical surfaces. These values are plotted for New
York City in Figure R·B·20.

The second example is a computation for variously tilted, south-facing collectors, and is somewhat different. Using the ratios of diffuse to daily total radiation, based on the ratio of measured radiation to extraterrestrial radiation,
determine the direct component of the mean daily radiation. Through trigonometric conversion, which takes into account the changing sun angle throughout
the day and the collector tilt, find the direct component for variously tilted south-
facing surfaces.

The diffuse component for a day is adjusted for the reduced portion of the sky

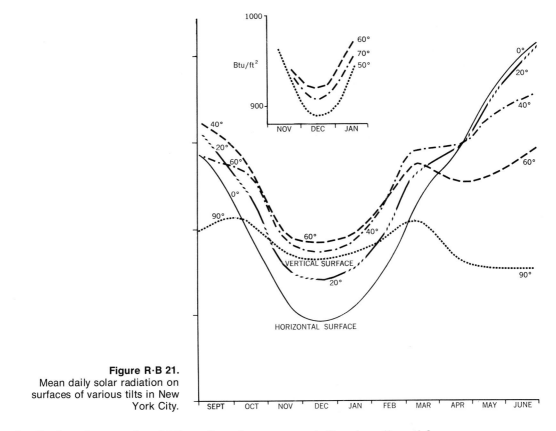

Figure R·B 21.
Mean daily solar radiation on surfaces of various tilts in New York City.

that the tilted surface can "see." The reflected component is likewise adjusted for the reduced portion of the ground affecting the collector. These are added to the adjusted direct component, giving mean daily insolation on the collector surfaces in question. The plots and tables of these data are given for New York City in Figure R·B·21.

All of these values may, with reasonable confidence, be multiplied by the number of days in a given month to obtain a mean expected solar radiation on the surface in question. Actual energy received can vary considerably, due to the extreme inconsistency of the weather, not only day to day, but year to year. In any case, these computations yield useable data.

RESOURCE SECTION R·C

Sol-Air Temperature

The heat flux, H, into an opaque sunlit surface is

$$H = \alpha I_t - \epsilon \Delta R + h_o(t_o - t_s)$$

H = total heat flux in Btu per square foot per hour of surface area (Btu/ft²/hr).

α = absorptance of the surface, in percent. This represents the ratio of the ability of the material to absorb solar radiation to the ability of a blackbody ($\alpha = 1$) to do the same. (See Appendix E on "Emittances and Absorptances" for values of some common materials.)

I_t = Incident solar irradiation, Btu per square foot per hour (Btu/ft²/hr).

ϵ = emittance of the surface, percent. This represents the ratio of the ability of the material to radiate its heat to the ability of a blackbody ($\epsilon = 1$) to do the same. (See Appendix E on "Emittances and Absorptances" for the values of some common materials.)

ΔR = the difference between the long-wave radiation incident on the surface from the sky and the surroundings, and the radiation emitted by a blackbody at outdoor air temperature, Btu per square foot per hour (Btu/ft²/hr).

h_o = coefficient of heat transfer by radiation and convection at the outer surface, Btu per degree Fahrenheit per square foot per hour (Btu/°F/ft²/hr).

t_o = outdoor air temperature, degrees Fahrenheit (°F).

t_s = outdoor surface temperature, degrees Fahrenheit (°F).

The "sol-air" temperature is an imaginary temperature of the outdoor air which combines the solar radiation terms (αI_t and $\epsilon \Delta R$) in the above equation with the convection term $h_o(t_o - t_s)$ so that the resultant sol-air temperature, t_e, is the one that the surface "sees" because of both convection and radiation. In terms of t_e, H becomes

$$H = h_o(t_e - t_s) = \alpha I_t - \epsilon \Delta R + h_o(t_o - t_s)$$

and t_e is therefore

$$t_e = t_o + (\alpha I_t / h_o) - (\epsilon \Delta R / h_o)$$

For example, Table 26 in the ASHRAE *Handbook of Fundamentals* (1972) gives sol-air temperatures for July 21 at 40°N latitude. The text in ASHRAE suggests that for horizontal surfaces, $\Delta R \approx 20$ Btu/ft²/hr, $\epsilon \approx 1$, $h_o \approx 3.0$ Btu/°F/ft²/hr; for vertical surfaces, $\Delta R \approx 0$. Plugging these values into the term $-\epsilon \Delta R / h_o$ results in the following:

$$\text{for horizontal surfaces: } -\epsilon \Delta R / h_o = -\frac{(1)(20)}{3} \approx -7.$$

$$\text{for vertical surfaces: } -\epsilon \Delta R / h_o = -\frac{\epsilon(0)}{h_o} = 0.$$

In Table 26 the parameter α / h_o has been given two values: 0.15 for light-colored surfaces and 0.30, the maximum likely value, for dark surfaces. See following table for the relative heat gain of some materials.

At noon on July 21, an average sunny day at 40°N latitude, approximately 126 Btu may be striking a south-facing surface each hour. If the air temperature is 90°F and the surface is light in color ($\alpha / h_o = 0.15$), the surface "sees" a sol-air temperature of 112°. Table 26 also shows that if the surface were dark ($\alpha / h_o = 0.30$), the sol-air temperature would be 134°.

As with the solar heat gain tables, this table is not a quick, useful design tool for reducing energy consumption for heating and cooling by taking advantage of the solar energy which hits walls and roofs. It should be possible for the designer to adjust surface orientation, size, color, and composition according to the effect each decision will have on the heating (or cooling) demand of the interior space.

RELATIVE HEAT GAIN OF SOME MATERIALS

$\alpha / h_o \approx 0.30$	4 —	Dark colored surfaces such as slate roofing
		tar roofing
		black paints
	3 —	Medium colored surfaces such as unpainted wood
		brick
		red tile
		dark cement
		red, grey, green paint
$\alpha / h_o \approx 0.15$	2 —	Light colored surfaces such as white stone
		light colored cement
		white paint
	1 —	

The primary source of information for this section is the ASHRAE *Handbook of Fundamentals*.

RESOURCE SECTION R·D

Solar Heat Gain Through Windows

The ASHRAE *Handbook of Fundamentals* (1972) explains that heat transfer through glass is affected by several factors. Among these are:

1. Solar irradiation intensity, I_t, and incident angle θ. (See Resource Sections R·B, "Insolation" and R·E, "Solar Angles and Shading" for help in determining the incident angle θ.) Figure R·D·1 indicates how transmittance, absorptance, and reflection vary according to incident angle.

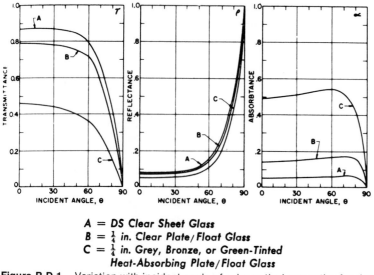

A = DS Clear Sheet Glass
B = $\frac{1}{4}$ in. Clear Plate/Float Glass
C = $\frac{1}{4}$ in. Grey, Bronze, or Green-Tinted
Heat-Absorbing Plate/Float Glass

Figure R·D·1. Variation with incident angle of solar-optical properties for double-strength sheet, clear and heat-absorbing plate/float glass. (Reprinted by permission from ASHRAE's *Handbook of Fundamentals*, 1974.)

2. Outdoor-indoor temperature difference. When the sun is not shining, heat flows according to the usual laws of conduction:

$$H = U (t_i - t_o).$$

This phenomenon is discussed in Part II and in the Resource Section R·A, "Heat Theory." Figure R·D·2 lists U-values for different types of glass.

3. Velocity and direction of air flow (wind) across the exterior fenestration surfaces. "Wind Control" is included in the discussions on conduction mentioned above.

4. Low temperature radiation exchange between the surface of the fenestration and the surroundings. This phenomenon is difficult to predict—it is assumed to be included in the outer and inner surface coefficients (air films), f_o and f_i. The glass becomes hot when the sun shines on it. Heat then flows by radiation and convection from its outer surface to the outside surrounding environment, and from its inner surface to the room. The rate of heat flow, H, inward by radiation and convection from an unshaded single pane of glass is:

RATE OF HEAT FLOW   ABSORBED SOLAR   CONDUCTION
THROUGH GLASS, H   =   HEAT GAIN   +   HEAT GAIN

where: ABSORBED SOLAR HEAT GAIN = N_i × absorbed radiation.

Type of Glass	U-Value	
	No Shading	Internal Shading[a]
Any Uncoated Single Glass[c]	1.06	0.81
Insulating Glass,[c] $\frac{3}{16}$ in. Air Space	0.66	0.54
uncoated $\frac{1}{4}$ in. Air Space	0.65	0.52
$\frac{3}{8}$ in. Air Space	0.61	0.50
$\frac{1}{2}$ in. Air Space	0.59	0.48
Prime Window Plus Storm Window, Air Space 1 in. or more	0.54[b]	0.47[b]
	No Supplementary Shading	
Double Glazing with Between-Glass Shading Louvered Sun Screen Separated by Air Space	0.63	
Venetian Blinds, Closed, in Air Space	0.44	
Glass Block Panels[d] Types I and II	0.56	
Types II, III, and IIIA	0.48	

Figure R·D·2.
Overall coefficients of heat transmission (U values) for fenestration under summer conditions (7.5 mph wind outdoors, still air indoors). (Reprinted by permission from ASHRAE's *Handbook of Fundamentals*, 1974.)

[a] Values apply to tightly closed Venetian and vertical blinds, draperies, and roller shades.
[b] Values apply to storm sash with a tight air space. Air leakage present in virtually all storm windows will, in effect, increase this value.
[c] *U*-values can be substantially reduced by low-emittance coatings applied to the inner surface of single or double glazing and to an air-space surface of insulating glass. Consult manufacturers for applicable *U*-values.
[d] Values listed are for $7\frac{3}{4} \times 7\frac{3}{4} \times 3\frac{7}{8}$ in. block. For $11\frac{3}{4} \times 11\frac{3}{4} \times 3\frac{7}{8}$ in. block, reduce the listed value by 0.04, and for $5\frac{3}{4} \times 5\frac{3}{4} \times 3\frac{7}{8}$ in. block, increase the listed value by 0.04. See Table 24 for definition of types.

Since absorbed radiation $= \alpha I_t$, then: ABSORBED SOLAR HEAT GAIN $= N_i(\alpha I_t)$.

N_i is the inward-flowing fraction of absorbed radiation. For unshaded single glazing, $N_i = U/f_o$, and: ABSORBED SOLAR HEAT GAIN $= \dfrac{U(\alpha I_t).}{f_o}$

Conduction heat gain is simply the U-value times the difference between the outdoor and indoor temperatures $(t_o - t_i)$. H, the rate of heat flow through the single glass, becomes

$$H = \begin{matrix} \text{ABSORBED SOLAR} \\ \text{HEAT GAIN} \end{matrix} + \begin{matrix} \text{CONDUCTION} \\ \text{HEAT GAIN} \end{matrix}$$

$$= \frac{U(\alpha I_t)}{f_o} + U(t_o - t_i)$$

$$= \frac{U(\alpha I_t) + f_o(t_o - t_i)}{f_o}$$

where:
   H = total heat flux in Btu per hour per square foot of surface area (Btu/ft²/hr).
   U = coefficient of heat transmission, Btu per degree Fahrenheit per square foot per hour (Btu/°F/ft²/hr).
   α = absorptance of the surface, in percent. This represents the ratio of the ability of the material to absorb solar radiation to the ability of a blackbody ($\alpha = 1$) to do the same. (See Appendix E, "Emittances and Absorptances," for values of some common materials.)
   I_t = incident solar irradiation, Btu per square foot per hour (Btu/ft²/hr).
   f_o = conductance of the outside air film, Btu per degree Fahrenheit per square foot per hour (Btu/°F/ft²/hr); usually taken as 6.0 Btu/°F/ft²/hr for a surface exposed to the weather; $1/f_o = 0.17$
   t_o = outdoor air temperature, degrees Fahrenheit (°F)
   t_i = indoor air temperature, degrees Fahrenheit (°F); usually taken as 65°F.

Since α, I_t and f_o vary greatly, primarily because of the continuously varying incident angle between the sun and the glass, the *Handbook* (1972) provides "Solar Heat Gain Tables" (pp. 388–392). Reference to these tables or to similar material is essential to designing for solar heat gain.

One of the most useful design tools for determining overall solar heat gain through windows is the chart (see Figure R·D·3) developed by F. W. Hutchinson at Purdue University. The example which accompanies the chart explains its use. Numbers particular to various cities and which can be used with this chart can be found in Part II (see Figure II·A·7). The normal outside air temperature and the fraction of possible sunshine are included there.

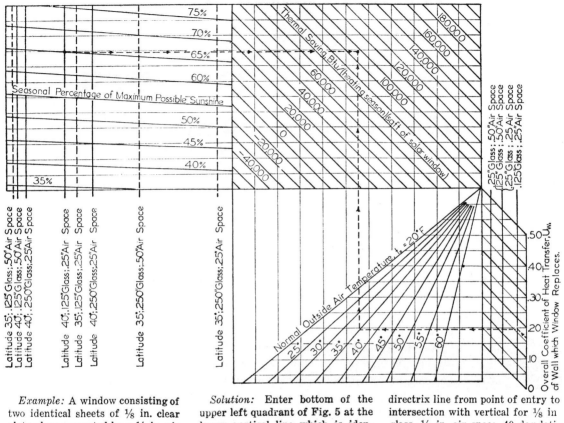

Example: A window consisting of two identical sheets of ⅛ in. clear plate glass separated by a ¼ in. air space is to be used in a south wall which has an overall coefficient of heat transfer of 0.165 Btu/(hr) (sq ft) (F). The normal outside temperature for the locality is 35 F and the sun shines for 65 per cent of the maximum possible hours between October 1 and May 1. Latitude is 40 degrees.

Solution: Enter bottom of the upper left quadrant of Fig. 5 at the heavy vertical line which is identified as applicable to ⅛ in. glass with ¼ in. air space at 40 deg latitude. Rise along this line to intersection with line for 65 per cent sunshine then move horizontally right (see dashed example line) to enter upper right quadrant. Now re-enter the figure at value of $U_w = 0.165$ on scale at right of lower right quadrant. Follow the

directrix line from point of entry to intersection with vertical for ⅛ in glass, ¼ in. air space, 40 deg latitude; then move horizontally left to intersect curve for $t_o = 35$ deg and from this point move vertically upward to intersect the horizontal line already established in the upper right quadrant. The point of intersection of these two lines gives the answer as 107,000 Btu saved per seven month heating season per square foot of window.

Figure R·D·3. Seasonal saving (or loss) attributable to 1 sq ft of double glass replacing an equal area of south wall with solar overhang of roof. (Note: the opaque wall area is not credited with any effective seasonal solar energy utilization.) Source: (HUT.)

RESOURCE SECTION R·E

Solar Angles And Shading

by Douglas Mahone,
Associate,
Total Environmental Action, Inc.

Most people realize that the sun's position in the sky changes from day to day and hour to hour. It is common knowledge that the sun is higher in the sky in the summer than in winter, and that the sun rises south of due east in the winter, and north of east in the summer. This is shown graphically by the accompanying sketch of the sun's path across the sky during the year; the encircled numbers indicate the time of day. In order to plan for the most effective use of shading, the sun's position must be determined. For instance, to decide on the size of a shading device to keep direct sun off a window between 10:00 am and 2:00 pm, it is necessary to know the angle from which the sunlight will be coming (the "angle of incidence"). Another situation requiring this knowledge is described in the Resource Section on "Insolation."

The sun's position in the sky is defined by two angular measurements: solar altitude and solar azimuth. Solar altitude (α) is measured up from the horizontal; solar azimuth (β) is measured from true south (see Figure R·E·1). These angles may be calculated or read from prepared tables or monographs.

The calculation depends upon three variables: latitude (L), declination (δ), and hour angle (H). Latitude is obvious—it can be read from any good map. Declination, a measure of how far north or south of the equator the sun has moved, varies from month to month (see Figure R·E·2). The hour angle

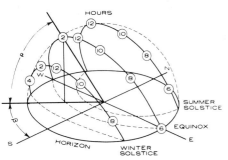

Figure R·E·1.
The sun's position in the sky.
Source: (RAM.)

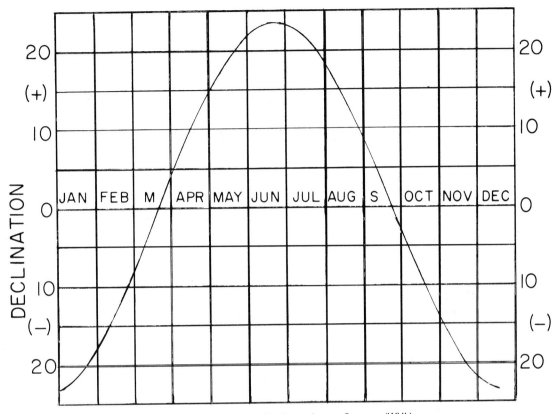

Figure R·E·2.   Graph to determine solar declination by time of year. Source: (WHI.)

depends on local solar time: H = 0.25 × (number of minutes from local solar noon). Solar time (the time read directly from a sundial) is based on solar noon when the sun is highest in the sky. Because of changes in the earth's velocity in orbit at different times during the year, the lengths of the days (measured from solar noon to solar noon) are slightly different from those of Mean Time days (measured by a clock running at a uniform rate). When local solar time is calculated, this difference is taken into account, along with a longitude correction if the observer is not on the standard time meridian for his time zone.

To correct local standard time (read from an accurate clock) to local solar time, several operations are necessary:

1. If daylight savings time is in effect, subtract one hour.
2. Determine the longitude meridian of the locality in question. Determine the standard time meridian for the locality (75° for EST, 90° for CST, 105° for MST, 120° for PST, 135° for YST, 150° for Alaska-Hawaii ST). Multiply the differences between the meridians by 4 minutes/degree. If the locality is east of the standard meridian, add the correction minutes to the standard time; if it is west, subtract them.
3. Add the equation of time (from Figure R·E·3) for the date in question to the corrected standard time. This is local solar time.

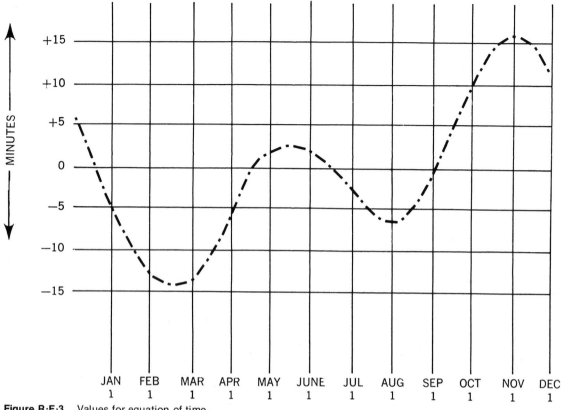

Figure R·E·3. Values for equation of time.

As an example, this procedure can be followed to determine the local solar time in Abilene, Texas, on December 1, at 1:30 pm (CST). Since it is not daylight savings time, no correction for that is necessary. On a map, we find that Abilene is on the 100°W meridian. Since the standard meridian for central time is 90°W, Abilene is 10° from the standard: 10° × 4 min/degree = 40 minutes. Since Abilene is west of the standard meridian, we subtract 40 minutes from local time: 1:30 − 40 = 12:50 pm. From the equation of time for December 1, we find we must add about 11 minutes: 12:50 + 11 = 1:01 local solar time, or 61 minutes from local solar noon.

From this, the hour angle (H) is determined as described above. Knowing latitude, declination, and hour angle, the solar altitude and azimuth follow:

$$\text{Solar altitude } \alpha = \cos L \cos \delta \cos H + \sin L \sin \delta$$
$$\text{Solar azimuth } \beta = \cos \delta \sin H / \cos \alpha$$

Solar altitude and azimuth can be determined for the 21st day of each month and for any hour of the day by use of sun path diagrams. A different diagram is required for each latitude, although interpolation between graphs is reasonably accurate. Reprinted here are eight diagrams which are adequate for the middle latitudes (see Figures R·E·4– 11).

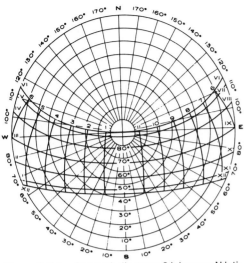

Figure R·E·4. Sun path diagram, 24 degrees N latitude. Source: (RAM.)

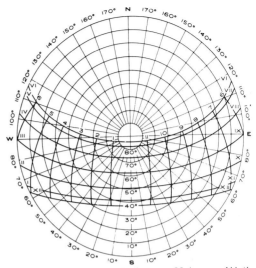

Figure R·E·5. Sun path diagram, 28 degrees N latitude. Source: (RAM.)

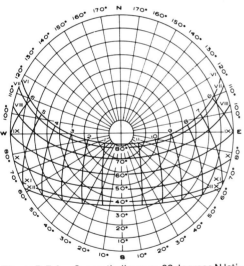

Figure R·E·6. Sun path diagram, 32 degrees N latitude. Source: (RAM.)

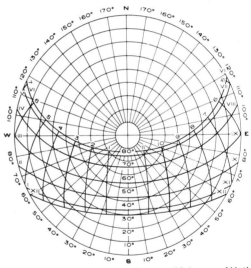

Figure R·E·7. Sun path diagram, 36 degrees N latitude. Source: (RAM.)

By using these diagrams, for example, the solar altitude and azimuth at 4:00 pm on April 21 in New York City (40°N) can be determined. Find the diagram for 40°N latitude (see Figure R·E·8) and locate the April line, the dark line running left-to-right numbered "IV" (April is the 4th month). Next locate the 4:00 pm line, the dark up-and-down line numbered "4." The intersection of these lines indicates the solar position. Solar altitude is read from the concentric circles; in this case it is 30°. The solar azimuth is read from the radial lines, S80°W

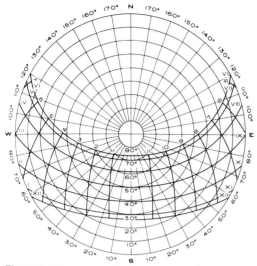

Figure R·E·8.   Sun path diagram, 40 degrees N latitude. Source: (RAM.)

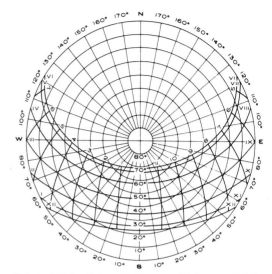

Figure R·E·9.   Sun path diagram, 44 degrees N latitude. Source: (RAM.)

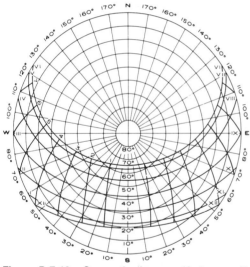

Figure R·E·10.   Sun path diagram, 48 degrees N latitude. Source: (RAM.)

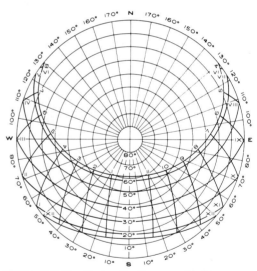

Figure R·E·11.   Sun path diagram, 52 degrees N latitude. Source: (RAM.)

in this case. When more accurate indications of the sun's position are required, Figures R·B·8 to R·B·14 may be used. They are not, however, as readily comprehensible as the sun path diagrams.

One of the primary applications of solar angle information is in the determination of shading angles for windows and collector surfaces, both to protect the surface from excessive sun and to ensure that a surface will not be shaded from useful solar energy.

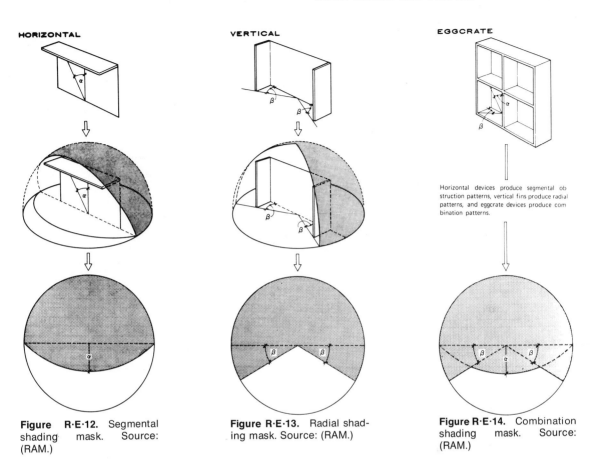

HORIZONTAL

VERTICAL

EGGCRATE

Horizontal devices produce segmental ob
struction patterns, vertical fins produce radial
patterns, and eggcrate devices produce com
bination patterns.

Figure R·E·12. Segmental shading mask. Source: (RAM.)

Figure R·E·13. Radial shading mask. Source: (RAM.)

Figure R·E·14. Combination shading mask. Source: (RAM.)

There are two basic ways of shading: by horizontal or vertical obstructions over the surface. Horizontal obstructions block out light from above (see Figure R·E·12). The extent of shading is determined by the relative geometries of the obstruction and of the surface: the broader the overhang, the more coverage; the higher the overhang, the less coverage. Vertical obstructions block out light from the sides (see Figure R·E·13). As with horizontal obstructions, the geometry of the shading angle controls the proportional sizes and proximities of the obstruction and the surface. These obstructions are represented in the figures as man-made shading devices, but they can just as well be trees, mountains, or buildings.

For any shading condition, a shading mask can be constructed to represent the amount and effectiveness of the shading on a given surface. A horizontal obstruction would result in a segmental shading mask, where the quantity α, indicated by the mask, corresponds to the angle α of the shading arrangement. A vertical obstruction would result in a radial shading mask, with the angle β corresponding to the same angle of the shading arrangement. For combinations of vertical and shading elements, a combination mask can be constructed (see Figure R·E·14).

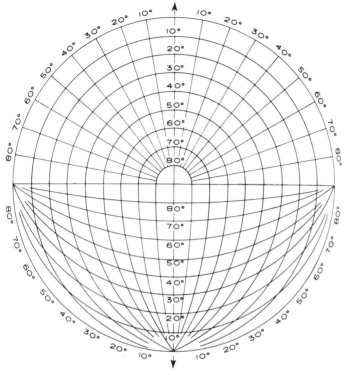

SHADING MASK PROTRACTOR

Figure R·E·15.
Shading mask protractor.
Source: (RAM.)

Shading masks are constructed and read with a shading mask protractor (Figure R·E·15). The bottom half of the protractor is used for studying the segmental shading effects of horizontal obstructions. The upper half, turned around so the 0° arrow indicates south, applies to the radial shading effect of vertical obstructions.

Shading masks may also be read with sun path diagrams. If a mask is superimposed over the appropriate diagram, the portions of the year when the surface will be shaded will be indicated. For example, in Figure R·E·16, if the shading mask for a horizontal obstruction ($\alpha = 60°$) is superimposed over the 40°N sun path diagram, the surface will be shaded by the obstruction from around March 21 to September 21. This process of relating the shading geometry to the yearly sun path can also work the other way around. If we define the times of the year when shading will be needed and plot these times on the sun path diagram, we will determine the shading mask of the obstruction needed to provide the shade. By superimposing the shading protractor over this mask, we can read off the angles (α and β) that are necessary. With these angles, we can figure the dimensions of the obstructions.

Sun path diagrams, shading masks, and the shading mask protractor are quick, convenient devices for studying and organizing the rather complicated geometries of solar angles. With their help, the designer can make maximum use of the light and heat (and absence thereof) that nature provides.

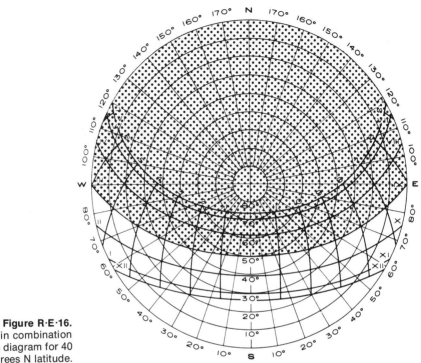

Figure R·E·16.
A shading mask in combination
with the sun path diagram for 40
degrees N latitude.

Consult the bibliography on "Designing for Direct Solar Radiation" for further information, especially Aladar and Victor Olgyay's book, *Solar Control and Shading*.

RESOURCE SECTION R·F

Absorption Cooling Principles

by Charles Michal,
Associate,
Total Environmental Action, Inc.

Solar collectors can provide the input heat necessary in an absorption refrigeration cycle, making solar air conditioning a logical and potentially effective use of solar energy.

The absorption cooling process and the more familiar compression air conditioner both use the vaporization of a liquid refrigerant to draw heat out of the air or water to be cooled. In conventional window air conditioners, electrical energy is used to compress the evaporated fluid so that it condenses and gives off the collected heat to the "outside" (see Figure R·F·1). The compressed, condensed

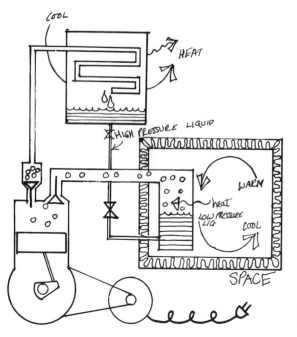

Figure R·F·1.
Compression cooling cycle.

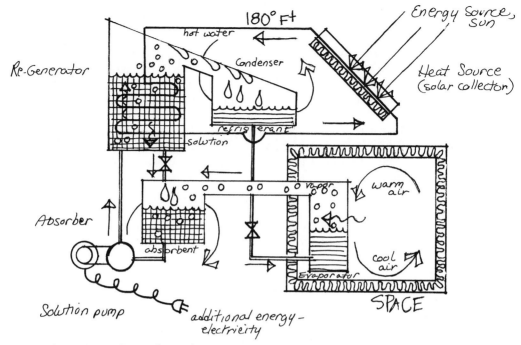

Figure R·F·2.   Absorption cooling cycle.

refrigerant is recycled and evaporates. This cools the space on the "inside" of the system (see Resource Section on "Heat Pumps").

An absorption refrigeration cycle uses two working fluids—a refrigerant and an absorbent—to accomplish the same thing (see Figure R·F·2). The evaporated refrigerant is absorbed from the cooling coils into the secondary fluid. The resulting solution is transferred by a low-power pump to a regenerator, where energy in the form of heat causes the refrigerant to be distilled out of the absorbent fluid. The refrigerant (now liquid) goes back to the evaporating coils, vaporizing and cooling the "inside" of the system. The absorbent is transferred back to the absorber, where the vaporized and heated refrigerant can be absorbed and carried away again.

The primary difficulty in using solar energy for cooling is due to the disparity between the relatively high temperatures which the absorption regenerator needs (250° − 350°F) and the comparatively low temperatures which solar collectors are able to provide most efficiently (150° − 200°F for the best flat plate collectors). The efficiency (COP) of the absorption cooler suffers with decreases in the operating (source) temperatures from the collector. On the other hand, collector efficiencies decrease as the absorber plate temperatures increase. For both components, the cooler and the collector, cost and complexity increase as they are redesigned to maintain their efficiencies at other than optimal temperatures. Fortunately, the disparity is somewhat ameliorated because of the frequent coincidence of large amounts of sunshine with a large cooling demand (desert areas for instance).

Absorption coolers for present solar powered projects are being designed to use temperatures as low as 180°F. In order to maintain reasonable efficiencies at those "high" temperatures, collectors must 1) be built to withstand potentially high pressures within the system; 2) have special transparent cover plates that admit the highest possible amount of incident radiation, yet be specially coated to reduce reradiation of the energy as heat; and 3) have absorber plates that are of high-grade metal, such as copper, with special selective coatings to increase absorptance and decrease emittance (reradiation of heat). In many cases, too, the collector must be made larger for the summer cooling mode than for the winter heating mode.

The selection of the refrigerant and absorbent fluids for an absorption cycle cooling system greatly influences the performance of the system. However, only a few fluid pairs are currently known to be practical for this application. Among those available, the parameters to be compared are basically two:

- the ability of the fluids to function at the selected temperatures; and
- the effect of the fluid properties on the pump power requirements of the system.

The following table includes the most competitive fluid pairs. Ammonia-water is generally regarded as the first choice, with water-lithium bromide as the second.

COMPARISON OF REFRIGERANT-ABSORBENT FLUID COMBINATIONS

Refrigerant	Absorbent	Remarks
Ammonia	Water	—Able to function at selected temperatures —Low pump power
R–21	DME–TEG	—Able to function at selected temperatures —High pump power
R–22	DME–TEG	—Able to function at selected temperatures —High pump power
Water	Lithium Bromide (and similar salts)	—Unable to function at selected temperatures (freezing occurs in the evaporator which operates below 32°F) —Low pump power

RESOURCE SECTION R·G

Heat Pump Principles

by Charles Michal,
Associate,
Total Environmental Action, Inc.

A heat pump is a combined heating and cooling device that can transfer heat out of conditioned spaces, making them cooler, or out of an outside source (such as the earth, air or water) into spaces, thereby warming them. Whether a heat pump is "heating" or "cooling" depends on which side of the refrigeration cycle the conditioned space is on. In the cooling mode, excess heat is pumped to a heat sink (air, water, or earth). The process is reversed during the heating season, when the outside environment (the heat source) is "cooled" in order to warm the building.

The transfer of heat (either to the heat sink or from the heat source) is accomplished by circulating a fluid, usually called a refrigerant, and forcing it to alternate between a liquid and a gaseous state. The latent heat involved in the phase change is drawn from the surrounding medium when the liquid refrigerant evaporates and is given off when the vapor condenses. Pumps, compressors, valves and other equipment cause the repeated evaporation and condensation to occur at appropriate points within the system.

The packaged, self-contained heat pump used in residential and small commercial applications generally reverses the direction of the refrigerant flow to change from one mode to the other. Figures R·G·1 and R·G·2 are diagrams of such a heat pump, first in the cooling mode and then in the heating mode. A four-way valve reverses the direction of fluid flow through the compressor so that the high pressure vapor condenses "inside" when heating is desired and low pressure liquid evaporates when cooling is desired.

Heat pumps may be classified according to the heat source or sink used, the type of fluid conditioned (air or water), or the operating cycle used. The heat pump in Figures R·G·1 and R·G·2 is an air-to-air pump with reversible refrigerant flow. Heat pumps which use air or water as the heat source/sink can have a fixed refrigerant flow by transporting the source/sink fluid to the condensing or evaporating coils, as required. ASHRAE's *Systems and Equipment* (1967) classifies various types of heat pumps.

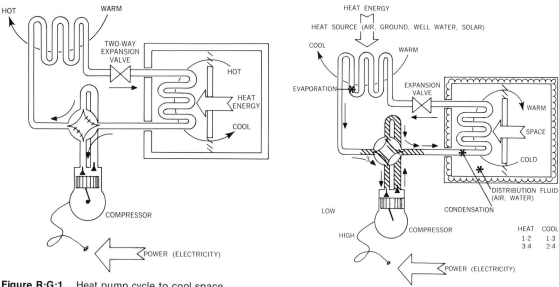

Figure R·G·1. Heat pump cycle to cool space.

Figure R·G·2. Heat pump cycle to heat space.

Heat pumps use electrical energy to manipulate heat distribution. The heat "produced" is a combination of the heat generated in compressing the refrigerant and the latent heat released by the condensing vapor. The effectiveness of heat pumps is indicated by a coefficient of performance (COP). This number is the ratio of the heat obtained or removed to the heat equivalent of the electrical energy used to operate the pump.

The COP of air-to-air heat pumps is usually lower than that of air-to-water or water-to-water heat pumps. Because the heat of compression is part of the heat output, the COP of a heat pump operating in the heating mode is often greater than that of the same pump operating in a cooling mode.

The dollar cost of a therm of heat produced with a heat pump can be obtained by dividing 10^5 Btu by the COP multiplied by 3413 Btu/kwh, and multiplying the result by the cost of electricity per kwh. The higher the COP, the lower the cost.

$$\$/\text{therm} = \frac{10^5 \text{ Btu/therm}}{\text{COP} \times 3413 \text{ Btu/kwh} \times \$/\text{kwh}}$$

$$= \frac{2.93 \text{ kwh/therm} \times \$/\text{kwh}}{\text{COP}}$$

The actual energy costs of heat pumps and considerations of their applicability to solar systems are discussed in Part V.

REFERENCES FOR RESOURCE SECTION

(BEN.) Bennett, Iven. "Monthly Maps of Mean Daily Insolation for the United States." *Solar Energy* 9(1965): 145–158.

(BUD.) Budyko, M. I. "Heat Balance of the Earth's Surface." USSR Central Geophysical Observatory, Leningrad, 1956.

(HUT.) Hutchinson, F. W. and Chapman, W. P. "A Rational Basis for Solar Heating Analysis." ASHRAE Journal Section in *Heating, Piping and Air Conditioning,* July 1946, pp. 109–117.

(JOR.) Jordan, Richard C., ed. *Low Temperature Engineering Applications of Solar Energy.* Technical Committee on Solar Energy Utilization of the American Society of Heating, Refrigeration & Air Conditioning Engineers, Inc. (ASHRAE), New York, N.Y., 1967.

(MOO.) Moorcraft, Colin. "Solar Energy in Housing." *Architectural Design,* October 1973, pp. 634–661.

(NOA.) U. S. Department of Commerce, Environmental Science Services Administration. *Climatic Atlas of the United States.* Washington, D.C., June 1968.

(OLG.) Olgyay, Victor. *Design with Climate: Bioclimatic Approach to Architectural Regionalism.* Princeton, N.J.: Princeton University Press, 1963.

(RAM.) Ramsey, Charles and Sleeper, Harold. *Architectural Graphic Standards.* New York: John Wiley & Sons, Inc., 1970.

(SEV.) Severns, W. H. and Fellows, J. R. *Air Conditioning and Refrigeration.* New York, N.Y.: John Wiley and Sons, Inc., 1958.

(WAG.) Building Research Institute. "Windows and Glass in the Exterior of Buildings." Publication No. 478. The National Academy of Sciences, Washington, D.C., 1957.

(WHI.) Whillier, Austin. "Solar Radiation Graphs." Printed under "Technical Note." *Solar Energy* 9(1965):164–165.

APPENDIXES

Building Design Information

APPENDIX A

Conversion Factors; Metric / English Equivalents

METRIC/ENGLISH EQUIVALENTS

English Measure	Metric Equivalent		Metric (S. I.) Measure	English Equivalent	
inch	2.54	centimeters	millimeter	.04	inch
foot	30.50	centimeters	centimeter	.39	inch
yard	.91	meter	meter	3.28	feet
mile (statute)	1.60	kilometers	meter	1.09	yards
			kilometer	.62	miles
square inch	6.45	square centimeters			
square foot	929.00	square centimeters	square centimeter	.16	square inch
square yard	.84	square meter	square meter	1.19	square yards
square mile	2.60	square kilometers	square kilometer	.38	square mile
ounce	28.30	grams	gram	.035	ounces
pound	.45	kilogram	kilogram	2.2	pounds
short ton	907.00	kilograms	ton (1,000 kg)	1.10	short tons
fluid ounce	29.60	milliliters	milliliters	.03	fluid ounce
pint	.47	liter	liter	2.1	pints
quart	.95	liter	liter	1.06	quarts
gallon	3.78	liters	liter	.26	gallon
cubic foot	.03	cubic meter	cubic meter	35.3	cubic feet
cubic yard	.76	cubic meter	cubic meter	1.3	cubic yards
Btu	251.98	calories	calorie	.004	Btu
pound (force)	4.45	newtons	newton	.225	pound (force)

Multiply:	by:	to obtain:
Acres	43,560	square feet
Acre-ft.	1,233.5	cubic meters
Angstrom units	1×10^{-8}	centimeters
Barrels, oil (crude)	5,800,000	Btu (energy)
Barrels, oil	5.615	cubic feet
Barrels, oil	42	gallons
Board feet	0.0833	cubic feet
Btu	0.55556	chu (centigrade heat units)
Btu	777.48	foot-pounds
Btu	1,055	joules
Btu	0.29305	watt-hours
Btu/hr/ft²/°F	5.682×10^4	watts/cm²/°C
Btu per square foot	0.271	langleys (calories per square centimeter)
Btu/hr/ft² (°F/in)	1	chu/hr/ft² (°C/in)
Calories	3.9685×10^{-3}	Btu
Calories	4.184	joules
Centigrade heat units (chu)	1.8	Btu
Common brick, number of	5.4	pounds
Cords	128	cubic feet
Cubic feet	0.037037	cubic yards
Cubic feet	7.48	gallons
Cubic feet of water	62.37	pounds at 60°F
Cubic feet of common brick	120	pounds
Cubic feet per second	448.83	gallons per minute
Cubic yards of sand	2,700	pounds
Feet of water (39.2°F)	0.4335	pounds per square inch
Feet of water	0.88265	inches of mercury at 0°C
Gallons	0.1337	cubic feet
Gallons of water	8.3453	pounds of water at 60°F
Horsepower	33,000	foot-pounds per minute
Horsepower	42.42	Btu per minute
Horsepower	2,546	Btu per hour
Horsepower	1.014	metric horsepower
Horsepower	0.7457	kilowatts
Horsepower, metric (chevalvapours)	0.9863	horsepower
Inches of mercury at 32°F	0.4912	pounds per square inch
Kilowatts	56.90	Btu per minute
Kilowatts	1.341	horsepower
Kilowatt-hours	3,413	Btu
Kilowatt-hours	2.66×10^6	foot-pounds
Langleys (cal/cm²)	3.69	Btu per square foot
Langleys per minute	0.0698	watts per square centimeter
Microns	1×10^{-4}	centimeters
Months (mean calendar)	730.1	hours
Newtons	0.22481	pounds (force)
Pounds of water	0.1198	gallons
Pounds of water evaporated at 212°F	970.3	Btu
Pounds per square inch	0.068046	standard atmospheres
Pounds per square inch	51.715	millimeters of Hg at 0°C
Standard atmosphere	14.696	pounds per square inch
Tons (long)	2,240	pounds
Tons (short)	2,000	pounds
Tons (short)	0.907185	metric tons
Tons (metric)	2,204.62	pounds
Tons of refrigeration	12,000	Btu per hour
Therms	1×10^5	Btu
Watts	3.413	Btu per hour
Watts	0.00134	horsepower

APPENDIX B

Degree Days And Design Temperatures

DEGREE DAY AND WINTER DESIGN TEMPERATURE DATA FOR 561 LOCALITIES IN THE UNITED STATES

Location	ASHRAE Winter Design Temp. (°F)	Winter Degree Days	Location	ASHRAE Winter Design Temp. (°F)	Winter Degree Days
ALABAMA			ARIZONA (cont.)		
Andalusia		1800	Chandler		1800
Anniston		2600	Douglas	18	2600
Bessemer		2600	Flagstaff	0	7200
Birmingham	19	2600	Gila Bend		1400
Decatur	15	3000	Mesa		1800
Dothan	23	1800	Phoenix	31	1800
Florence	13	3000	Prescott	15	4600
Huntsville	16	3000	Scottsdale		1800
Mobile	26	1600	Tombstone		2400
Montgomery	22	2200	Tucson	29	1800
Muscle Shoals		3000	Winslow	9	4800
Phenix City		2400	Yuma	37	1000
Selma	23	2200	ARKANSAS		
Sheffield		3000	Benton		3000
Talladega	15	2600	Blytheville	12	3400
Tuscaloosa	19	2600	Dumas		2400
			El Dorado	19	2200
ALASKA			Fayetteville	9	3400
Anchorage	− 25	10800	Fort Smith	15	3200
			Hot Springs	18	3000
ARIZONA			Jonesboro	14	3600
Bisbee		2600			

Location	ASHRAE Winter Design Temp. (°F)	Winter Degree Days
ARKANSAS (cont.)		
Little Rock	19	3200
Mountain Home		3600
Pine Bluff	20	2800
Searcy		3400
Stuttgart		3000
Texarkana	22	2600
Walnut Ridge		3400
CALIFORNIA		
Alameda		2800
Anaheim		2000
Bakersfield	31	2200
Berkeley		3000
Eureka	32	4600
Fresno	28	2600
Lafayette		3000
Long Beach	36	1800
Los Angeles	41	2000
Los Gatos		3000
Merced	30	2400
Modesto	32	2400
Oakland	35	2800
Pasadena	36	2000
Redding	31	4600
Sacramento	30	2600
San Bernardino	31	2000
San Diego	42	1400
San Francisco	35	3000
San Jose	34	3000
San Leandro		2800
San Mateo		2800
San Rafael		3000
Santa Ana	33	2000
Santa Barbara	34	2000
Santa Clara		3000
Santa Cruz	32	2800
Santa Maria	32	3000
Santa Monica	43	2200
Santa Rosa	29	3000
Stockton	30	2600
Yreka	13	5400
COLORADO		
Boulder	4	5600
Colo. Springs	−1	6400
Denver	−2	6200
Fort Collins	−9	7000
Grand Junction	8	5600
Leadville	−9	10600
Pueblo	−5	5400
Sheridan		6200

Location	ASHRAE Winter Design Temp. (°F)	Winter Degree Days
COLORADO (cont.)		
Trinidad	1	5400
CONNECTICUT		
Bridgeport	4	5600
Danbury		6000
Greenwich		5400
Hartford	1	6200
Meriden		6000
New Britain		6000
New Haven	5	5800
Norwalk	0	5400
Stamford		5400
DELAWARE		
Dover	13	4600
Newark		5000
Wilmington	12	5000
DIST. OF COLUMBIA		
Washington	16	4200
FLORIDA		
Cocoa Beach		600
Daytona Beach	32	800
Deland		800
Fort Lauderdale	41	200
Fort Meade		600
Fort Myers	38	400
Fort Pierce	37	400
Ft. Walton Beach		1200
Gainesville	28	1000
Jacksonville	29	1200
Lakeland	35	600
Lake Wales		600
Lake Worth		200
Melbourne	37	600
Miami	42	200
Ocala	29	800
Orlando	33	800
Panama City	32	1400
Pensacola	29	1400
Punta Gorda		400
St. Augustine	31	1000
St. Petersburg	39	600
Sarasota	35	600
Tallahassee	25	1400
Tampa	36	600
Titusville		600
Venice		600
West Palm Beach	40	200
Winter Haven		600
Winter Park		800

Location	ASHRAE Winter Design Temp. (°F)	Winter Degree Days
GEORGIA		
Albany	26	1800
Athens	17	3000
Atlanta	18	3000
Augusta	20	2400
Brunswick	27	1200
Columbus	23	2400
Dalton	15	3400
Gainesville	16	3200
La Grange	16	2600
Macon	23	2200
Marietta	17	3200
Moultrie	26	1600
Rome	16	3400
Savannah	24	1800
Valdosta	28	1600
HAWAII		
Honolulu	60	(a)
IDAHO		
Aberdeen		7000
Boise	4	5800
Burley	4	5800
Coeur D'Alene	2	6600
Idaho Falls	−12	7200
Lewiston	6	5600
Moscow	−3	5800
Mountain Home	2	6400
Nampa		5800
Pocatello	−8	7000
Twin Falls	4	5400
ILLINOIS		
Alton		5000
Aurora	−7	6600
Belleville	6	4800
Bloomington	−1	5600
Cairo		3800
Carbondale	7	4200
Champaign	0	5800
Chicago	−3	6600
Danville	−1	5400
Decatur	0	5400
De Kalb		6600
Elgin	−8	6600
Evanston		6600
Galesburg	−4	6200
Joliet	−5	6600
Kankakee	−4	6000
Molino	−7	6400
Ottawa		6400
Peoria	−2	6000

Location	ASHRAE Winter Design Temp. (°F)	Winter Degree Days
ILLINOIS (cont.)		
Quincy	−2	5400
Rantoul	−1	5800
Rockford	−7	6800
Rock Island		6400
Springfield	−1	5400
Urbana	0	5800
INDIANA		
Anderson	0	5400
Bloomington	3	4800
Columbus	3	5400
Connersville		5400
Elkhart		6400
Evansville	6	4400
Fort Wayne	0	6200
Gary		6200
Greensburg		5200
Hammond		6200
Indianapolis	0	5600
Kokomo	0	5600
Lafayette	−1	5800
Marion	−2	5800
Muncie	−2	5600
New Albany		4600
Richmond	−1	5400
Seymour		4800
South Bend	−2	6400
Terre Haute	3	5400
IOWA		
Algona		7400
Ames	−11	6800
Burlington	−4	6200
Cedar Falls		7400
Cedar Rapids	−8	6600
Clinton	−7	6800
Council Bluffs	−7	6600
Davenport		6400
Des Moines	−7	6600
Dubuque	−11	7400
Fort Dodge	−12	7400
Iowa City	−8	6400
Keokuk	−3	5600
Marshalltown	−10	6800
Mason City	−13	7600
Muscatine		6400
Ottumwa	−6	6400
Sioux City	−10	7000
Waterloo	−12	7400
KANSAS		
Coffeyville		4000

Location	ASHRAE Winter Design Temp. (°F)	Winter Degree Days	Location	ASHRAE Winter Design Temp. (°F)	Winter Degree Days
KANSAS (cont.)			MASS. (cont.)		
Dodge City	3	5000	Lowell	−1	6800
Garden City	−1	5200	Pittsfield	−5	7600
Hutchinson	2	4600	Springfield	−3	6600
Kansas City		4800	Worcester	−3	7000
Leavenworth		5000	MICHIGAN		
Salina	3	5000	Adrian	0	6400
Topeka	3	5200	Alpena	−5	8600
Wichita	5	4600	Ann Arbor		6800
			Battle Creek	1	6600
KENTUCKY			Benton Harbor	−1	6200
Covington	3	5200	Coldwater		6400
Hopkinsville	4	4000	Detroit	4	6200
Lexington	6	4600	Escanaba	−7	8600
Louisville	8	4600	Flint	−1	7400
Murray		4000	Grand Rapids	2	6800
Owensboro	6	4400	Holland	2	6400
Paducah	10	4200	Jackson	0	6400
			Kalamazoo	1	6600
LOUISIANA			Lansing	2	7000
Alexandria	25	2000	Ludington		7400
Baton Rouge	25	1600	Marquette	−8	8400
Bossier City		2200	Midland		7200
Lafayette	28	1400	Monroe		6200
Lake Charles	29	1400	Mt. Pleasant	−8	7200
Monroe	23	2200	Muskegon	4	6600
New Orleans	32	1400	Port Huron	−1	7200
Shreveport	22	2200	Saginaw	−1	7000
			Sault Ste. Marie	−12	9400
MAINE			MINNESOTA		
Bangor	−8	8000	Austin		8000
Lewiston	−8	7800	Bloomington		8400
Portland	−5	7600	Duluth	−19	10000
			Minneapolis	−14	8400
MARYLAND			Moorhead		9400
Annapolis		4400	Rochester	−17	8200
Baltimore	12	4600	St. Cloud	−20	8800
Cambridge		4200	St. Paul	−14	8400
College Park		4400			
Cumberland	5	5200	MISSISSIPPI		
Frederick	7	5000	Biloxi	30	1600
Hagerstown	6	5200	Columbus	18	2600
Rockville		4400	Greenville	21	2400
Silver Spring		4400	Gulfport		1600
Wheaton		4400	Hattiesburg	22	1800
			Jackson	21	2200
MASSACHUSETTS			Laurel	22	2000
Attleboro		5800	Meridian	20	2200
Boston	6	5600	Natchez	22	1800
Cape Cod		6000			
Fall River	5	5800			
Lawrence	−3	6800			

Location	ASHRAE Winter Design Temp. (°F)	Winter Degree Days	Location	ASHRAE Winter Design Temp. (°F)	Winter Degree Days
MISS. (cont.)			N. JERSEY (cont.)		
Pascagoula		1600	Edison		5000
Picayune		1600	Flemington		5200
Vicksburg	23	2000	Lakewood		5400
MISSOURI			Morristown		5600
Cape Girardeau	8	4200	Newark	11	5000
Columbia	2	5000	New Brunswick	8	5000
Hannibal	−1	5400	New Milford		5400
Jefferson City	2	4800	Paterson	8	5400
Joplin	7	4000	Trenton	12	5000
Kansas City	4	4800	NEW MEXICO		
St. Joseph	−1	5400	Alamagordo	18	3000
St. Louis	4	5000	Albuquerque	14	4400
Sedalia	4	5000	Carlsbad	17	2600
Springfield	5	5000	Clovis	14	4200
MONTANA			Hobbs	15	2600
Billings	−10	7000	Las Cruces	19	2800
Butte	−24	9800	Roswell	16	3800
Great Falls	−20	7800	Santa Fe	7	6200
Helena	−17	8200	NEW YORK		
Missoula	−7	8200	Albany	−5	6800
NEBRASKA			Binghamton	−2	7200
Columbus	−7	6600	Buffalo	3	7000
Grand Island	−6	6600	Elmira	1	6400
Hastings	−3	6200	Glens Falls	−11	7200
Lincoln	−4	5800	Hempstead		5200
Norfolk	−11	7000	New City-Nyack		5800
North Platte	−6	6600	New York	12	5000
Omaha	−5	6600	Niagara Falls	4	7000
Scottsbluff	−8	6600	Poughkeepsie	−1	6200
NEVADA			Rochester	2	6800
Elko	−13	7400	Rome	−7	7400
Las Vegas	23	2800	Schenectady	−5	6800
Reno	2	6400	Syracuse	−2	6800
NEW HAMPSHIRE			Watertown	−14	7200
Concord	−11	7400	White Plains		5600
Dover		7200	NORTH CAROLINA		
Keene	−12	7400	Asheville	13	4000
Laconia		7800	Burlington		3800
Manchester	−5	7200	Charlotte	18	3200
Nashua		7000	Durham	15	3400
Portsmouth	−2	7200	Elizabeth City	18	3400
NEW JERSEY			Fayetteville	17	3000
Atlantic City	14	4800	Gastonia		3200
Bernardsville		5400	Greensboro	14	3800
Camden		4600	Jacksonville	21	2600
Cape May		4800	Kinston		2800
Clinton		5200	Raleigh	16	3400
			Rocky Mount	16	3600

Location	ASHRAE Winter Design Temp. (°F)	Winter Degree Days	Location	ASHRAE Winter Design Temp. (°F)	Winter Degree Days
N. CAROLINA (cont.)			OREGON		
Salisbury		3200	Baker	− 3	7000
Wilmington	23	2400	Corvallis		4800
Winston-Salem	14	3600	Eugene	22	4800
			Grants Pass	22	5000
NORTH DAKOTA			Klamath Falls	1	6400
Bismarck	−24	8800	Medford	21	5000
Fargo	−22	9200	Pendleton	3	5200
Grand Forks	−26	9800	Portland	21	4600
Minot	−24	9600	Salem	21	4800
OHIO			PENNSYLVANIA		
Akron	1	6000	Allentown	3	5800
Ashland		5800	Altoona	1	6200
Ashtabula	3	6400	Chambersburg	5	5400
Canton	1	6000	Erie	7	6400
Chillicothe	5	5000	Gettysburg		5200
Cincinnati	8	4400	Harrisburg	9	5200
Cleveland	2	6400	Johnstown	1	5600
Columbus	2	5600	Kingston		6200
Dayton	0	5600	Lancaster	2	5400
Defiance	−1	6200	Philadelphia	11	4400
Dover		6000	Phila. Suburban		5200
Elyria		6400	Pittsburgh	5	6000
Findlay	0	6200	Reading	6	5000
Greenville		5800	Scranton	2	6200
Lancaster	1	5200	State College	2	6200
Lima	0	6000	Sunbury	3	5600
Mansfield	1	6400	Uniontown	4	5200
Marietta		4800	Washington		5400
Newark	−1	5600	Wilkes-Barre	2	6200
Norwalk	−1	6200	Williamsport	1	5800
Sandusky	4	5800	York	4	5400
Sidney		5800			
Springfield	3	5600	RHODE ISLAND		
Toledo	1	5800	Newport	5	5800
Troy		5600	Providence	6	6000
Wintersville		5800	Westerly		5800
Youngstown	1	6400			
			SOUTH CAROLINA		
OKLAHOMA			Charleston	23	2000
Altus	14	3000	Columbia	20	2400
Bartlesville	5	4000	Florence	21	2600
Enid	10	3800	Greenville	19	3000
Lawton	13	3000	Myrtle Beach		(b)
Moore		3200	Spartanburg	18	3000
Norman	11	3200	Walhalla		3200
Oklahoma City	11	3200			
Ponca City	8	4000	SOUTH DAKOTA		
Shawnee		3200	Aberdeen	−22	8600
Stillwater	9	3800	Huron	−16	8200
			Rapid City	−9	7400

Location	ASHRAE Winter Design Temp. (°F)	Winter Degree Days	Location	ASHRAE Winter Design Temp. (°F)	Winter Degree Days
S. DAKOTA (cont.)			**VIRGINIA (cont.)**		
Sioux Falls	−14	7800	Charlottesville	11	4200
Watertown	−20	8400	Danville	13	3600
			Lynchburg	15	4200
TENNESSEE			Martinsville		3800
Chattanooga	15	3200	Newport News		3400
Jackson	14	3400	Norfolk	20	3400
Johnson City		4000	Petersburg	15	3600
Knoxville	13	3400	Richmond	14	3800
Memphis	17	3200	Roanoke	15	4200
Nashville		3600			
Oak Ridge		3800	**WASHINGTON**		
			Bellingham	8	5400
TEXAS			Kennewick	4	5200
Abilene	17	2600	Longview	14	5200
Amarillo	8	4000	Olympia	21	5200
Austin	25	1800	Poulsbo		5200
Beaumont	29	1000	Seattle	28	5200
Brownsville	36	600	Spokane	−2	6600
Brownwood	20	2200	Tacoma	20	5200
Corpus Christi	32	1000	Walla Walla	12	4800
Dallas	19	2400	Yakima	6	6000
El Paso	21	2800			
Fort Worth	20	2400	**WEST VIRGINIA**		
Galveston	32	1200	Charleston	9	4400
Houston	28	1400	Clarksburg	3	5000
Killeen	22	2000	Fairmont	3	5000
Longview	21	2400	Huntington	10	4400
Lubbock	11	3600	Parkersburg	8	4800
Lufkin	24	1800	Wheeling	5	5200
Midland	19	2600			
Paris		2600	**WISCONSIN**		
San Angelo	20	2200	Appleton	−10	7800
San Antonio	25	1600	Beloit	−7	6800
Sherman	18	2600	Eau Claire	−15	8000
Texarkana		2600	Fond du Lac	−11	7600
Victoria	28	1200	Green Bay	−12	8000
Waco	21	2000	La Crosse	−12	7600
Wichita Falls	15	2800	Madison	−9	7800
			Milwaukee	−6	7600
UTAH			Racine	−4	7400
Ogden	7	5600	Superior		9800
Salt Lake City	5	6000	Watertown		7800
			Wausau	−18	8400
VERMONT					
Burlington	−12	8200	**WYOMING**		
Montpelier		8800	Casper	−11	7400
Rutland	−12	8000	Cheyenne	−6	7400
			Lander	−16	7800
VIRGINIA			Sheridan	−12	7600
Alexandria		4200			

NOTES

WINTER DESIGN TEMPERATURES are taken from the ASHRAE *Handbook of Fundamentals.* They are needed to determine heating load. Where no temperature is listed, interpolation between figures for nearby localities will be reasonably accurate.

WINTER DEGREE DAYS are taken from the ASHRAE *Handbook of Fundamentals* and other reference sources. Values have been rounded to the nearest 200 degree days.

This table is a summarization of information in *Insulation Manual* by NAHB Research Foundation, Inc.

(a) No heating required.

(b) Heating data unavailable.

APPENDIX C

Insulating Values Of Building Materials

The Resource Section on "Heat Theory" includes a complete explanation of how to use the information in this table. In summary, the total heat transmitted by conduction per year is

H = (24 hours per day) × (number of Degree Days per year) × (U-value of the construction) × (area of the construction)

The number of Degree Days for a particular locale can be found in Appendix B, "Degree Days and Design Temperatures," or by asking local oil or gas distributors. The U-value is determined as shown below.

Example calculations to determine the U-value of an exterior wall and relative yearly heat loss:

Wall Construction	Uninsulated Wall Resistance, R	Insulated Wall Resistance, R
Outside surface (film), 15 mph wind	0.17	0.17
Wood bevel siding, lapped	0.81	0.81
½ in sheathing, reg. density	1.32	1.32
3½ in air space	1.01	
R-11 insulation		11.00
½ in gypsumboard	0.45	0.45
Inside surface (film)	0.68	0.68
Total R	4.44	14.43

For the uninsulated wall, U = 1/R = 1/4.44 = 0.23 Btu/hr/ft²/°F. Therefore, the heat loss for 100 sq ft of the uninsulated wall section in a 6,000 Degree Day area is equal to 24 hours/day × 6000 × 0.22 × 100 sq ft, or 310,000 Btu per year.

For the insulated wall, U = 1/R = 1/14.45 = 0.07 Btu/hr/ft²/°F. Under the same conditions, therefore, the heat loss for the uninsulated wall is equal to 24 hours/day × 6000 DD × 0.07 × 100 sq ft or 100,000 Btu per year, a saving of over 200,000 Btu per year over the 100 sq ft of uninsulated wall. The resultant fuel savings is two gallons of oil (or 60 kwh of electric resistance heating).

WOOD DOORS	1/C	C Without Storm Door	C With Storm Door	
			Wood	Metal
1 in.	1.56	0.64	0.30	0.39
1¼ in.	1.82	0.55	0.28	0.34
1½ in.	2.04	0.49	0.27	0.33
1¾ in.—solid wood core	1.96	.51	—	—
2 in.	2.33	0.43	0.24	0.29

AIR SURFACES

Position of Surface	Direction of Heat Flow	Type of Surface		
		Non-Reflective Materials Resistance (R)	Reflective Aluminum Coated Paper Resistance (R)	Highly Reflective Foil Resistance (R)
STILL AIR				
Horizontal	Upward	0.61	1.10	1.32
45° slope	Upward	0.62	1.14	1.37
Vertical	Horizontal	0.68	1.35	1.70
45° slope	Down	0.76	1.67	2.22
Horizontal	Down	0.92	2.70	4.55
MOVING AIR (any position)				
15 mph wind	Any	0.17 (Winter)	—	—
7½ mph wind	Any	0.25 (Summer)	—	—

AIR SPACES

Position of Air Space and Thickness (inches)		Heat Flow Dir.	Season	Types of Surfaces on Opposite Sides		
				Both Surfaces Non-Reflective Materials Resistance (R)	Aluminum Coated Paper/Non-Reflective Materials Resistance (R)	Foil/Non-Reflective Materials Resistance(R)
Horizontal	¾	Up	W	0.87	1.71	2.23
	¾		S	0.76	1.63	2.26
	4		W	0.94	1.99	2.73
	4		S	0.80	1.87	2.75
45° slope	¾	Up	W	0.94	2.02	2.78
	¾		S	0.81	1.90	2.81
	4		W	0.96	2.13	3.00
	4		S	0.82	1.98	3.00
Vertical	¾	Down	W	1.01	2.36	3.48
	¾		S	0.84	2.10	3.28
	4		W	1.01	2.34	3.45
	4		S	0.91	2.16	3.44
45° slope	¾	Down	W	1.02	2.40	3.57
	¾		S	0.84	2.09	3.24
	4		W	1.08	2.75	4.41
	4		S	0.90	2.50	4.36
Horizontal	¾	Down	W	1.02	2.39	3.55
	1½		W	1.14	3.21	5.74
	4		W	1.23	4.02	8.94
	¾		S	0.84	2.08	3.25
	1½		S	0.93	2.76	5.24
	4		S	0.99	3.38	8.03

Material & Description	Density lb/ft^3	Resistance (R)[a] — Per inch thickness 1/k	Resistance (R)[a] — For thickness listed 1/C	Conductance, Conductivity k	Conductance, Conductivity C
BUILDING BOARDS, PANELS, FLOORING, ETC.					
Asbestos-cement board	120	0.25	—	4.0	—
Asbestos-cement board ⅛ in.	120	—	0.033	—	30.30
Gypsum or plaster board ⅜ in.	50	—	0.32	—	3.13
Gypsum or plaster board ½ in.	50	—	0.45	—	2.22
Plywood (See Siding Materials)	34	—	—	—	—
Sheathing, wood fiber (impregnated or coated) 25/32 in.	20	—	2.06	—	.49
Wood fiber board, laminated or homogeneous	22	2.44	—	.41	—
	25	2.27	—	.44	—
	26	2.38	—	.42	—
	33	1.82	—	.55	—
Wood fiber, hardboard type ¼ in.	65	—	0.18	—	5.56
Wood fiber, hardboard type	65	0.72	—	1.39	—
Wood subfloor 25/32 in.	—	—	0.98	—	1.02
Wood, hardwood finish ¾ in.	—	—	0.68	—	1.47
BUILDING PAPER					
Vapor-permeable felt	—	—	0.06	—	16.67
Vapor-seal, 2 layers of mopped 15 lb felt	—	—	0.12	—	8.33
Vapor-seal plastic film	—	—	Negl.	—	—
INSULATING MATERIALS					
BLANKET AND BATT[b]					
Mineral wool, fibrous form processed from rock, slag, or glass	0.5	3.12	—	.32	—
Wood fiber	1.5–4.0	3.70	—	.27	—
BOARDS AND SLABS	3.2–3.6	4.00	—	.25	—
Cellular glass 90°F	9	2.44	—	.41	—
60°F		2.56	—	.39	—
30°F		2.70	—	.37	—
0°F		2.86	—	.35	—
−30°F		3.00	—	.33	—
Corkboard 90°F	6.5–8.0	3.57	—	.28	—
60°F		3.70	—	.27	—
30°F		3.85	—	.26	—
0°F		4.00	—	.25	—
90°F	12	3.22	—	.31	—
60°F		3.33	—	.30	—
30°F		3.45	—	.29	—
0°F		3.57	—	.28	—
Glass Fiber 90°F	4–9	3.85	—	.26	—
60°F		4.17	—	.24	—
30°F		4.55	—	.22	—
0°F		4.76	—	.21	—
−30°F		5.26	—	.19	—
Expanded rubber (rigid) 75°F	4.5	4.55	—	.22	—

Material	Density	Temp	Conductivity k	Conductance C	Resistance $1/k$	Resistance $1/C$
Expanded polyurethane (R-11 blown) (Thickness 1 in. & greater)	1.5–2.5	100°F	.18	—	5.56	—
		75°F	.17	—	5.88	—
		50°F	.16	—	6.25	—
		25°F	.17	—	5.88	—
		0°F	.17	—	5.88	—
Expanded polystyrene, extruded	1.9	75°F	.26	—	3.85	—
		60°F	.25	—	4.00	—
		30°F	.24	—	4.17	—
		0°F	.22	—	4.55	—
		−60°F	.19	—	5.26	—
Expanded polystyrene, molded beads	1.0	75°F	.28	—	3.57	—
		30°F	.26	—	3.85	—
		0°F	.24	—	4.17	—
Mineral wool with resin binder	15	90°F	.29	—	3.45	—
		60°F	.28	—	3.57	—
		30°F	.27	—	3.70	—
		0°F	.25	—	4.00	—
Mineral fiberboard, wet felted						
Core or roof insulation	16–17		.34	—	2.94	—
Acoustical tile	18		.35	—	2.86	—
Acoustical tile	21		.37	—	2.73	—
Mineral fiberboard, wet molded						
Acoustical tile[g]	23		.42	—	2.38	—
Wood or cane fiberboard						
Acoustical tile[c]	½ in.		—	.84	—	1.19
Acoustical tile[c]	¾ in.		—	.56	—	1.78
Interior finish (plant, tile)	15		.35	—	2.86	—
Insulating roof deck[h]						
Approximately	1½ in.		—	.24	—	4.17
Approximately	2 in.		—	.18	—	5.56
Approximately	3 in.		—	.12	—	8.33
Wood shredded (cemented, preformed slabs)	22		.60	—	1.67	—
LOOSE FILL						
Macerated paper or pulp product	2.5–3.5		.28	—	3.57	—
Mineral wool (glass, slag, or rock)	2.0–5.0	90°F	.30	—	3.33	—
		60°F	.27	—	3.70	—
		30°F	.25	—	4.00	—
		0°F	.23	—	4.35	—
Perlite (expanded)	5.0–8.0	90°F	.38	—	2.63	—
		60°F	.36	—	2.78	—
		30°F	.34	—	2.94	—
		0°F	.32	—	3.12	—
Vermiculite (expanded)	7.0–8.2	90°F	.48	—	2.08	—
		60°F	.46	—	2.18	—
		30°F	.44	—	2.27	—
		0°F	.42	—	2.38	—
	4.0–6.0	90°F	.45	—	2.22	—
		60°F	.43	—	2.33	—
		30°F	.40	—	2.50	—
		0°F	.38	—	2.63	—

MATERIAL & DESCRIPTION		DENSITY lb/ft³	RESISTANCE (R)[a] Per inch thickness 1/k	RESISTANCE (R)[a] For thickness listed 1/C	CONDUCTANCE, CONDUCTIVITY k	CONDUCTANCE, CONDUCTIVITY C
INSULATING MATERIALS						
Sawdust or Shavings						
ROOF INSULATION[d]		0.8–15	2.22	—	.45	—
Preformed, for use above deck						
Approximately	½ in.	—	—	1.39	—	.72
Approximately	1 in.	—	—	2.78	—	.36
Approximately	1½ in.	—	—	4.17	—	.24
Approximately	2 in.	—	—	5.26	—	.19
Approximately	2½ in.	—	—	6.67	—	.15
Approximately	3 in.	—	—	8.33	—	.12
Cellular glass		—	2.56	—	.39	—
MASONRY MATERIALS—CONCRETES						
Cement mortar		116	0.20	—	5.00	—
Gypsum-fiber concrete, 87½% gypsum, 12½% wood chips		51	0.60	—	1.67	—
Lightweight aggregates including:						
expanded shale, clay or slate;		120	0.19	—	5.26	—
expanded slags; cinders; pumice;		100	0.28	—	3.57	—
		80	0.40	—	2.50	—
perlite; vermiculite; also		60	0.59	—	1.70	—
cellular concretes		40	0.86	—	1.62	—
		30	1.11	—	.90	—
		20	1.43	—	.70	—
Sand & gravel or stone aggregate (oven dried)		140	0.11	—	.90	—
Sand & gravel or stone aggregate (not dried)		140	0.08	—	12.50	—
Stucco		116	0.20	—	5.00	—
MASONRY UNITS						
Brick, common[i]		120	0.20	—	5.00	—
Brick, face[i]		130	0.11	—	9.09	—
Clay tile, hollow						
1 cell deep	3 in.	—	—	0.80	—	1.25
1 cell deep	4 in.	—	—	1.11	—	.90
2 cells deep	6 in.	—	—	1.52	—	.66
2 cells deep	8 in.	—	—	1.85	—	.54
2 cells deep	10 in.	—	—	2.22	—	.45
3 cells deep	12 in.	—	—	2.50	—	.40
Concrete blocks, three oval core:						
Sand and gravel aggregate	4 in.	—	—	0.71	—	1.40
	8 in.	—	—	1.11	—	0.90
	12 in.	—	—	1.28	—	0.78
Cinder aggregate	3 in.	—	—	0.86	—	1.16
	4 in.	—	—	1.11	—	0.90
	8 in.	—	—	1.72	—	0.58
	12 in.	—	—	1.89	—	0.53
Lightweight aggregate	3 in.	—	—	1.27	—	0.79
(expanded shale, clay,	4 in.	—	—	1.50	—	0.67
slate or slag; pumice	8 in.	—	—	2.00	—	0.50

Material	Thickness	Density (lb/ft³)	k	C	R (1/k)	R (1/C)
Concrete blocks, rectangular core						
Sand and gravel aggregate	12 in.			0.44		2.27
2 core, 8 in., 36 lb				0.96		1.04
Same with filled cores				0.52		1.93
Lightweight aggregate (expanded shale, clay, slate or slag, pumice):						
3 core, 6 in., 19 lb				0.61		1.65
Same with filled cores				0.33		2.99
2 core, 8 in., 24 lb				0.46		2.18
Same with filled cores				0.20		5.03
3 core 12 in., 38 lb				0.40		2.48
Same with filled cores				0.17		5.82
Stone, lime or sand			12.5		0.08	
Gypsum partition tile:						
3 × 12 × 30 in., solid				0.79		1.26
3 × 12 × 30 in., 4-cell				0.74		1.35
4 × 12 × 30 in., 3-cell				0.60		1.67
Granite, Marble		150–175			.05	
METALS						
Aluminum		159–175	1416		0.0007	
Brass, red		524–542			0.0014	
Brass, yellow		524–542			0.0014	
Copper, cast rolled		550–555	2640		0.0004	
Iron, gray cast		438–445	336		0.0030	
Iron, pure		474–493			0.0023	
Lead		704	240		0.0040	
Steel, cold drawn		490	312		0.0032	
Steel						
Stainless, type 304					0.0055	
Zinc, cast					0.0013	
PLASTERING MATERIALS						
Cement plaster, sand aggregate		116	5.0		0.20	
Sand aggregate	½ in.			10.00		0.10
Sand aggregate	¾ in.			6.66		0.15
Gypsum plaster:						
Lightweight aggregate	½ in.	45		3.12		0.32
Lightweight aggregate	⅜ in.	45		2.67		0.39
Lightweight aggregate on metal lath	¾ in.			2.13		0.47
Perlite aggregate		45	1.5		0.67	
Sand aggregate	½ in.	105	5.6	11.10	0.18	0.09
Sand aggregate	⅜ in.	105		9.10		0.11
Sand aggregate	¾ in.	105		7.70		0.1
Sand aggregate on metal lath	¾ in.			2.50		0.40
Sand aggregate on wood lath						
Vermiculite aggregate		45	1.7		0.59	
ROOFING						
Asbestos-cement shingles		120		4.76		0.21
Asphalt roll roofing		70		6.50		0.15
Asphalt shingles		70		2.27		0.44

MATERIAL & DESCRIPTION		DENSITY lb/ft³	RESISTANCE (R)[a] Per inch thickness 1/k	RESISTANCE (R)[a] For thickness listed 1/C	CONDUCTANCE, CONDUCTIVITY k	CONDUCTANCE, CONDUCTIVITY C
ROOFING						
Built-up roofing	⅜ in.	70	—	0.33	—	3.00
Slate	½ in.	—	—	0.05	—	20.00
Wood shingles		—	—	0.94	—	1.06
WOODS						
Maple, oak, and similar hardwoods		45	0.91	—	1.10	—
Fir, pine, and similar softwoods		32	1.25	—	0.80	—
Fir, pine, and similar softwoods	23 in.	32	—	0.98	—	1.02
	1¾ in.	32	—	2.03	—	0.49
	2⅜ in.	32	—	3.28	—	0.30
	3⅜ in.	32	—	4.55	—	0.22
SIDING MATERIALS (ON FLAT SURFACE)						
Shingles						
Asbestos-cement		120	—	0.21	—	4.76
Wood, 16 in.; 7½ in. exposure		—	—	0.8	—	1.15
Wood, double 16 in.; 12 in. exposure		—	—	1.19	—	0.84
Wood, plus insulating backer board	5/16 in.	—	—	1.40	—	0.71
Siding						
Asbestos-cement, ¼ in. lapped		—	—	0.21	—	4.76
Asphalt roll siding		—	—	0.15	—	6.50
Asphalt insulating siding (½ in. bd.)		—	—	1.46	—	0.69
Wood, drop, 1 × 8 in.		—	—	0.79	—	1.27
Wood, bevel, ½ × 8 in., lapped		—	—	0.81	—	1.23
Wood, bevel, ¾ × 10 in., lapped		—	—	1.05	—	0.95
Plywood, ¾ in, lapped		—	—	0.39	—	1.59
¼ in. plywood		—	—	0.31	—	3.23
⅜ in. plywood		—	—	0.47	—	2.13
½ in. plywood		—	—	0.62	—	1.62
⅝ in. plywood		—	—	0.78	—	1.23
¼ in. hardboard		—	—	0.18	—	5.56
Stucco, per inch		—	—	0.20	—	5.00
Building paper		—	—	0.06	—	16.67
Insulating board sheathing, ½ in. nail-base		—	—	1.14	—	.88
Insulating board sheathing, ½ in. regular density		—	—	1.32	—	.76
Insulating board sheathing, 25/32 in. regular density		—	—	2.04	—	.49
FINISH MATERIALS						
Carpet and fibrous pad		—	—	2.08	—	.48
Carpet and rubber pad		—	—	1.23	—	.81
Cork tile	⅛ in.	—	—	0.28	—	3.57
Terrazzo	1 in.	—	—	0.08	—	12.50
Tile-asphalt, linoleum, vinyl, rubber		—	—	0.05	—	20.00
½ in. gypsumboard				0.45	—	2.22
⅝ in. gypsumboard				0.56	—	1.79
25/32 in. hardwood flooring				0.68	—	1.47

[a]Representative values for dry materials are selected by the ASHRAE Technical Committee 2.4 on Insulation. They are intended as design (not specification) values for materials of building construction in normal use. For conductivity of a particular product, the user may obtain the value supplied by the manufacturer or secure the results of unbiased tests.

[b]Resistance values are the reciprocals of C before rounding off C to two decimal places.

[c]See also Insulating Materials, Board.

[d]Includes paper backing and facing if any.

[e]Conductivity varies also with fiber diameter.

[f]These are values for aged board stock.

[g]Insulating values of acoustical tile vary depending on the board and on the type, size, and depth of the perforations. An average conductivity value is 0.42.

[h]The U.S. Department of Commerce, *Simplified Practice Recommendation for Thermal Conductances Factors for Preformed Above-Deck Roof Insulation*, No. R 257.55, recognizes the specification of roof insulation on the basis of the C values shown. Roof insulation is made in thicknesses to meet these values. Therefore, thickness supplied by different manufacturers may vary depending on the conductivity k value of the particular material.

[i]Face brick and common brick do not always have these specific densities. When the density is different from that shown, there will be a change in the thermal conductivity.

[j]Data on rectangular core concrete blocks differ from the above data on oval core blocks due to core configuration, different mean temperatures and possibly differences in unit weights. Weight data on the oval core blocks tested is not available.

[k]Weights of units approximately $7\frac{5}{8}$ in. high and $15\frac{5}{8}$ in. long. These weights are given as a means of describing the blocks tested, but conductance values are all for one square foot of area.

[l]Vermiculite, perlite or mineral wool insulation. Where insulation is used, vapor barriers or other precautions must be considered to keep insulation dry.

Copyright by the American Society of Heating, Refrigerating and Air-Conditioning Engineers, Inc. Reprinted by permission from ASHRAE *Handbook of Fundamentals*, 1967

COEFFICIENTS OF TRANSMISSION (U) OF WINDOWS, SKYLIGHTS, AND LIGHT TRANSMITTING PARTITIONS
Part A—Vertical Panels (Exterior Windows and Partitions)— Flat Glass, Glass Block, and Plastic Sheet

Description	Exterior[a]		Interior
	Winter	Summer	
Flat Glass			
Single glass	1.13	1.06	0.73
Insulating glass—double[b]			
3/16 in. air space	0.69	0.64	0.51
1/4 in. air space	0.65	0.61	0.49
1/2 in. air space	0.58	0.56	0.46
Insulating glass—triple[b]			
1/4 in. air spaces	0.47	0.45	0.38
1/2 in. air spaces	0.36	0.35	0.30
Storm windows			
1 in.–4 in. air space	0.56	0.54	0.44
Glass Block[c]			
6 × 6 × 4 in. thick	0.60	0.57	0.46
8 × 8 × 4 in. thick	0.56	0.54	0.44
—with cavity divider	0.48	0.46	0.38
12 × 12 × 4 in. thick	0.52	0.50	0.41
—with cavity divider	0.44	0.42	0.36
12 × 12 × 2 in. thick	0.60	0.57	0.46
Single Plastic Sheet	1.09	1.00	0.70

Part B—Horizontal Panels (Skylights)—Flat Glass, Glass Block, and Plastic Bubbles

Description	Exterior		Interior
	Winter[d]	Summer[e]	
Flat Glass			
Single glass	1.22	0.83	0.96
Insulating glass—double[b]—			
3/16 in. air space	0.75	0.49	0.62
1/4 in. air space	0.70	0.46	0.59
1/2 in. air space	0.66	0.44	0.56
Glass Block[c]			
11 × 11 × 3 in. thick with cavity divider	0.53	0.35	0.44
12 × 12 × 4 in. thick with cavity divider	0.51	0.34	0.42
Plastic Bubbles[f]			
Single walled	1.15	0.80	—
Double walled	0.70	0.46	—

Part C—Adjustment Factors for Various Window Types (Multiply U-values in Parts A and B by These Factors)

Window Description	Single Glass	Double or Triple Glass	Storm Windows
All Glass[g]	1.00	1.00	1.00
Wood Sash—80% Glass	0.90	0.95	0.90
Wood Sash—60% Glass	0.80	0.85	0.80
Metal Sash—80% Glass	1.00	1.20	1.20[h]

[a]See Part C for adjustment for various window types.

[b]Double and triple refer to the number of lights of glass.

[c]Dimensions are nominal.

[d]For heat flow up.

[e]For heat flow down.

[f]Based on area of opening, not total surface area.

[g]Refers to windows with negligible opaque area.

[h]Value becomes 1.00 when storm sash is separated from prime window by a thermal break.

APPENDIX D

Specific Heats And Densities Of Heat Storage Materials

SPECIFIC HEAT

The thermal capacity or specific heat of a substance is the amount of heat added to or taken from a unit of weight of a material to produce a change of 1°F in its temperature.

All specific heats vary with a change of temperature, and distinctions must be made between the true or instantaneous and the mean specific heat. The instantaneous specific heat is the amount of heat that must be added or removed at any definite temperature to produce a temperature change of 1°F per unit weight of the material. The mean specific heat, over a given temperature range, is the average amount of heat required to produce a change of 1°F of temperature per unit weight of the substance.

Specific heats of gases and vapors are also dependent on the conditions of maintenance, i.e., either constant pressure or constant volume. The specific heat at constant pressure is greater than the specific heat at constant volume. When volume changes are produced at constant pressure, by temperature changes, work is done, and the heat equivalent of the work is reflected in the specific heat at constant pressure.

PROPERTIES OF SOLIDS
(Values are for room temperature unless otherwise noted in brackets)

Material Description	Specific Heat Btu/lb/°F		Density lb/ft³
Aluminum (alloy 1100)	0.214		171
Aluminum Bronze			
(76% Cu, 22% Zn, 2% Al)	0.09		517
Alundum (aluminum oxide)	0.186		
Asbestos: fiber	0.25		150
insulation	0.20		36
Ashes, wood	0.20		40
Asphalt	0.22		132
Bakelite	0.35		81
Bell metal	0.086	[122]	
Bismuth tin	0.040		
Brick, building	0.2		123
Brass:			
red (85% Cu, 15% Zn)	0.09		548
yellow (65% Cu, 35% Zn)	0.09		519
Bronze	0.104		530
Cadmium	0.055		540
Carbon (gas retort)	0.17		
Cardboard			
Cellulose	0.32		3.4
Cement (Portland clinker)	0.16		120
Chalk	0.215		143
Charcoal (wood)	0.20		15
Chrome Brick	0.17		200
Clay	0.22		63
Coal	0.3		90
Coal Tars	0.35	[104]	75
Coke (petroleum, powdered)	0.36	[752]	62
Concrete (stone)	0.156	[392]	144
Copper (electrolytic)	0.092		556
Cork (granulated)	0.485		5.4
Cotton (fiber)	0.319		95
Cryolite (AlF₃ 3NaF)	0.253		181
Diamond	0.147		151
Earth (dry and packed)			95
Felt			20.6
Fireclay brick	0.198	[212]	112
Flourspar (CaF₂)	0.21		199
German Silver (nickel silver)	0.09		545
Glass:			
crown (soda-lime)	0.18		154
flint (lead)	0.117		267
pyrex	0.20		139
"wool"	0.157		3.25
Gold	0.0312		1208
Graphite:			
powder	0.165		
"Karbate" (impervious)	0.16		117
Gypsum	0.259		78
Hemp (fiber)	0.323		93
Ice: [32°F]	0.487		57.5
[−4°F]	0.465		
Iron:			
cast	0.12	[212]	450
wrought			485
Lead	0.0309		707
Leather (sole)			62.4
Limestone	0.217		103
Linen			

Material Description	Specific Heat Btu/lb/°F		Density lb/ft³
Litharge (lead monoxide)	0.055		490
Magnesia:			
powdered	0.234	[212]	49.7
light carbonate			13
Magnesite brick	0.222	[212]	158
Magnesium	0.241		108
Marble	0.21		162
Nickel	0.105		555
Paints:			
White lacquer			
White enamel			
Black lacquer			
Black shellac			63
Flat black lacquer			
Aluminum lacquer			
Paper	0.32		58
Paraffin	0.69		56
Plaster			132
Platinum	0.032		1340
Porcelain	0.18		162
Pyrites (Copper)	0.131		262
Pyrites (Iron)	0.136	[156]	310
Rock Salt	0.219		136
Rubber:			
Vulcanized (soft)	0.48		68.6
(hard)			74.3
Sand	0.191		94.6
Sawdust			12
Silica	0.316		140
Silver	0.0560		654
Snow (freshly fallen)			7
(at 32°F)			31
Steel (mild)	0.12		489
Stone (quarried)	0.2		95
Tar:			
pitch	0.59		67
bituminous			75
Tin	0.0556		455
Tungsten	0.032		1210
Wood:			
Hardwoods:	0.45/0.65		23/70
Ash, white			43
Elm, American			36
Hickory			50
Mahogany			34
Maple, sugar			45
Oak, white	0.570		47
Walnut, black			39
Softwoods:			22/46
Fir, white	0.65		27
Pine, white	0.67		27
Spruce			26
Wool:			
Fiber	0.325		82
Fabric			6.9/20.6
Zinc:			
Cast	0.092		445
Hot-rolled	0.094		445
Galvanizing			

PROPERTIES OF LIQUIDS

Name or Description	Specific Heat, c_p		Specific Gravity or Density, ρ	
	Btu/lb/°F, c_p	Temperature, °F	lb/ft³, ρ	Temperature, °F
Acetaldehyde			48.9	64.4
Acetic Acid	0.522	79–203	65.49	68
Acetone	0.514	37–73	49.4	68
Air	0.24	68	0.075	68
Allyl Alcohol	0.655	70–205	53.31	68
n-Amyl Alcohol			51.06	59
Ammonia	1.099	32	43.50	−50
Alcohol-ethyl	0.680	32–208	49.27	68
Alcohol-methyl	0.601	59–68	49.40	68
Aniline	0.512	46–180	63.77	68
Benzene	0.412	68	54.9	68
Bromine	0.107	68	194.7	68
n-Butyl Alcohol	0.563	68	50.6	68
n-Butryic acid	0.515	68	60.2	68
Calcium Chloride Brine				
(20% by wt)	0.744	68	73.8	68
Carbon Disulfide	0.240	68	78.9	68
Carbon Tetrachloride	0.201	68	99.5	68
Chloroform	0.234	68	92.96	68
Decane-n	0.50	68	45.6	68
Ethyl Ether	0.541	68	44.61	68
Ethyl Acetate	0.468	68	52.3	68
Ethyl Chloride	0.368	32	56.05	68
Ethyl Iodide	0.368	32	120.85	68
Ethylene Bromide	0.174	68	136.05	68
Ethylene Chloride	0.301	68	77.10	68
Ethylene Glycol			69.22	68
Formic Acid	0.526	68	76.16	68
Glycerine (glycerol)			78.72	68
Heptane	0.532	68	42.7	68
Hexane	0.538	68	41.1	68
Hydrogen Chloride			74.6	b.p.
Isobutyl Alcohol	0.116	68	50.0	68
Kerosine	0.50	68	51.2	68
Linseed Oil			58	68
Methyl Acetate	0.468	68	60.6	68
Methyl Iodide			142	68
Naphthalene	0.402	m.p.	60.9	m.p.
Nitric Acid	0.42	68	94.45	68
Nitrobenzene	0.348	68	75.2	68
Octane	0.51	68	43.9	68
Petroleum	0.4–0.6	68	40–66	68
n-Pentane	0.558	68	39.1	68
Propionic acid	0.473	68	61.9	68
Sodium Chloride Brine				
20% by wt	0.745	68	71.8	68
10% by wt	0.865	68	66.9	68
Sodium Hydroxide and Water				
15% by wt	0.864	68	72.4	68
Sulfuric Acid and Water				
100% by wt	0.335	68	114.4	68

Name or Description	Specific Heat, c_p		Specific Gravity or Density, ρ	
	Btu/lb/°F, c_p	Temperature, °F	lb/ft³, ρ	Temperature, °F
Sulfuric Acid and Water				
95% by wt	0.35	68	114.6	68
90% by wt	0.39	68	113.4	68
Toluene				
($C_5H_5CH_2$)	0.404	68	54.1	68
Turpentine	0.42	68	53.9	68
Water	0.999	68	62.32	68
Xylene				
$C_6H_4(CH_3)_2$				
Ortho	0.411	68	55.0	68
Mata	0.400	68	54.1	68
Para	0.393	68	53.8	68
Zinc Sulfate and Water				
10% by wt	0.90	68	69.2	68
1% by wt	0.80	68	63.0	68

APPENDIX E

Emittances And Absorptances Of Materials

Heat is disseminated, or moved, from one point in a substance to another point or between substances in three ways. Two of these, conduction and convection, are utilized in all conventional space heating systems. The third, radiation, is equally important for the successful application of solar energy to space heating and cooling.

Heat is energy and can take the form of long-wave electromagnetic radiation. All radiation travels in straight lines at the same speed (that of light: 186,000 miles per second) but at differing wavelengths.

The amount of energy transmitted by radiation is inversely proportional to its wavelengths (i.e., the shorter the wavelength, the higher the energy content). Radiant heat is a long-wave, low energy form of radiation. When radiation strikes any substance, it is either reflected from, transmitted through, or absorbed by that substance. Each substance reflects, transmits, and absorbs incident radiation differently, according to its absolute temperature and its physical and chemical characteristics and the wavelengths of the incident radiation. Glass, for instance, transmits a major portion of the visible light that strikes it, but very little infrared radiation.

Numerical ratings can be assigned to each material, indicating the reflectance, transmittance, and absorptance of that material within a certain temperature range and for a certain portion of the electromagnetic spectrum. The sum of the absorptance, reflectance, and transmittance of a material is 1.0, indicating that 100% of the incident energy is accounted for. For most opaque solids, the energy transmitted is effectively zero, so that the sum of absorptance and reflectance is considered to be 1.0.

Radiant energy, once absorbed, is transformed into heat. This heat may be conducted away, or reradiated, or emitted as long-wave radiation from the material. The emittance, ϵ, of a material is a numerical indicator of that material's capability of giving off long-wave radiation. Emittances are ratios of the radiating power of substances to the radiating power of a theoretical "blackbody" (i.e., blackbody, $\epsilon = 1.0$; black paint, $\epsilon = .95$; selective black, $\epsilon = .05$). These figures are useful because they indicate the *relative* performance of various materials.

For instance, masonry, brick, and concrete, which have emittances around 0.9, make better radiators of heat than brass or aluminum, which at best have emittances of 0.22. Asphalt paving, which has an absorptance greater than 0.9, transforms much more of the incident solar radiation into heat than sand (absorptance between 0.6 and 0.75), as anyone who has walked in bare feet from parking lot to beach can testify.

The ratio between the absorptance of short-wave radiation and the emittance of long-wave radiation by any one material has special importance for the designer of a solar collector. Materials with high ratios, called "selective blacks," can be used to coat the surfaces of the collector plate, so that the maximum amount of energy will be absorbed and a minimum amount of energy will be lost due to reradiation, or emittance.

The following tables list the absorptances and emittances of various materials. Except as noted, the figures are for short-wave absorption and long-wave emission. Likewise, the temperature of the material is assumed to be between 0 and 212°F. The materials are listed alphabetically in five categories, each of which includes materials with similar figures.

EMITTANCES AND ABSORPTANCES OF MATERIALS

Class I Substances: Absorptance to Emittance Ratios (α/ϵ) Less than 0.5

Substance	Short-wave Absorptance	Long-wave Emittance	α/ϵ
Magnesium carbonate, $MgCO_3$	0.025–.04	0.79	0.03–.05
White plaster	.07	0.91	.08
Snow, fine particles, fresh	.13	0.82	.16
White paint, .017 in. on aluminum	.20	0.91	.22
Whitewash on galvanized iron	.22	0.90	.24
White paper	.25–.28	0.95	.26–.29
White enamel on iron	.25–.45	0.9	.28–.5
Ice, with sparse snow cover	.31	0.96–0.97	.32
Snow, ice granules	.33	0.89	.37
Aluminum oil base paint	.45	0.90	.50
White powdered sand	.45	0.84	.54

Class II Substances: Absorptance to Emittance Ratios (α/ϵ) Between 0.5 and 0.9

Substance	Short-wave Absorptance	Long-wave Emittance	α/ϵ
Asbestos felt	.25	0.50	.50
Green oil base paint	.5	0.9	.56
Bricks, red	.55	0.92	.60
Asbestos cement board, white	.59	0.96	.61
Marble, polished	.5–.6	0.9	.61
Wood, planed oak	—	0.9	—
Rough concrete	.60	0.97	.62
Concrete	.60	0.88	.68
Grass, green, after rain	.67	0.98	.68
Grass, high and dry	.67–.69	0.9	.76
Vegetable fields and shrubs, wilted	.70	0.9	.78
Oak leaves	.71–.78	0.91–.95	.78–.82
Frozen soil	—	0.93–.94	—
Desert surface	.75	0.9	.83
Common vegetable fields and shrubs	.72–.76	0.9	.82
Ground, dry plowed	.75–.80	0.9	.83–.89
Oak woodland	.82	0.9	.91
Pine forest	.86	0.9	.96
Earth surface as a whole (land and sea, no clouds)	.83	—	—

EMITTANCES AND ABSORPTANCES OF MATERIALS

Class III Substances: Absorptance to Emittance Ratios (α/ϵ) Between 0.8 and 1.0

Substance	Short-wave Absorptance	Long-wave Emittance	α/ϵ
Grey paint	.75	.95	.79
Red oil base paint	.74	.90	.82
Asbestos, slate	.81	.96	.84
Asbestos, paper		.93–.96	—
Linoleum, red-brown	.84	.92	.91
Dry sand	.82	.90	.91
Green roll roofing	.88	.91–.97	.93
Slate, dark grey	.89	—	—
Old grey rubber	—	.86	—
Hard black rubber	—	.90–.95	—
Asphalt pavement	.93	—	—
Black cupric oxide on copper	.91	.96	.95
Bare moist ground	.9	.95	.95
Wet sand	.91	.95	.96
Water	.94	.95–.96	.98
Black tar paper	.93	.93	1.0
Black gloss paint	.90	.90	1.0
Small hole in large box, furnace or enclosure	.99	.99	1.0
"Hohlraum," theoretically perfect black body	1.0	1.0	1.0

Class IV Substances: Absorptance to Emittance Ratios (α/ϵ) Greater than 1.0

Substance	Short-wave Absorptance	Long-wave Emittance	α/ϵ
Black silk velvet	.99	.97	1.02
Alfalfa, dark green	.97	.95	1.02
Lamp black	.98	.95	1.03
Black paint, 0.017 in. on aluminum	.94–.98	.88	1.07–1.11
Granite	.55	.44	1.25
Graphite	.78	.41	1.90

High Ratios, but Absorptances Less Than .80

Substance	Short-wave Absorptance	Long-wave Emittance	α/ϵ
Dull brass, copper, lead	.2–.4	.4–.65	1.63–2.0
Galvanized sheet iron, oxidized	.8	.28	2.86
Galvanized iron, clean, new	.65	.13	5.00
Aluminum foil	.15	.05	3.00
Magnesium	.3	.07	4.3
Chromium	.49	.08	6.13
Polished zinc	.46	.02	23.0
Deposited silver (optical reflector) untarnished	.07	.01	

EMITTANCES AND ABSORPTANCES OF MATERIALS

Class V Substances: Selective Surfaces[1]

Substance	Short-wave Absorptance	Long-wave Emittance	α/ϵ
Plated metals:[2]			
Black sulfide on metal	.92	.10	9.2
Black cupric oxide on sheet aluminum	.08–.93	.09–.21	
Copper (5 × 10⁻⁵ cm thick) on nickel or silver-plated metal			
Cobalt oxide on platinum			
Cobalt oxide on polished nickel	.93–.94	.24–.40	3.9
Black nickel oxide on aluminum	.85–.93	.06–.1	14.5–15.5
Black chrome	.87	.09	9.8
Particulate coatings:			
Lampblack on metal			
Black iron oxide, 47 micron grain size, on aluminum			
Geometrically enhanced surfaces:[3]			
Optimally corrugated greys	.89	.77	1.2
Optimally corrugated selectives	.95	.16	5.9
Stainless steel wire mesh	.63–.86	.23–.28	2.7–3.0
Copper, treated with NaClO₂ and NaOH	.87	.13	6.69

[1]Selective Surfaces absorb most of the solar radiation between 0.3 microns and 1.9 microns, and emit very little in the 5 to 15 micron range—the infrared.

[2]For a discussion of Plated Selective Surfaces, see Daniels, *Direct Use of the Sun's Energy*, especially chapter 12.

[3]For a discussion of how surface selectivity can be enhanced through surface geometry, see K. G. T. Hollands, "Directional Selectivity Emittance and Absorptance Properties of Vee Corrugated Specular Surfaces," *The Journal of Solar Energy Science and Engineering*, 3 (July, 1963).

Information in these tables was gathered from several sources, including:
ASHRAE. *Handbook of Fundamentals.* 1972.
Bowden. "Heat Theory." *Alternative Sources of Energy,* July 1973.
McAdams. *Heat Transmission.* 1954.
Severns and Fellows. *Air Conditioning and Refrigeration.* 1966.
Souders. *The Engineer's Companion.* 1966.
McDonald. "Spectral Reflectance Properties of Black Chrome for Use as a Solar Selective Coating." NASA Technical Memorandum THX-71596.

APPENDIX F

Heat Equivalents Of Fuels For Solar Back-up

HEAT EQUIVALENTS OF FUELS (and other energy sources)

MATERIAL	HEATING VALUE[a]	SOURCE[b]	HEAT OBTAINABLE[c]
SOLIDS	**(Btu/lb)**		**(Btu/lb)**
Anthracite coal	12,700–13,600	(1)	6,800–10,150
Bituminous coal	11,000–14,350	(1)	4,400–10,045
Subbituminous coal	9,000	(1)	
"Good Illinois" coal	8,500	(2)	
Lignite coal	6,900	(1)	
Coke	11,000–12,000	(3)	
Newspaper	8,500	(2)	
Brown paper	7,670	(2)	
Corrugated board	7,400	(2)	
Food cartons	7,700	(2)	
Pulp trays	8,300	(2)	
Waxed milk cartons	11,680	(2)	
Plastic film	13,780	(2)	
Polystyrene	15,730	(2)	
Polyethylene	14,890	(2)	
Typical urban refuse	5,000	(5)	
Wood—general	8,000–10,000		
—green		(4)	3,000–4,600
—dry		(4)	5,300–6,000
LIQUIDS	**(Btu/gal)**		**(Btu/gal)**
Distillate fuel oils			
—Grade 1	132,900–137,000	(1)	94,000
—Grade 2	137,000–141,800	(1)	97,300
—Grade 4	143,100–148,100	(1)	102,200
Residual fuel oils			
—Grade 5L	146,800–150,000	(1)	
—Grade 5H	149,500–152,000	(1)	
—Grade 6	151,300–155,900	(1)	
Alcohol			
Kerosene	133,000		
Gasoline	111,000		

HEAT EQUIVALENTS OF FUELS (and other energy sources)

MATERIAL	HEATING VALUE[a]	SOURCE[b]	HEAT OBTAINABLE[c]
GASES	**(Btu/ft³)**		**(Btu/ft³)**
Natural gas	1,000–1,050	(1)	780
Commercial propane	2,500	(1)	1,870
Commercial butane	3,200	(1)	2,400
Propane-air or butane-air	500–1,800	(1)	350–1,250
Acetylene	1,500	(3)	
Bio-gas	550		
Methane	950–1,050		
Manufactured gas (from coal)	450		

OTHER SOURCES	**Potential Maximum**		**Heat Obtainable**
Electricity			
—resistance heating	3,413 Btu/kwh		3,413 Btu/kwh
—heat pumps			
Water/gravity			
—per foot of heat	60 kwh/acre/ft		36 kwh/acre/ft
	1.4 kwh/1,000 ft³		.8 kwh/1,000 ft³
Wind (per sq ft collector)			
— 5 mph avg			.5 kwh/month
—10 mph avg			4.0 kwh/month
—15 mph avg			8.0 kwh/month
Sun (per sq ft collector)	432 Btu/hr		150 Btu/hr
	(solar constant, outer atmosphere)		

[a]Heat of combustion or califoric values. The heat produced by complete combustion of the specific fuel. This value also includes the latent heat generated by the condensation of the water vapor content of the fuel.

[b]Sources for the values found in column 2 are:

(1) ASHRAE. *Handbook of Fundamentals,* 1972.

(2) MIT. *Technology Review,* February, 1972.

(3) Ram Bux Singh. *Biogas Plant.* Gobar Gas Research Station, India, 1971.

(4) Peter Allen. *Firewood for Heat.* Department of Resources and Economic Development, New Hampshire. Bulletin #17.

(5) *Power Generation Alternatives.* City of Seattle, 1972.

[c]Heat obtainable, or useful heat, is equal to the heat of combustion minus heat losses due to incomplete combustion, waste flue gases, water vapor in fuels, equipment limitations, etc. These losses vary between 20% of the heat of combustion for a well-engineered gas or oil unit and 50% for a hand-fired, uncontrolled coal burning unit.

[d]Energy received from the sun and wind varies widely with time and place. These figures are illustrative only.

BIBLIOGRAPHY

CHARLES HAVERSTICK

Books on Solar Energy of General Interest

Anderson, Bruce. "Solar Energy and Shelter Design." M.A. thesis, Massachusetts Institute of Technology, 1973.
———. *The Solar Home Book-Heating, Cooling, and Designing with the Sun.* Harrisville, New Hampshire. Cheshire Books, 1976.

Association for Applied Solar Energy. *A Directory of World Activities and Bibliography of Significant Literature.* 2nd ed. Special supplement to *Solar Energy.* Edited by Jean Smith Jensen. Menlo Park, Calif.: AFASE and Stanford Research Institute, 1959.

Association for Applied Solar Energy, Stanford Research Institute, and University of Arizona. *World Symposium on Applied Solar Energy.* Phoenix, Arizona, November 1–5, 1955; reprint ed. New York: Johnson Reprint Corp., 1964.

Brinkworth, Brian Joseph. *Solar Energy for Man.* Halsted Press Book. New York: John Wiley and Sons, Inc., 1972.

Crowther, Richard L. *Sun Earth,* Denver, Colorado. Crowther/Solar Group/Architects. 1976.

Daniels, Farrington. *Direct Use of the Sun's Energy.* New Haven: Yale University Press, 1964.

Duffie, J. A. and Beckman, W. A. *Solar Energy Thermal Processes.* New York: John Wiley and Sons, Inc., 1974.

Halacy, Daniel S., Jr. *The Coming Age of Solar Energy.* New York: Harper and Row, 1964.

Hamilton, Richard, ed. *Space Heating with Solar Energy.* Proceedings of a course symposium at MIT, August 20–26, 1950. Massachusetts Institute of Technology, 1954.

Herdeg, Walter. *The Sun in Art.* Zurich: The Graphic Press, 1968.

International Solar Energy Society. "The Sun in the Service of Mankind." *Proceedings of the Paris Solar Conference,* July 1973.

Kreider, Jan F. and Kreith, Frank. *Solar Heating and Cooling.* New York. McGraw-Hill Book Co. 1976.

National Science Foundation and National Aeronautics and Space Administration. *An Assessment of Solar Energy as a National Energy Resource.* NSF/NASA Solar Energy Panel, December 1972.

Rau, Hans. *Solar Energy.* New York: Macmillan, 1964.

United Nations. *Proceedings of the United Nations Conference on New Sources of Energy.* Vols. 4–6. Rome, 1961.

University of Maryland Department of Mechanical Engineering. *Proceedings of the Solar Heating and Cooling for Building Workshop.* Washington, D.C., March 21–23, 1973.

Zarem, A. M. and Erway, Duane D. *Introduction to the Utilization of Solar Energy.* New York City: McGraw-Hill, 1963.

General Solar Energy

"Albuquerque Solar Building Tests Radiant-Heat Systems." *Popular Mechanics* 108 (November 1957): 157. Describes Bridgers and Paxton solar-heated office building in Albuquerque, New Mexico; heated by means of south-facing collector panels and a 6000-gallon water tank.

Alhadithi, A. E. A. "Solar Air Heater for Space Heating." Paper presented at the Paris Solar Conference, July 1973.

Allcut, E. A. and Hopper, F. C. "Solar Energy in Canada." *Proceedings of the United Nations Conference on New Sources of Energy.* Rome, 1961.

Altman, Manfred. "Conservation and Better Utilization of Electric Power by Means of Thermal Energy Storage and Solar Heating." Interim Report, University of Pennsylvania, October 1971.
———. "Conservation and Better Utilization of Electric Power by Means of Thermal Energy Storage and Solar Heating." Phase II Progress Report, University of Pennsylvania, December 1972.

Ambrose, E. R. "Heat Pump System Employing Storage and Using Auxiliary Heat." Paper presented at the Space Heating with Solar Energy Conference, Massachusetts Institute of Technology, August 21–26, 1950.

American Gas and Electric Service Corporation and

E. R. Ambrose Air Conditioning. "The Heat Pump and Solar Energy." *Proceedings of the World Symposium on Applied Solar Energy,* Phoenix, Arizona, 1955.

Anderson, Lawrence B. "Architectural Problems." Abstract of a paper delivered at the Space Heating with Solar Energy Conference, Massachusetts Institute of Technology, August 21–26, 1950.

Anderson, Lawrence B.; Hottel, Hoyt C.; and Whillier, Austin. "Solar Heating Design Problems." Publication No. 36. Godfrey L. Cabot Solar Energy Conversion Research Project, Massachusetts Institute of Technology.

———. "Solar Heating Design Problems." In *Solar Energy Research.* Madison: University of Wisconsin Press, 1961.

Baer, Steve. "Solar House." *Alternative Sources of Energy* 10 (March 10, 1973).

Baumer, H. "Solar Design." *Ohio State University Experimental Station News* 19 (June 1947): 38–40.

Beckman, W. A. "Solution of Heat Transfer Problems on a Digital Computer." *Solar Energy* 13 (1972): 293–300.

Bennett, Iven. "Monthly Maps of Mean Daily Insolation for the United States." *Solar Energy* 9 (1965): 145–158.

Bhardwaj, R. K.; Gupta, B. K.; Prakash, R. "Performance of a Flat-Plate Solar Collector." *Solar Energy* 2 (1967): 160–162.

Black, James F. "Weather Control: Use of Asphalt Coatings to Tap Solar Energy." *Science* 139 (January 1963): 226–227.

Bliss, Raymond W., Jr. "Atmospheric Radiation Near the Surface of the Ground: A Summary for Engineers." *Solar Energy* 5 (1961): 103–120.

———. "The Derivations of Several 'Plate-Efficiency Factors' Useful in the Design of Flat-Plate Solar Heat Collectors." *Solar Energy* 3 (1959): 55–64.

———. "Solar Energy Utilization in Chile." A Report to the National Academy of Sciences, Tucson, Arizona, June 15, 1961.

Böer, K. W. "A Description of the Energy Conversion System of the University of Delaware Solar House." University of Delaware, Institute of Energy Conversion, July 20, 1973.

———. "Solar Heating and Cooling of Buildings." Paper presented to NSF/RANN Symposium, November 1973.

———. "A Solar House System Providing Supplemental Energy for Consumers and Peak-Shaving with Power-On-Demand Capability for Utilities." Publication No. 166. University of Delaware, Institute of Energy Conversion.

Buelow, F. H. and Boyd, J. S. "Heating Air by Solar Energy." *Architectural Engineering* 38 (January 1957): 28–30.

Charters, W. W. S. and Lawand, T. A. "The Measurement of Temperature in Solar Energy Apparatus." Technical Report No. T-64, April 1970. Brace Research Institute, McGill University.

Charters, W. W. S. and Peterson, L. F. "Free Convection Suppression Using Honeycomb Cellular Materials." *Solar Energy* 13 (1972): 353–361.

Chiou, J. P.; El-Wakil, M. M.; and Duffie, J. A. "A Slit-and-Expanded Aluminum-Foil Matrix Solar Collector." *Solar Energy* 9 (1965): 73–80. Proposes the use of porous matrices as the heat-absorbing media in air-cooled solar collectors.

Chung, R.; Löf, G. O. G.; and Duffie, J. A. "Design of a Solar Absorption Space Cooler." Paper Presented at the A.I. Chemical Engineers Meeting, Salt Lake City, September 1958. Instructions for building solar water heater with materials readily available in Belgian Congo. Tables give size of absorber for various locations, price of electricity, initial cost, yearly cost and yearly savings.

Close, D. J. "Rock Pile Thermal Storage for Comfort Air Conditioning." *Mechanical and Chemical Engineering Transactions of the Institute of Engineers, Australia* (May 1965): 11–22.

———. "Solar Air Heaters for Low and Moderate Temperature Applications." *Solar Energy* 7 (1963): 117–124.

Cooper, P. I. "Some Factors Affecting the Absorption of Solar Radiation in Solar Stills." *Solar Energy* 13 (1972): 373–381.

Cox, John-Robertson. "Architectural Planning and Design Analysis of Energy Conservation in Housing Through Thermal Energy Storage and Solar Heating." NSF/RANN GI 27976.

Cox, S. K.; VonderHaar, T. H.; Hanson, K. J.; and Suomi, V. E. "Measurements of Absorbed Shortwave Energy in a Tropical Atmosphere." *Solar Energy* 14 (1973): 169–173.

Daniels, Farrington. "The Solar Era, Part 2—Power Production with Small Solar Engines." *Mechanical Engineering* 94 (September 1972): 16–19.

Dietz, Albert G. H. "Diathermanous Materials and Properties of Surfaces." Abstract of paper delivered at the Space Heating with Solar Energy Conference, Massachusetts Institute of Technology, August 21–26, 1950.

———. "Weather and Heat Storage Factors." Abstract of paper delivered at the Space Heating with Solar Energy Conference, Massachusetts Institute of Technology, August 21–26, 1950.

Donovan, Mary and Bliss, Ray. *Calculated Performance of a Buried Rockpile for Solar Heat Storage.* Technical Report No. 8, October 1, 1953. Heat capacity, internal heat transfer rates, air pressure drop, and approximate loss rates.

——. "Calculated Performance of Four Air-Heating Flat-Plate Solar Heat Collectors." Technical Report No. 3, October 1952. Calculates typical mid-winter clear-day useful heat collection of four types of air-heating collectors operating under identical conditions similar to space-heating use.

——. "Necessary Capacities of Solar-Heat Collector and Heat Storage Unit as Indicated by Tucson Weather Records." Technical Report No. 5, September 1953. Analyzes Tucson weather data over a 19 year period from the standpoint of probable collector performance as related to winter space-heating load. Calculates approximate winter collector performance and heating load on a day-to-day basis for a 6-year period with a view of estimating sizing of collector and heat storage unit to supply all needed heat load requirements.

Douglas, Roger. "Available Solar Heat." *Alternative Sources of Energy* 13 (February 1974): 19.

Dudley, James C. "Thermal Energy Storage Unit for Air Conditioning Systems Using Phase Change Material." Research Project of the University of Pennsylvania, August 1972.

Dudley, James C. and Freedman, Steve I. "Off-Peak Air Conditioning Using Thermal Energy Storage." Research Project of the University of Pennsylvania, August 1972.

Duffie, John A. "New Materials in Solar Energy Utilization." *Proceedings of the United Nations Conference on New Sources of Energy.* Rome, 1961.

Eibling, J. A. "Battelle Institute Studies New Concept for Solar Heating of Homes. *ASHRAE Journal* 17 (February 1975): 44.

Eisenstadt, M.; Flanigan, F.; and Farber, E. "Solar Air Conditioning with An Ammonia-Water Absorption Refrigeration System." Paper No. 59-A-276. American Society of Mechanical Engineers.

——. "Tests Prove Feasibility of Solar Air Conditioning." Leaflet No. 127. *Engineering Progress,* University of Florida.

Eldridge, Frank R. *Solar Energy Systems.* Mitre Corporation, 1973. Alternative solar energy systems which can be used to gather electricity and to produce hydrogen fuels.

"The Energy Crisis." Special issue of *Technology Review,* December 1973.

Erlandson, Paul M. "Direct Conversion of Solar Energy." *Proceedings of the World Symposium on Applied Solar Energy.* Phoenix, Arizona, 1955.

Fairbanks, John W. and Morse, Frederick H. "Passive Solar Array Orientation Devices for Terrestrial Applications." *Solar Energy* 14 (1972): 67–69.

Farber, Erich A. "Design and Performance of A Compact Solar Refrigeration System." Paper presented at the International Solar Energy Society Conference, Melbourne, Australia, 1970.

——. "The Direct Use of Solar Energy to Operate Refrigeration and Air-Conditioning Systems." *Engineering Progress,* November 1965.

——. "Selective Surfaces and Solar Absorbers." *Journal of Solar Energy Science and Engineering* 3 (April 1959).

——. "Solar Energy Conversion Research and Development at the University of Florida Solar Energy and Energy Conversion Laboratory." *Building Systems Design,* June 1972.

——. "Solar Energy: Conversion and Utilization." *Proceedings of 21st Annual Air Conditioning Conference,* February 24–25, 1972.

Farber, Erich A.; Bussell, W. H.; and Bennett, J. D. "Solar Energy Used to Supply Service Hot Water." *Air Conditioning, Heating and Ventilating* 54 (October 1957): 75–79. Data for a system that uses solar energy for heating water.

Farber, Erich A. and Morse, Gordon L. "Combining the Collector and Generator of a Solar Refrigeration System." Technical Paper No. 426. Florida Engineering and Industrial Experiment Station, University of Florida, Gainesville, Florida.

Farber, Erich A. and Reed, J. C. "Practical Application of Solar Energy." Leaflet Series No. 83. *Engineering Progress,* November 1956.

"French Switch on to Sun Power." *Business Week,* May 9, 1970, p. 126.

Fukuo, Nobuhei; Kozuka, Takeshi; Iida, Shozo; Fujishiro, Ikuya; Irisawa, Toshiaki; Yoshida, Masaie; and Mii, Hisao. "Installations for Solar Space Heating in Girin." *Proceedings of the United Nations Conference on New Sources of Energy.* Rome, 1961.

Gardenshire, Lawrence. "Solar Energy Heats New Mexico Home." *SAW* 2 (March 1957):6. Operating data on the collector and storage system in author's solar-heated house in New Mexico.

Garg, H. P. "Effect of Dirt on Transparent Covers in Flat-Plate Solar Energy Collectors." *Solar Energy* 15 (1974): 299–302.

Gaucher, Leon P. "The Solar Era, Part I: The Practical Promise." *Mechanical Engineering* 94 (August 1972): 9–12.

Ghaswala, S. K. "Solar Heating of Buildings." *Industrialized Building* 2 (December 1954): 33–38, 43. Early solar energy experiments, principles and planning of buildings in India.

Gillette, R. B. "Analysis of the Performance of a Solar Heated House." M.S. Thesis in Mechanical Engineering, University of Wisconsin, 1959.

Goldberg, B. and Klein, W. H. "Comparison of Normal Incident Solar Energy Measurements at Washington, D.C." *Solar Energy* 13 (1971): 311–321.

Grillo, Paul Jacques. "Sun-Heated Ski Lodge Slit Into Mountain Slope." *Interiors,* January 1951, pp. 114–115.

Gupta, C. I. "On Generalizing the Dynamic Performance of Solar Energy Systems." *Solar Energy* 13 (1971): 301–310.

Gutierrez, G.; Hincapie, F.; Duffie, J. A.; and Beckman, W. A. "Simulation of Forced Circulation Water Heaters; Effects of Auxiliary Energy Supply, Load, Type and Storage Capacity." *Solar Energy* 15 (1974): 282–298.

Hammond, Allen L. "Solar Energy: The Largest Resource." *Science* 177 (September 22, 1972): 1088–1090.

Hand, I. F. "Solar Energy for House Heating: An Application of Solar Data to Heating Through Windows Facing South." *Heating and Ventilation* 44 (December 1947): 80–94. Weather Bureau radiation data for planning houses.

Harrison, P. L. "A Device for Finding True North." *Solar Energy* 15 (1974): 303–308.

Hay, H. R. "The California Solarchitecture House." Los Angeles, California: Sky Therm Processes and Engineering Company.

———. "Energy, Technology, and Solarchitecture." Paper presented to American Society of Mechanical Engineers, August 16, 1972.

———. "New Roofs for Hot Dry Regions." *Ekistics* 31 (February 1971): 158–164.

———. "Solar Energy, Solar Power, and Pollution." Paper presented at the Paris Solar Conference, July 1973.

———. "Some Solar Radiation Adaptations and Implications." Paper presented to American Society of Mechanical Engineers, July 26, 1971.

———. "The Solar Era, Part 3. Solar Radiation: Some Implications and Adaptations." *Mechanical Engineering* 94 (October 1972): 24–29.

Hay, H. R. and Yellott, J. I. "A Naturally Air-Conditioned Building." *Mechanical Engineering* 92 (January 1970): 19–25.

"Heat From the Sun." *Architectural Forum* 104 (January 1951): 148–149.

Hesselschwerdt, August L. "Energy Transport to Storage." Paper presented at the Space Heating with Solar Energy Conference, Massachusetts Institute of Technology, August 21–26, 1950.

———. "Engineering and Control Problems." Paper presented at the Space Heating with Solar Energy Conference, Massachusetts Institute of Technology, August 21–26, 1950.

Hollands, K. G. T. "Directional Selectivity: Emittance and Absorptance Properties of Vee Corrugated

Specular Surfaces." *Solar Energy* 7 (1963): 108–116.

———. "Honeycomb Devices in Flat-Plate Solar Collectors." *Solar Energy* 9 (1965): 159–164.

Hottel, Hoyt C. "The Performance of Flat-Plate Solar Energy Collectors." Paper presented at the Space Heating with Solar Energy Conference, Massachusetts Institute of Technology, August 21–26, 1950.

———. "Residential Uses of Solar Energy." *Proceedings of the World Symposium on Applied Solar Energy,* Phoenix, Arizona, 1955.

Hottel, Hoyt C. and Woertz, B. B. "The Performance of Flat Plate Solar Heat Collectors." Publication No. 3. Godfrey L. Cabot Solar Energy Conversion Research Project, Massachusetts Institute of Technology.

Howe, Bob. "Some Notes on Solar House Heating." *Alternative Sources of Energy* 13 (1974): 14.

Hunley, N. R. "Solar Energy Evaluation for Modular Integrated Utility Systems." Research sponsored by the U.S. Department of Housing and Urban Development under HUD Interagency Agreement No. IAA-H-40-72, February 1973.

Hunter, James M. "The Architectural Problem of Solar Collectors—a Roundtable Discussion." *Proceedings of the World Symposium on Applied Solar Energy,* Phoenix, Arizona, 1955.

Johnson, L. E. "How to Combine Solar Heating, Radiant Heating, and the Heat Pump." *Heating and Ventilation* 45 (December 1948): 86–90.

Johnson, Timothy E. and Wellesley-Miller, S. "Space Conditioning with Variable Membranes." Department of Architecture, Massachusetts Institute of Technology, February 1973.

Jordan, Richard C., ed. *Low Temperature Engineering Application.* American Society of Heating, Refrigeration and Air Conditioning Engineers, Inc., 1967.

Jordan, Richard C. and Ibele, Warren E. "Mechanical Energy From Solar Energy." *Proceedings of the World Symposium on Applied Solar Energy,* Phoenix, Arizona, 1955.

Jordan, Richard C. and Threlkeld, J. L. "Solar Heat Collectors Used with Heat Pumps." *Heating and Ventilation* 51 (April 1954): 96–98. University of Minnesota study on solar heat collectors with heat pumps. Heat loss and the collector area are calculated for one model of a solar house in four different cities; advantages and disadvantages involved in heating with a solar energy pump system.

Kelly, Bruce P.; Eckert, John A.; and Berman, Elliot. "Investigation of Photovoltaic Applications." Paper presented at the Paris Solar Conference, July 1973.

Klass, P. J. "Gains Cited in Thin-Film Solar-Cell Efforts." *Aviation Week* 89 (August 26, 1968): 74.

Klein, S. A.; Duffie, J. A.; and Beckman, W. A. "Transient Considerations of Flat-Plate Collectors." Paper presented at ASME Meeting, November 11–15, 1973.

Kobayashi, Takao and Sargent, Stephen. "A Survey of Breakage-Resistant Materials for Flat-Plate Solar Collector Covers." Paper presented to the U.S. Section of International Solar Energy Society, Colorado State University, Ft. Collins, Colorado, August 20–23, 1974.

Lawand, T. A. "Simple Solar Still for the Production of Distilled Water." Technical Report No. T-17, October 1965. Brace Research Institute, McGill University.

———. "Preliminary Report on Solar Air Heater Tests." Technical Report No. T-3, March 1963. Brace Research Institute, McGill University.

Löf, G. O. G. "Cooling with Solar Energy." *Proceedings of the World Symposium on Applied Solar Energy,* Phoenix, Arizona, 1955.

———. "Round-Up: Recent Solar Heating Installations." *Progressive Architecture,* March 1959.

———. "Performance of Solar Energy Collectors of Overlapped-Glass-Plate Type." Paper presented at the Space Heating with Solar Energy Conference, Massachusetts Institute of Technology, August 21–26, 1950.

———. "Selected Results, from Design Study, Solar Heated House in Denver, Colorado." Paper presented at Space Heating with Solar Energy Conference, Massachusetts Institute of Technology, August 21–26, 1950.

———. "Use of Solar Energy for Heating Purposes: Space Heating." *Proceedings of the United Nations Conference on New Sources of Energy.* Rome, 1961.

Löf, G. O. G.; Close, D. J.; and Duffie, J. A. "A Philosophy for Solar Energy Development." *Solar Energy* 12 (1968): 243–250.

Löf, G. O. G.; El-Wakil, M. M.; and Chiou, J. P. "Design and Performance of Domestic Heating System Employing Solar Heated Air—The Colorado Solar House." *Proceedings of the United Nations Conference on New Sources of Energy.* Rome, 1961.

Löf, G. O. G. and Hawley, R. W. "Unsteady-State Heat Transfer Between Air and Loose Solids." *Industrial and Engineering Chemistry* 40 (June 1948): 1061–1070.

Löf, G. O. G. and Tybout, R. A. "Cost of House Heating with Solar Energy." *Solar Energy* 14 (1973): 253–277.

———. "The Design and Cost of Optimized Systems for Residential Heating and Cooling by Solar Energy." *Solar Energy* 16 (1974): 9–18.

Longmore, J. and Ne'Eman, E. "The Availability of Sunshine and Human Requirements for Sunlight in Buildings." Paper presented at the Conference on Environmental Research in Real Buildings, NIC Committee TC-3.3 (Fundamentals of the Physical Environment), Cardiff, September 1973.

Mills, Clarence, A. "Experimental Cooling-Heating System." *Architectural Forum,* November 1950, pp. 127–131. Using reflective interior surfaces to keep heat in and to act as radiators.

Minardi, John E. and Chuang, Henry N. "Performance of a Black Liquid Flat-Plate Solar Collector." Paper presented at the International Solar Energy Society, U.S. Section Annual Meeting, August 19–23, 1974.

Moore, Gordon, L. "Combining the Collector and Generator of a Solar Refrigeration System." Paper presented to the Solar Energy Applications Group, Winter Annual Meeting and Energy Systems Exposition, American Society of Mechanical Engineers, Pittsburgh, Pennsylvania, November 12–17, 1967.

Moorcraft, Colin. "Solar Energy in Housing." *Architectural Design,* October 1973, pp. 634–661.

Moore, S. W.; Balcomb, J. D.; and Hedstrom, J. C. "Design and Testing of a Structurally Integrated Steel Solar Collector Unit Based on Expanded Flat Metal Plates." Los Alamos Scientific Laboratory, Los Alamos, New Mexico.

Morrison, C. A. and Farber, E. A. "Development and Use of Solar Insolation Data for South Facing Surfaces in Northern Latitudes." University of Florida, Department of Mechanical Engineering.

Morrow, Walter E., Jr. "Solar Energy: Its Time is Near." *Technology Review,* December 1973, pp. 31–43. The potential utilization of solar energy for homes, total energy systems for commercial and industrial buildings, and large scale electric power generation; costs and overall energy savings.

NSF/NASA Solar Energy Panel. "An Assessment of Solar Energy as a National Energy Resource." December 1972. Available from National Technical Information Service, U.S. Department of Commerce, Springfield, Virginia.

Nevins, R. G. and McNall, P. E., Jr. "A High Flux, Low-Temperature Solar Collector." *Heating, Piping and Air Conditioning* 29 (November 1957): 171–176.

Newton, A. B. "Solar Energy Heats, Cools Maryland School." *ASHRAE Journal* 17 (February 1975): 34.

O'Connor, Egan. "Solar Energy—How Soon?" Environmental Action Reprint Service, University of

Colorado. Excellent on *why* solar energy.

Olgyay, Aladar. "Design Criteria of Solar Heated Houses." *Proceedings of the United Nations Conference on New Sources of Energy.* Rome, 1961.

Olgyay, Aladar and Telkes, Maria. "Solar Heating for Houses." *Progressive Architecture,* March 1959, pp. 195–207. Basic design considerations for solar-heated houses: heating loss, solar collectors and heat storage systems, the winter-summer balance, and design features which differ from those in conventional residences. Economic comparison of solar heating with conventional systems. Three recent solar houses are described: the Löf house in Denver, MIT house in Lexington, and AFASE house in Phoenix.

O'Neill, Mark J.; McDanal, A. J.; and Sims, W. H. *The Development of a Residential Heating and Cooling System Using NASA-Derived Technology.* National Aeronautics and Space Administration, Marshall Space Flight Center, Alabama. Available from National Technical Information Service (NTIS).

Pelletier, Robert J. "Solar Energy: Present and Foreseeable Uses." *Agricultural Engineering* 40 (March 1959): 142–144, 151.

Pellette, P.; Cobble, M.; and Smith P. "Honeycomb Thermal Trap." *Solar Energy* 12 (1968): 263–265.

Pleijel, Gunnar and Lindstrom, Bert. "A Swedish Solar-Heated House at Capri." *Proceedings of the United Nations Conference on New Sources of Energy.* Rome, 1961.

Proctor, D. "The Use of Waste Heat in a Solar Still." *Solar Energy* 14 (1973) 433–449.

Rabl, Ari and Nielsen, Carl E. "Solar Ponds for Space Heating." *Chemical Technology* 6 (1975), 608–616.

Read, W. R.; Choda, A.; and Copper, P. I. "Technical Note: A Solar Timber Kiln." *Solar Energy* 15 (1974); 309–316.

Reddy, S. Jeevananda. "An Empirical Method for Estimating Sunshine from Total Cloud Amount." *Solar Energy* 15 (1974): 281–285.

Rogers, Benjamin T. "Using Nature to Heat and Cool." *Building System Design,* October/November 1973.

Safwat, H. H. and Souka, A. F. "Design of a New Solar-Heated House Using Double-Exposure Flat-Plate Collectors." *Solar Energy* 13 (1970): 105–119.

Sargent, S. L. and Beckman, W. A. "Theoretical Performance of an Ammonia-Sodium Thiocyanate Intermittent Absorption Refrigeration Cycle." *Solar Energy* 12 (1968): 137–146.

Satcunanathan, Suppramaniam and Deonarine, Stanley. "A Two-Pass Solar Air Heater." *Solar Energy* 15 (1973): 41–49.

Schaeper, H. R. A. and Farber, Erich A. "The Solar Era, Part 4—The University of Florida 'Electric.'" *Mechanical Engineering* 94 (November 1972).

Schönholzer, Ernest. "Hygienic Clean Winter Space Heating with Solar and Hydroelectric Energy Accumulated During the Summer and Stored in Insulated Reservoirs." *Solar Energy* 12 (1969): 379–385.

Sheridan, Norman R. "Performance of the Brisbane Solar House." *Solar Energy* 13 (1972): 395–401.

Shoemaker, M. J. "Notes on a Solar Collector with Unique Air Permeable Media." Research Projects Corporation, Madison, Wisconsin.

Siple, Paul A. "Climatic Considerations of Solar Energy for Space Heating." Paper presented at the Space Heating with Solar Energy Conference, Massachusetts Institute of Technology, August 21–26, 1950.

Smith, Gerry E. "Economics of Solar Collectors, Heat Pumps and Wind Generators." Working Paper No. 3. University of Cambridge, Department of Architecture. Report on the economics of solar collection, wind and ambient energy utilization in Britain.

Solar Age. SolarVision, Inc. Rt. 515 Box 288, Vernon, New Jersey. 07462. Monthly.

Solar Energy Digest. William B. Edmondson, PO Box 17776, San Diego, California 92117. Monthly.

Solar Energy, The Journal of Solar Energy Science and Technology. Published by Pergamon Press for the International Solar Energy Society.

Solar Engineering. Solar Engineering Publishers, Inc., 8435 N. Stemmons Freeway, Suite 880, Dallas, Texas 75247. Monthly.

"Solar Heating Test." *Architectural Forum* 108 (April 1959): 135.

"Solar Space Heating for Houses." *Architectural Forum* 18 (April 1955): 227–30. Various solar houses in the United States: Donovan and Bliss, Hunter-Löf, Donald F. Monell, Austin Whillier, American Gas & Electric Service Corporation.

Soliman, S. H. "Effect of Wind on Solar Distillation." *Solar Energy* 13 (1972): 403–415.

Speyer, E. "Solar Energy Collection With Evacuated Tubes." *Journal of Engineering for Power* 87 (July 1965): 270.

Stonier, Thomas. "An International Solar Energy Development Decade: A Proposal for Global Cooperation." *Bulletin of the Atomic Scientists,* May 1972, pp. 31–34.

Swartman, R. K.; Swaminathan, C.; and Robertson, J. G. "Effects of Changes in the Atmosphere on Solar Insolation." *Solar Energy* 14 (1973): 197–202.

Swartman, R. K.; Ha, Vinh; Michel, Julien; and Whitney, D. J. "The Solar Era Part 5—The Pollution of Our Solar Energy." *Mechanical Engineering,* December 1972.

Telkes, Maria. "Future Uses of Solar Energy." *Bulletin of the Atomic Scientists,* August 1951.

———. "A Review of Solar House Heating." *Heating and Ventilation* 46 (September 1949): 68–74. Solar house heating, heat storage, and collection.

———. "Solar Heat Storage." In *Solar Energy Research,* pp. 57–61. Edited by Farrington Daniels and John A. Duffie. Madison: University of Wisconsin Press, 1955. Comparison of specific-heat type and heat-of-fusion type of heat storage. The second method provides maximum heat storage capacity in a given volume, therefore it should be useful for the storage of solar heat. Further tests are needed to obtain valid results.

———. "Solar House Heating—A Problem of Heat Storage." Publication No. 19, MIT Solar Energy Conversion Project.

———. "Solar Stills." *Proceedings of the World Symposium on Applied Solar Energy,* Phoenix, Arizona, 1955.

———. "Space Heating with Solar Energy." *Science Monthly* 69 (December 1949): 394–97. Based on an address presented at UN Scientific Conference on the Conservation and Utilization of Resources, Lake Success, New York, August 17–September 6, 1949. Experiments with solar heating, heat collection, heat storage; Dover, Mass., experimental house.

———. "Storage of Heating and Cooling." Paper presented at the Annual meeting of ASHRAE, Montreal, June 23, 1974. Report of Dr. Telkes' phase-changing eutectic salt heat storage and the apparent solution to the problems of super cooling, thickening, and nucleating.

Thomason, Harry E. and Thomason, Harry Jack Lee, Jr. "Solar House Heating and Air-Conditioning Systems: Comparisons and Limitations." Barrington, New Jersey: Edmund Scientific Company, 1974.

Threlkeld, J. L. and Jordan, R. C. "Utilization of Solar Energy for House Heating." *Heating, Piping and Air Conditioning,* January 1954. Methods of heating with solar energy; solar energy heat pump systems; method of calculating daily collector efficiency on clear and partly cloudy days; optimum number of glass panes for a flat-plate type collector; storage of solar energy by specific heat and heat effusion methods; application of heat pumps to year-round air-conditioning. Result of research sponsored by the American Society of Heating and Ventilating Engineers in cooperation with the Department of Mechanical Engineering of the University of Minnesota.

Tybout, Richard A. and Löf, G. O. G. "Solar House Heating." *Natural Resources Journal* 10 (April 1970): 268–326. Significant economic analysis for using solar energy.

Twarowski, M. *Soleil et Architecture.* Paris: Dunod Editeur, Co., 1967. Limited data; some design ideas.

U.S. Department of Commerce, Weather Bureau. *Sunshine and Cloudiness at Selected Stations in the United States, Alaska, Hawaii, and Puerto Rico.* Technical Paper No. 12, Washington, 1951. Available from Superintendent of Documents, Asheville, N.C.

U.S. Department of Commerce, Weather Bureau. *Weekly Mean Values of Daily Total Solar and Sky Radiation.* Technical Paper No. 11, Washington, 1949. Includes Supplement No. 1, Washington, 1955. Weekly means (up to 30 year period of record) graphed over course of year for over 30 stations. Available from Department of Documents, Asheville, N.C.

U.S. House of Representatives, 92nd Congress. Committee of Science and Astronautics. "Solar Energy Research: A Multi-Disciplinary Approach." Staff report of Second Session, December 1972. Available from U.S. Government Printing Office.

University of Maryland, Department of Mechanical Engineering. *Proceedings of the Solar Heating and Cooling for Buildings Workshop,* Washington, D.C., March 21–23, 1973. Available from National Technical Information Service (NTIS).

University of Wisconsin. "Modeling of Solar Heating and Air Conditioning." Project Report to the National Science Foundation from the Solar Energy Laboratory at the University of Wisconsin, January 31, 1973.

Ward, G. T. "Calculation of Direct Insolation on Plane Surface Over Specified Time Intervals." Technical Report No. T-35, August 1961. Brace Research Institute, McGill University.

"Warm Winter Behind Glass." *Life,* December 16, 1956, pp. 41–71. Bridgers and Paxton solar-heated office building, Albuquerque, New Mexico; heated by means of south-facing collector panels and a 6000 gallon water storage tank.

Weaver, Kenneth and Krist, Emory. "The Search for Tomorrow's Power." *National Geographic,* November 1972.

Weingart, Jerome. "Everything You've Always Wanted to Know About Solar Energy, But Were Never Charged Up Enough to Ask." *Environmental Quality Magazine,* December 1972, pp. 39–43.

Whillier, Austin. "Black-Painted Solar Air Heaters of Conventional Design." *Solar Energy* 8 (1964): 31–37. Excellent analysis of solar air-type collector performance.

———. "Principles of Solar House Design." *Progressive Architecture,* May 1955, pp. 122–126.

———. "Solar Energy Collection and Its Utilization for House Heating." Ph.D. dissertation, Massachusetts Institute of Technology, 1953.

———. "Solar House Heating—A Panel." *Proceedings of the World Symposium on Applied Solar Energy,* Phoenix, Arizona, 1955. Designing a solar-heated house for the New England region of the U.S.

———. "Solar Radiation Graphs." *Solar Energy* 9 (1965): 164–165.

———. "Thermal Resistance of the Tube-Plate Bond in Solar Heat Collectors." *Solar Energy* 8 (1964): 95–98.

Wilson, J. L. "Analysis of Solar Heating and Cooling Buildings." *ASHRAE Journal* 17 (January 1975): 72.

Wright, Henry Niccolis. "Solar Radiation as Related to Summer Cooling and Winter Radiation in Residences." Preliminary study for John B. Pierce Foundation, New York City.

Yanagimachi, Masanosuke. "On the Study of Utilization of Solar Energy as a Heat Source for House Heating, Cooling and Hot Water Supply by Heat Pump." *Journal of the Japanese Association of Domestic and Sanitary Engineers,* September 1947, pp. 158–79.

Yeh, Hsuan. "Conservation and Better Utilization of Electric Power by Means of Thermal Energy Storage and Solar Heating." Research Project of the University of Pennsylvania. Phase III Progress Report, March 31, 1973.

Yellott, John I. "Utilization of Sun and Sky Radiation for Heating and Cooling of Buildings." *ASHRAE Journal,* December 1973, pp. 31–42.

Solar Domestic Water Heating and Swimming Pool Heating

Abou-Hussein, M. S. M. "Ten Years' Experience with Solar Water Heaters in the United Arab Republic." *Proceedings of the United Nations Conference on New Sources of Energy.* Rome, 1961.

American Society of Heating and Ventilating Engineers. "Solar Water Heaters." *Heating, Ventilating, Air-Conditioning Guide* 28 (1950): 995–98. Design and operation of solar water heating devices.

Andrassy, Stella. "Solar Water Heaters." *Proceedings of the United Nations Conference on New Sources of Energy.* Rome, 1961.

Bhardwaj, R. K. "Investigation on Three Closed Chamber Solar Water Heaters." Paper presented at the Paris Solar Conference, July 1973.

Brace Research Institute, McGill University. "How to Build a Solar Water Heater." Do-it-Yourself Leaflet No. L-4. February 1965; rev. February 1973.

Brooks, F. A. "Solar Energy and its Use for Heating Water in California." Bulletin 602, November 1932. University of California at Berkeley.

———. "Use of Solar Energy for Heating Water." In *Solar Energy Research,* pp. 75–77. Edited by Farrington Daniels and John A. Duffie. Madison: University of Wisconsin Press, 1955. Use of solar water heating in California, for water heaters in rural areas, mountain cabins and locations remote from gas or liquid-fuel, in swimming pools, in agriculture, and in the industrial production of salt from tideland pools.

Chinnery, D. N. W. "Extending the Swimming Season with Solar Energy." *Heating, Air Conditioning and Refrigeration* 5 (May 1973).

Close, D. J. "The Performance of Solar Hot Water Heaters with Natural Circulation." *Solar Energy* 6 (1962): 33–40.

Commonwealth Scientific and Industrial Research Organization (CSIRO). *Solar Water Heaters.* Circular #2 of the Division of Mechanical Engineering, Melbourne, Australia, 1964. Principles of design, construction, and installation.

Czarnecki, J. T. "Performance of Experimental Solar Water Heaters in Australia." *Solar Energy* 2 (July–October 1958): 2–6. Seven experimental solar water heaters were installed at CSIRO laboratories throughout Australia in order to gain field experience and performance data for various localities. Each heater included an insulated 70 gallon hot water storage tank with a built-in electric booster and two solar absorbers with a total active area of 45 square feet. Average monthly values for a 12 month period are given for the daily electric power consumption and the solar contribution.

———. "Solar Energy for Domestic Hot Water Service." *Electrical Engineer and Merchandiser* 32 (February 1956): 345–48. The system consists of (1) flat heat absorber which receives the radiant energy from the sun and converts it into heat energy; (2) storage water tank of 70 gallon capacity insulated to keep heat losses to outside at a minimum; (3) auxiliary electric heater together with a thermostat, both mounted inside the storage tank. Tests carried out on such an arrangement in Melbourne, using active absorber areas of 45 square feet and daily water usage of 45 gallons at 135°F, have shown that the average annual saving of fuel can be 60% to 70%.

Davey, E. T. "Solar Water Heating." *Building Materials* 8 (October–November 1966): 57–61.

de Winter, Francis. *How to Design and Build a Solar Swimming Pool Heater.* New York: Copper Development Association, 1974. Excellent publication from theory and physical design to construction plans.

de Winter, Francis and Lyman, W. S. "Home-Built Solar Water Heaters for Swimming Pools." Paper presented at the Paris Solar Conference, July 1973.

"Domestic Water Heating by the Sun." *Mechanics,* June 17, 1955, pp. 161–63; and June 24, 1955, pp. 181–83, 191. Interview with Dr. Harold Heywood. How domestic water can be heated by the sun; constructional details for an absorber that can be made at home.

Farber, E. A. "Solar Water Heating." *Air Conditioning, Heating and Ventilating* 56 (July 1959): 53–55.

———. "Solar Water Heating: Present Practices and Installations." Paper presented at the ASME Semi-Annual Meeting, San Francisco, June 9–13, 1957. Use of solar energy for water heating; present-day practices, designs and considerations in installing a solar hot-water system; availability of sunshine; the features of solar hot-water systems, types of absorber, storage requirements, absorber size, location and position of absorber and storage tank, materials and their properties; design and economic considerations; installations which have performed satisfactorily for some time.

———. "The Use of Solar Energy for Heating Water." *Proceedings of the United Nations Conference on New Sources of Energy.* Rome, 1961.

Farber, E. A. and Bussell, W. H. "Solar Energy Used to Supply Service Hot Water." *Air Conditioning, Heating, and Ventilating* 54 (October 1957): 75–79. Data for designer of a system that uses solar energy for heating water; availability of sunshine in the United States; what can be expected from a good solar water heating installation; the effect of multiple glass covers on the absorber; heat losses; how much energy the water can be expected to absorb; comparison of the costs of heating water by solar energy, gas and electric.

Farber, E. A. and Triandafyllis, J. "Solar Swimming Pool Heating: Plans." Paper presented at the Paris Solar Conference, July 1973.

Geoffroy, J. "Use of Solar Energy for Water Heating." *Proceedings of the United Nations Conference on New Sources of Energy.* Rome, 1961.

Gutierrez, G.; Hincapie, F.; Duffie, J. A.; and Beck-man, W. A. "Simulation of Forced Circulation Water Heaters: Effects of Auxiliary Energy Supply, Load Type and Storage Capacity." Paper presented at the Paris Solar Conference, July 1973.

"How to Construct Solar Water Heaters." *Domestic Engineering* 172 (July 1948): 120–22, 146–47. How a solar water heater works; its costs and advantages; details of construction.

Kasuda, T. "Solar Water Heating in Japan." Paper presented at University of Maryland Workshop on Solar Heating and Cooling of Buildings, March 21–23, 1973.

Kemp, C. M. *Apparatus for Utilizing the Sun's Rays for Heating Water.* U.S. Patent 451, 384, April 28, 1891. A glass-covered box is used as a collector for the heat of the sun and is furnished with water-filled pipes.

Khanna, M. L. and Mathur, K. N. "Some Observations on a Domestic Solar Water Heater." *Proceedings of the New Delhi Symposium on Wind and Solar Energy,* 1954. Experiments and results obtained with a domestic water heating system developed by the authors for average-size families in India in the winter season. The heater utilizes a flat-plate absorber arranged so that water flows directly in contact with two copper sheets. Efficiency varies from 35%–40% in the early and late day time, to 80% when the sun is at nearly normal incidence on the absorbing surfaces.

Mathur, K. N. and Khanna, M. L. "Solar Water Heaters." *Proceedings of the United Nations Conference on New Sources of Energy.* Rome, 1961.

"Miromit Solar Heaters." *Business Daily,* August 11, 1955. Commercial production of solar water heaters designed by L. F. Yissar.

Morse, R. N. "The Design and Construction of Solar Water Heaters." Report ED1, Commonwealth Scientific and Industrial Research Organization (CSIRO), April 1954. Factors to be considered in designing a solar water heater; importance of the flow conditions in the connecting piping. Sufficient data given to design solar absorbers for Melbourne conditions based on observations over a 21 month period. Detailed drawings are provided for a standard absorber being installed in a number of localities throughout Australia.

———. "Solar Water Heaters." *Proceedings of the World Symposium on Applied Solar Energy,* Phoenix, 1955. Recent work carried out in solar water heaters; account of some Australian work; factors affecting the design of solar water heaters from the point of view of performance and economy.

———. "Solar Water Heaters for Domestic and Farm

Use. Report ED3, Commonwealth Scientific and Industrial Research Organization (CSIRO), July 1956. Solar water heater has an output of 48 gallons of hot water per day. Drawings, materials lists, and photographs; problems met in actual installations; performance and operating costs under Melbourne conditions.

———. "Water Heating by Solar Energy." *Proceedings of the United Nations Conference on New Sources of Energy.* Rome, 1961.

Morse, R. N.; Davey, E. T.; and Welch, L. W. "High Temperature Solar Water Heating." Paper presented at the Paris Solar Conference, July 1973.

Savornin, J. "Study of Solar Water Heating in Algeria." *Proceedings of the United Nations Conference on New Sources of Energy.* Rome, 1961.

Sobotka, R. *Solar Water Heaters.* Tel-Aviv: Miromit Sun Heaters, Ltd.

"Solar Heating for Domestic Hot Water." *Domestic Engineering* 171 (1948): 124–25, 261–66.

Tanishita, Ichimatsu. "Recent Development of Solar Water Heaters in Japan." *Proceedings of the United Nations Conference on New Sources of Energy.* Rome, 1961.

Whillier, A. and Richards, S. J. "Hot Water From the Sun." *Farmers Weekly,* November 19, 1958. Details for constructing a solar water heater developed at the National Building Research Institute, Pretoria, South Africa.

———. "A Standard Test for Solar Water Heaters." *Proceedings of the United Nations Conference on New Sources of Energy.* Rome, 1961.

Whillier, A. and Saluja, G. "Effect of Materials and Construction Details on the Thermal Performance of Solar Water Heaters." *Solar Energy* 9 (January 1965): 21.

Designing for Direct Solar Radiation

Danz, Ernst. *Architecture and the Sun.* London: Thames & Hudson. An international survey of sun protection methods.

Farber, E. A.; Smith, W. A.; Pennington, C. W.; and Reed, J. C. "Theoretical Analysis of Solar Heat Gain Through Insulating Glass with Inside Shading." *ASHRAE Journal* 5 (1963).

Hutchinson, F. W. "The Solar House." *Heating and Ventilation* 44 (March 1947): 55–9.

———. "Solar House: Analysis and Research." *Progressive Architecture* 28 (May 1947): 90–94. An analysis of solar house design based on research being conducted at Purdue University. Conclu-
sions are that the available solar gain for double windows in south walls in most cities in the U.S.A. is more than sufficient to offset the excess transmission losses through the glass. In evaluating solar designs for most American localities, it is possible to determine the theoretically exact amount of south glazing which will result in maximum benefit.

Hutchinson, F. W. and Chapman, W. P. "A Rational Basis for Solar Heating Analysis." *Heating, Piping and Air Conditioning* 18 (July 1946): 109–117.

Olgyay, Aladar and Olgyay, Victor. *Solar Control and Shading Devices.* Princeton: Princeton University Press, 1967.

Pennington, Clark W. "ASHRAE Solar Calorimeter and the Shading of Sunlit Glass." *ASHRAE Journal* 8 (March 1966).

Pennington, Clark W. and Moore, Gordon L. "Measurement of Solar-Optical Properties of Glazing Materials." *ASHRAE Journal* 13 (July 1971): 55–58.

Pennington, C. W. and Smith, W. A. "Solar Heat Gain Through Double Glass with Between-Glass Shading." *ASHRAE Journal* 6 (October 1964): 50–53.

Pennington, C. W.; Smith, W. A.; Farber, E. A.; and Reed, John C. "Experimental Analysis of Solar Heat Gain Through Insulating Glass with Indoor Shading." *ASHRAE Journal* 6 (February 1964): 27–29.

Thomas, Wendell. "The Self-Heating, Self-Cooling House." *The Mother Earth News,* No. 10, pp. 76–79.

Vild, Donald J. *ASHRAE Research and Principles of Heat Transfer Through Glass Fenestration.* Publication No. 478. Building Research Institute, National Academy of Sciences, 1957.

Yellott, John I. "When Sunshine Falls on Roofs and Walls." Paper presented at Heating, Piping, Air-Conditioning Conference on Controlling the Industrial Environment, Chicago, November 2–4, 1970.

Energy Conservation

Aronin, Jeffrey Ellis. *Climate and Architecture.* New York: Reinhold Publishing Corporation, 1953.

ASHRAE. *Handbook on Fundamentals.* New York: American Society of Heating, Refrigerating and Air Conditioning Engineers, 1967.

ASHRAE. *Handbook on Fundamentals.* New York: American Society of Heating, Refrigerating, and

Air Conditioning Engineers, 1972.

Becker, H. P. "Energy Conservation Analysis of Pumping Systems." *ASHRAE Journal* 17 (April 1975): 43.

Berg, Charles. "Energy Conservation Through Effective Utilization." *Science*, July 13, 1973.

Bird, R. B.; Stewart, W. E.; and Lightfoot, E. N. *Transport Phenomena*. New York: John Wiley and Sons, Inc., 1960.

Brooks, D. B. "Energy Conservation: How Big a Target?" *ASHRAE Journal* 16 (August 1974): 7.

Cannon, James. "Steel: The Recyclable Material." *Environmental Magazine*, November 1973, p. 11. Chart p. 17: "Energy Costs for Production of Common Materials."

Carrier Air Conditioning Company. *Handbook of Air Conditioning System Design*. New York City: McGraw-Hill Company, 1965. About cooling only.

Caudill, W. W.; Lawyer, Frank D.; and Bullock, Thomas. *A Bucket of Oil*. Boston: Cahners Publishing Co., Inc., 1974.

Conklin, Groff. *The Weather Conditioned House*. New York: Reinhold Publishing Corporation, 1958.

Council on Environmental Quality. "Energy and the Environment: Electric Power." Paper No. 0-523-292, August 1973. Available from Superintendent of Documents, Washington, D.C.

Davis, Albert J. and Shubert, Robert P. "Natural Energy Sources in Building Design." Blacksburg, Virginia: Virginia Polytechnic Institute, Department of Architecture, c. 1974.

Diamant, R. M. E. *Internal Environment of Dwellings*. London: Hutchinson Educations, Ltd.

———. *Total Energy*. Vol. 6. International Series of Heating, Ventilation and Refrigeration. N. S. Billington and E. Ower, eds. New York: Pergamon Press, 1970.

Dubin, Fred S. "Energy for Architects." Reported by Margot Villecco in *Architecture Plus*, July 1973, pp. 39–49.

———. "Energy Conservation Through Building Design and a Wiser Use of Electricity." Prepared for the Annual Conference of the American Public Power Association, San Francisco, California, June 26, 1972.

———. "Total Energy for Mass Housing." *Actual Specifying Engineer*, February 1973. How total energy and housing can be integrated.

Eccli, Sandy, ed. *Alternative Sources of Energy: Practical Technology and Philosophy for Decentralized Society*. Alternative Sources of Energy, Milaca, Minn., 1974.

Emerik, Robert. *Heating Handbook: A Manual of Standards, Codes and Methods*. New York: McGraw Hill Book Company, 1964.

"Flows of Energy." *Scientific American,* September, 1971.

Forest Products Laboratory. "Fuel Value of Wood." Technical Note No. 98, Madison, Wisconsin.

Forest Products Laboratory. "Thermal Insulation From Woods for Buildings: Effects of Moisture and Its Control." Madison, Wisconsin, July 1968.

Geiger, Rudolf. *The Climate Near the Ground*. Cambridge, Massachusetts: Harvard University Press, 1950. A detailed technical analysis of microclimate, not directly related to shelter design.

General Electric (Lighting Development/Nelo Park). *Interior Lighting Design Workbook*. TPC-42, April 1972. Zonal cavity method of lighting design, data tables, example problems and work sheets.

Givoni, B. *Man, Climate, and Architecture*. Barking, Essex, U.K.: Applied Science Publishers, 1969.

Hammond, Jonathan; Hunter, Marshall; Cramer, Richard; and Neubauer, Loren. "A Strategy for Energy Conservation: Proposed Energy Conservation and Solar Utilization Ordinance for the City of Davis, California." Prepared for the City of Davis with the support of The Case Institute, August 1974.

Hannon, Bruce. "Options for Energy Conservation." *Technology Review,* February 1974.

Harvey, D. G. and Kudrick, J. A. "Minimization of Residential Energy Consumption." Paper presented at the 7th Intersociety Energy Conversion Conference, September 1972.

Henke, K. C., Jr. "Pending State Legislation on Energy Conservation." *ASHRAE Journal* 17 (April 1975): 41.

Hirst, Eric and Moyers, John C. "Efficiency of Energy Use in the United States." *Science* 179 (March 30, 1973): 1299–1304.

Hottel, H. C. and Howard, J. B. *New Energy Technology: Some Facts and Assessments*. Cambridge, Massachusetts: The MIT Press, 1971.

MacKillop, Andrew. "Low Energy Housing." *Ecologist Magazine* 2 (December 1972): 4–10. Excellent summary of energy consumption and conservation issues in housing.

Malarky, John T. "High Performance Glasses for Energy Efficient Buildings." *Professional Engineer,* February 1974.

McAdams, W. H. *Heat Transmission*. 3rd ed. New York City: McGraw-Hill, 1954. Useful for rock thermal storage design.

Mother Earth News. *Handbook of Homemade Power*. New York: Bantam Books, Inc., 1974.

Mueller, Robert F. "Energy Conservation Alternatives to Nuclear Power: A Case Study." Goddard Space Flight Center, July 1973. Charts of energy consumption.

NAHB Research Foundation, Inc. *Insulation Manual.* Rockville, Maryland, September 1971.

National Bureau of Standards. "Technical Options for Energy Conservation in Buildings." NBS Technical Note 789. Institution for Applied Technology, Washington, D.C., July 1973.

Nevins, R. G. "Energy Conservation Strategies and Human Comfort." *ASHRAE Journal* 17 (April 1975): 33.

Olgyay, Victor. *Design with Climate.* Princeton: Princeton University Press, 1963. Required for anyone interested in designing with energy.

Severns, William H. and Fellows, Julian R. *Air Conditioning and Refrigeration.* New York: John Wiley & Sons, Inc., 1958, rev. 1966.

Spethman, D. H. "The Importance of Control in Energy Conservation." *ASHRAE Journal* 17 (February 1975): 35.

Stanford Research Institute. "Patterns of Energy Consumption in the United States." Paper No. 4106-0034. Available from the Superintendent of Documents, U.S. Government Printing Office, Washington, D.C.

Stein, Richard G. "A Matter of Design." *Environment Magazine,* October 1972, pp. 17ff. Methods of reducing energy consumption of buildings.

Stoner, Carol Hupping, ed. *Producing Your Own Power.* Emmaus, Pennsylvania: Rodale Press, Inc., 1974.

Szczelkun, Stefan A. *Survival Scrapbook #3: Energy.* New York: Schocken Books, Inc. Ways to decrease energy dependence: solar, wind, tidal, bio-gas, animal power.

United Nations. *Proceedings of the United Nations Conference on New Sources of Energy.* Vols. 1–7, Rome, 1961. Technical information on wind, solar, geothermal energy utilization.

U.S. Department of Agriculture. *Selecting and Growing Shade Tree #205.* U.S. Department of Agriculture, Washington, D.C., December 1973. Shade trees: characteristics, types, choice, heights, planting.

U.S. Department of Commerce. *Climatological Data.* Superintendent of Documents, Government Printing Office, Washington, D.C. State and regional weather information by year.

U.S. Department of Commerce, Environmental Science Services Administration. *Climatic Atlas of the United States.* Washington, D.C., June 1968.

U.S. Department of Commerce, Environmental Science Services Administration. *Selected Climatic Maps.* Office of Data Information, Superintendent of Documents, 1966.

U.S. Deparment of Commerce. *Selective Guide to Climatic Data Sources.* Doc. No. 4.11, Environmental Data Service, Washington, D.C., Superintendent of Documents, 1969.

U.S. Department of Housing and Urban Development. *Heat Loss Calculations, Minimum Design Standards for HUD Handbook.* No. 4940.6. Washington, D.C., March 13, 1973.

U.S. Division of Housing Research. *Applications of Climatic Data to House Design.* U.S. Superintendent of Documents, Washington, D.C., 1954.

U.S. Superintendent of Documents. *Engineering Weather Data.* Air Force, Army, Navy Manual, June 15, 1967. Out of print.

INDEX